Emergence, Complexity and Computation

Volume 29

The Emergence, Complexity and Computation (ECC) series publishes new developments, advancements and selected topics in the fields of complexity, computation and emergence. The series focuses on all aspects of reality-based computation approaches from an interdisciplinary point of view especially from applied sciences, biology, physics, or chemistry. It presents new ideas and interdisciplinary insight on the mutual intersection of subareas of computation, complexity and emergence and its impact and limits to any computing based on physical limits (thermodynamic and quantum limits, Bremermann's limit, Seth Lloyd limits…) as well as algorithmic limits (Gödel's proof and its impact on calculation, algorithmic complexity, the Chaitin's Omega number and Kolmogorov complexity, non-traditional calculations like Turing machine process and its consequences,…) and limitations arising in artificial intelligence field. The topics are (but not limited to) membrane computing, DNA computing, immune computing, quantum computing, swarm computing, analogic computing, chaos computing and computing on the edge of chaos, computational aspects of dynamics of complex systems (systems with self-organization, multiagent systems, cellular automata, artificial life,…), emergence of complex systems and its computational aspects, and agent based computation. The main aim of this series it to discuss the above mentioned topics from an interdisciplinary point of view and present new ideas coming from mutual intersection of classical as well as modern methods of computation. Within the scope of the series are monographs, lecture notes, selected contributions from specialized conferences and workshops, special contribution from international experts.

More information about this series at http://www.springer.com/series/10624

Juan C. Burguillo

Self-organizing Coalitions for Managing Complexity

Agent-based Simulation of Evolutionary Game Theory Models using Dynamic Social Networks for Interdisciplinary Applications

Springer

Juan C. Burguillo
Department of Telematics Engineering,
School of Telecommunications Engineering
University of Vigo
Vigo
Spain

ISSN 2194-7287 ISSN 2194-7295 (electronic)
Emergence, Complexity and Computation
ISBN 978-3-319-69896-0 ISBN 978-3-319-69898-4 (eBook)
https://doi.org/10.1007/978-3-319-69898-4

Library of Congress Control Number: 2017959900

Printed on acid-free paper

This Springer imprint is published by Springer Nature
The registered company is Springer International Publishing AG
The registered company address is: Gewerbestrasse 11, 6330 Cham, Switzerland

To my family:
Alejandra, Sofía and Álex.
They are my motivation,
and the coauthors of my Life!

Preface

This book mainly deals with the paradigm of how using self-organized coalitions, in games played by selfish agents, can support the emergence of cooperation in complex scenarios. The main idea was to convert pure competitive games into a mix of competition and cooperation that provides a high profit to the selfish players and also to the whole population. At the same time, the whole game outcome should increase the efficiency in the resource assignment from the game designer perspective.

The use of self-organized coalitions is presented in this book from several perspectives, considering combinatorial optimization problems, time series prediction, smart grid infrastructures, game theory competitions, indirect reciprocity and even the popular Conway's Game of Life. My interest in these topics, and the use of mixed competition/cooperation scenarios, came to me fifteen years ago when reading the seminal works from Robert Axelrod and Martin Nowak about how cooperation may emerge in complex scenarios played by selfish agents. Both authors used game theory and evolutionary games as tools to model and simulate complex scenarios with a reduced set of simple rules. In fact, along last decades, the use of evolutional game theory has become a relevant technic to simulate them, and to perform experiments in many different scientific fields as evolutionary biology, sociology, economics, computer science and networked computing among others.

The first part of this book gives an introductory background on several disciplines that are strongly interconnected and related to the research chapters presented afterwards. These disciplines are complex systems, complex networks, cellular automata, multi-agent systems, self-organization and game theory. After the background part, the next two parts of this book provide some research scenarios and deal with my interest in exploring, together with several coauthors, the use of self-organizing coalitions to allow the emergence of cooperation in competitive games played by selfish agents. The second part describes the use of self-organized coalitions in algorithms dealing with optimization problems, time series prediction or managing energy distribution in smart grid environments. The third part is more oriented to game scenarios, considering a possession-trading game, the evolution of indirect reputation or the classical and popular Game of Life from John Conway.

Partial and more focused versions of most of the chapters of the research part were published since 2010 in journals, book chapters and international conferences mainly related to artificial intelligence topics like optimization, multi-agent systems, computer science, parallel computing. Nevertheless, the present chapters have been reorganized integrating several publications, re-simulated again in more complex scenarios, rewritten and deeper extended than the ones provided in the original versions. The last chapter was not published before, and it is a direct application of the ideas presented in the previous ones into the complex scenario provided by the Game of Life. Publishing all these works together offers a wider perspective of the potential of self-organizing coalitions for finding and evaluating optimal solutions to complex problems.

Almost all the research chapters presented in this book, except the one about smart grids, have been simulated using CellNet: a Java simulating framework released with this book as free and open-source software, to support a hands-on approach to the scenarios and experiments described along the book. An appendix describes how to download CellNet, execute it, navigate through its user-friendly interface and to access or modify its source code.

Considering the introductory chapters, the research ones and the CellNet simulating framework provided as supplementary material; this book should be accessible for graduated and master students that want to enhance their knowledge and perform their thesis related to the topics discussed in this book. However, this book is particularly interesting for Ph.D. and postdoctoral students that are familiar with some of these topics, but want to have a more general perspective to explore new ideas and to design and create their own new models, games and scenarios.

Finally, I would like to thank my research colleagues from the Telematics Department at the University of Vigo for discussions, opinions, research proposals and mainly fun. Funding support for several works presented here has been provided by the Atlantic Research Center for Information and Communication Technologies (AtlantTIC). Some of these ideas, and many more not included in this volume, came from stays in relevant institutions, whose researchers quizzed me with interesting problems to solve from an interdisciplinary point of view. I would like to thank Prof. Ivan Zelinka, who invited me years ago to meet his research group in Czech Republic and was the catalyzer of my closer approach to complex systems. I also want to thank Dr. Djamel Kadrahoui for inviting me to stay at the Public Research Center Henri Tudor in Luxembourg that allowed me to meet new colleagues that have also influenced this book. My special thanks to Prof. Michael Luck at Kings College London, where I enjoyed a nice working atmosphere that helped me to finish some of the introductory chapters of this book. I also want to thank the coauthors of some of the chapters for their generous support and collaboration. Finally, I also want to thank Ms. Varsha Prabakaran and Dr. Tom Ditzinger from Springer, for their patience and support while completing this endeavour.

Vigo, Spain
September 2017

Juan C. Burguillo

Contents

Part II Self-Organizing Algorithms

Part III Evolutionary Games

Chapter 1
Introduction

Many problems that challenge us today can be traced back to a profound tension between what is good and desirable for society as a whole and what is good and desirable for an individual.

Martin Nowak (preface from [4])

This book aims to enhance the knowledge on the use of coalitions, which self-organize themselves, in order to manage problems that may appear in multiple disciplines. On the one hand we consider coalitions among members of a team that self-organize internally in order to maximize the group goal, but also keeping in mind the individual member goals. On the other hand, such coalitions can be also used to model, manage or evaluate complex problems modeled by disciplines like complex networks, cellular automata, multi-agent systems and game theory. In fact, one of the main goals of this volume is also to address the use of cooperative scenarios introducing coalitions in competitive games.

Complexity and its underlying theory refers to the study of many agents and their interactions. Here, the concept of agent is very broad and describe autonomous entities including animals, people, teams, organizations, etc. The system resulting from such interactions among the agents is denoted as a Multi-agent System (MAS).

Trying to model a large number of entities, and their non-linear interactions in a continuously changing environment, by means of classical mathematical tools can become a very hard task and many times impossible. Within this context, computer simulations become the natural tool for performing such kind of analysis and evaluation. In the last two decades, these interactions have been usually modeled by means of Agent-based Models (ABM), and simulated using Agent-based Simulations (ABS). The aim of these agent-based models and simulations is to better

J. C. Burguillo, *Self-organizing Coalitions for Managing Complexity*, Emergence, Complexity and Computation 29, https://doi.org/10.1007/978-3-319-69898-4_1

understand the real scenarios, analyzing their properties, strengths, weaknesses and limitations; but also to consider alternative worlds or agent societies that could have different configurations, rules or properties than the ones available in the real world. Summarizing Axelrod in the introduction to his book *The Complexity of Cooperation* [1]: *the use of simulations can be considered a third method of doing science, contrasted with the two standard mathematical methods of induction and deduction. Induction aims to discover patterns in empirical data, while classical deduction involves a set of axioms (assumptions about the model) and the proving of propositions and theorems that can be derived from those assumptions. On the one hand, the use of agent-based modeling shares conceptual elements with the previous two methods, as it starts with a simple set of assumptions to design and run a simulation that afterwards produces data that can be analyzed inductively. On the other hand, it does not prove theorems, and the data is produced from simulations and not from real world measurements.* Therefore, agent-based modeling and simulation aid intuition to analyze real or artificial models, and also can be used to discard certain models or assumptions that try to describe reality.

Simulations can be divided in two main categories [1]. In the first one, we can include those simulations that need to be very precise, providing a detailed image of reality; for instance, security simulations, aeronautics, army battles or evacuation scenarios usually demand a fine-grained well tuned simulation. In a second category, we find those simulations where the goal is to enhance our understanding of fundamental processes. In this case the use of simple assumptions is relevant in order to abstract unnecessary details that complicate the model but do not throw light over the system behavior. The simulations presented in this book belong to this second category, and the models run in the experiments try to be simple enough to abstract unnecessary details, but at the same time enhance our understanding of what happens at the microlevel, to explain the effects emerging at the macrolevel. In these simulation scenarios, even if the assumptions are simple, the results may be not; in fact, in some cases from a very simple set of rules can emerge a set of behaviors that can be extremely complex to analyze. Conway's Game of Life is a good example of how very simple rules can generate very complex behavior, and universal computation in the ultimate term, using a simple cellular automata model.

Cellular automata are a class of systems that usually are described by a spatial discrete lattice of homogeneous cells, that take a finite number of discrete states, and also use discrete dynamics to update each cell state taking into account also the states of its local neighbors. Basically, they have been traditionally used for: parallel computation, to simulate discrete dynamical systems, to study pattern formation, to model fundamental physics and, of course, to study complexity. Due to this, cellular automata is a basic framework for several models described in this book.

Cellular automata can be basic model to describe homogeneous spatial interactions among agents, and the modeling of agent-based systems usually consider a collection of agents interacting in some specified way. To model such system interactions it is necessary to specify its topology (who interacts with whom) and the system dynamics (how the individual entities interact). Most complex systems have complicated non-regular topologies, that require a complex network framework for their representation. A complex network is a structure made up of nodes connected by one or more specific types of interdependency. Nodes represent individuals, groups, or organizations, while connections (links, edges or ties) represent relations such as friendship, economic deals, internet paths, neuron interactions, etc. The resulting graph-based structures are often very complex, being social networks the most popular application, but the analysis of these structures has carried out a great number of research papers in fields like: Economics, Telecommunications, Biology, Artificial Intelligence, Bioinformatics, Anthropology, Information Science, Social Psychology, Sociolinguistics, among others. Network Theory has emerged as a key technique to be applied in order to model, analyze, simulate and understand those complex network topologies; from a static and a dynamic point of view. In this book we will consider the static and dynamic perspectives allowing agents to interact with a certain neighborhood, and even to modify their neighbor set dynamically by means of partner switching.

Nowadays, the growing complexity of the ICT ecosystem and the appearance of concepts like sensor networks, traffic management, autonomic computing, ubiquitous computing, ambient intelligence, internet of things, etc.; needs new solutions to support the design and the analysis for autonomous, adaptive and robust complex distributed systems. It is unrealistic to be able to achieve distributed optimal control of such systems, and even more from a centralized point of view. This is not feasible because of the huge size of those systems, the unpredictability of their dynamic organization, their interactions with the environment, and the diversity of the goals pursued by the different devices. Self-organized models are potentially good candidates to understand such complex behavior, where emergent phenomena may appear from their numerous interacting components, and where self-organization can be a powerful tool to manage their complex behavior. Self-organization is a process that consist of a huge number of autonomous entities distributed over space, and connected locally or using a network topology, but with a limited communication range. The building blocks are autonomous entities inherently dynamic, which work distributed and decentralized in a loosely coupled model over a continuously changing environment. Due to these characteristics multi-agent systems and agent-based simulation have been a reference model to design and engineer self-organizing systems.

The interactions among the agents in a multi-agent system may be of multiple types, but we are interested in those ones that usually convey some kind of competition and cooperation among them. In this framework, agents usually have selfish goals, that not necessarily coincide with the goals of the multi-agent system as a whole. Therefore, their actions or long term strategies may be selected according to a certain *utility*. The utility concept has been studied for years in the framework of game theory, that analyzes how rational agents should play games. Rationality is a central assumption in classical game theory and some other disciplines like economics. A rational player always plays to maximize his own payoff, assuming certain actions that the other players will do. The goal of game-theoretic analysis is to give advice on how to play a rational game against other rational opponents. The assumption that most of the humans only use rational choice when taking decisions is clearly unrealistic, but it usually allows to apply classical mathematical tools and particularly deduction. This leads to find nice properties of the system under equilibrium conditions, which is one of the main outcomes in neoclassical economic models. Unfortunately, when agents use other types of adaptive behavior, instead of rationality, or when the system is out of equilibria; then deduction becomes hard to apply due to multiple non-linear effects, and we enter in the realm of simulation.

The rationality assumption can be relaxed or bounded, and the resulting models have been more recently applied to the analysis of real observed behavior. When trying to model human behavior, behavioral game theory is a branch of game theory that analyzes how humans play, and it is based in three main ingredients: (i) how moral obligation and vengeance affect the way people behave, (ii) the brain has limited resources to reason about the game and how other players behave, and (iii) people usually act depending on their personal experiences (learning), but also imitating other players (memetics [2]) to take popular decisions. In fact, social interactions are strongly related with concepts like altruism that has puzzled biologist and anthropologist for decades. Martin Nowak, in his article titled *Five Rules for the Evolution of Cooperation* [3] explains the emergence of altruism in human societies from a game theoretical point of view.

In the models present in this book, the agents use some kind of adaptability instead of full rationality, as they play with limited resources and the only requirement is that players learn by trial and error; incorporate what they have learnt into their future behavior, and disappear or somehow change if they do not. This is coherent with the evolutionary game theory approach (EGT). EGT does not need that players act rationally, instead, the notion of rationality is replaced with the much weaker concept of reproductive success. Players have a strategy, and the evolutionary game will show how good it is. In fact, evolution by natural selection tests alternative strategies for their ability to survive and reproduce. In Biology, strategies are genetically inherited and control individual's actions, just like a computer program, by means of their genes [2]. The success of a strategy does not depend only on how good it is isolated, it also depends how good such strategy plays against other alternative strategies, considering their relative frequencies within a competing population. Thus, EGT differs from classical game theory by focusing more on the strategy dynamics, and also on the effect of the frequency of those competing strategies over the whole population. Under this model, the payoff utility is measured in terms of fitness units, that describe the reproductive success: strategies that are successful on average will be used more frequently, and prevail in the end.

The first part of the book gives an introductory background on the disciplines mentioned in the last paragraphs, that are strongly interconnected and related with the research chapters included at the end of the book. Initially the background part started as an unique chapter to support and unify the concepts managed at the rest of the chapters. With time, the background introduction grew up to a level that it was much more coherent to divide it into several chapters specifically dedicated to those topics related with complex systems. These background topics are introduced in the next order: complex systems, complex networks, cellular automata, multi-agent systems, self-organization and game theory. The main idea behind their contents was to provide a clear introduction to those research areas with some key references to explain them in a simple but complete manner.

The second part of the book describes the use of self-organized coalitions in algorithms dealing with optimization problems, time series prediction or managing energy distribution in smart grid environments.

In 2011 I enjoyed a one month stay at CRP Henri Tudor's in Luxembourg, invited by Dr. Djamel Kadrahoui and supported by FNR from Luxembourg. One of the unexpected outcomes of my visit was meeting some researchers from the outstanding research group leaded by Prof. Pascal Bouvry. As a result, I ended studying with one of his research members, Bernabé Dorronsoro, how the use of self-organized coalitions under a evolutionary game theoretic scenario can be combined with cellular automata and genetic algorithms (cGAs) in optimization problems. cGAs are a kind of genetic algorithms with decentralized population in which interactions among individuals are restricted to the closest ones. The use of decentralized populations in GAs allows to keep the population diversity for longer, usually resulting in a better exploration of the search space and, therefore in a better performance of the algorithm. However, the use of decentralized populations supposes the need of several parameters that have a major impact on the behavior of the algorithm. In the case of cGAs, these parameters are the population and neighborhood shapes. Chapter 8 presents a parameter free algorithm based in cellular automata and coalitions to manage dynamically and adaptively the cell neighborhoods in combinatorial optimization problems.

One year later, and thanks to an invitation from Prof. Ivan Zelinka to visit and collaborate with his new team at VSO-TUO in Czech Republic, I became interested about the possible connections among cellular automata, self-organized maps (SOM, a competitive learning neural network architecture) and self-organized coalitions. Chapter 9 explores, combines and enhances a set of several research works combining such research areas, and applying them for Time Series Prediction (TSP). In this work I have collaborated with Juan García-Rois, a present Ph.D. student with an outstanding research future, that helped me to evaluate in deep and to validate my results with an alternative model written in Matlab. The main outcome of this chapter is the redesign and evaluation of a Coalitional Algorithm for SOM (CASOM), as a way to create dynamic neighborhoods, by means of coalitions. The CASOM algorithm is mainly parameter free, provides better results than SOM and has lower training needs in multiple scenarios, including static or dynamic neuron network topologies.

The use of self-organized coalitions to find simple but powerful heuristic approaches to simplify the coalition structure generation problem leads me to collaborate with the remarkable Ph.D. student Gabriel de Oliveira-Ramos and his advisor Prof. Ana L. Bazzan. Chapter 10 is the result of our common work within the nowadays relevant smart grid scenario. Through Vehicle-To-Grid (V2G) sessions, plug-in electric vehicles (PEVs) can sell their surplus energy to the grid. However, profiting from V2G sessions is not trivial for singletons. Thereby, the formation of coalitions among PEVs becomes a relevant approach to tackle this issue. We present here a complete study of how to model such problem from different perspectives, also describing how the use of self-organized coalitions can help to solve efficiently several V2G scenarios.

The third part of the book is more oriented to game scenarios, considering a possession-trading game, the evolution of indirect reputation or the classical and popular Game of Life from John Conway.

Chapter 11 explores an evolutionary extension of the popular Hawk–Dove game from classical game theory. In particular, a networked version of the Possessor's and Trader's game is revised, and extended with multiple scenarios and evolutionary approaches. In such game, among the two basic strategies used in the Hawks–Dove game, hawks (H) and doves (D), we also consider two other strategies based on the property of resources: Possession (P), as the right to occupy or possess what one owns; and Trade (T), as the right to buy and sell ownership. The simulations presented in this chapter describe how evolutionary forces, depending on the simulation parameters, allow the emergence of the different type of populations (D, C, P or T) over several complex topologies. An informal trading social network is also introduced, where traders seek for trading mates to buy or sell resources; avoiding the hawks in their own neighborhoods. However, in this game scenario we have found that static topologies are more successful in general as the emergence of trading needs neighborhood stability, and we also have found a strong dependence on the initial conditions.

Around 2014 I explored together with Ana Peleteiro, a brilliant Ph.D. student I advised, how cooperation can be achieved in complex real-world scenarios not limited to direct interactions. In particular, cooperation can consider prior indirect interactions among other players, i.e., indirect reciprocity. Such work was based on previous collaborations done with Siang Yew Chong, a colleague from the University of Nottingham at the Malaysia Campus. In such model, agents play against each other the donation game, an indirect reciprocity game, where they can create coalitions to share information about agents' reputation or change their personal network of social contacts. Chapter 12 revises and extends from many perspectives such work to improve cooperation among self-interested agents placed in a complex network.

Chapter 13 has not been published before, and introduces the use of coalitions into the popular Conway's Game of Life. The new game, called CoaLife is built as an upper layer, keeping the basic rules used by Life, and adding new ones to enable to create, modify or release coalitions along the game execution. This allows coalitions to evolve along the game and serve as a basis for creating new layers over the two previous ones. As an example, an Iterated Prisoner's Dilemma (IPD) extension for CoaLife (IPD-CoaLife) is presented, as an upper layer for CoaLife, where alive cells play the IPD with their alive neighbors. The aim of this last chapter was to follow a similar approach to the one shown in the previous chapters, to create a simple IPD model of self-organized coalitions over the Life game. The combination of coalitions and the IPD gives the possibility to explore competition and cooperation among cells in the rich and challenging environment provided by Conway's Game of Life.

Finally, an appendix describes how to access, download and run the self-contained framework CellNet, written in Java and released for free. CellNet includes the executables and the source code to support a hands-on approach for most of the models, scenarios and experiments presented in this book.

References

1. Axelrod, R.M.: The Complexity of Cooperation: Agent-Based Models of Competition and Collaboration. Princeton University Press, Princeton (1997)
2. Dawkins, R.: The Selfish Gene. Oxford University Press, Oxford (1976)
3. Nowak, M.A.: Five rules for the evolution of cooperation. Science **314**, 1560–1563 (2006)
4. Nowak, M., Highfield, R.: Supercooperators: Altruism, Evolution, and Why We Need Each Other to Succeed. Simon and Schuster, New York (2011)

Part I
Background

Chapter 2
Complex Systems

Ivan Zelinka and Juan C. Burguillo

A complex system is an entity composed of interconnected parts, such that the collective behavior of those parts is *more than the sum* of the individual components. Those collective behaviors that appear as the interaction of the interconnected parts are usually called emergent.

Complex systems and their dynamics are a fundamental part of the world around us, and are also present in many branches of science and even in our daily life [26]. The term complex system is often used in a broad sense encompassing a research approach to problems in many diverse disciplines including anthropology, artificial intelligence, artificial life, chemistry, computer science, economics, evolutionary computation, earthquake prediction, meteorology, molecular biology, neuroscience, physics, psychology and sociology. Examples of complex systems include natural as well as artificial systems, p.e., colonies of social insects, flocking or schooling behavior in birds or fish, ecosystems, the economy, financial markets, social structures, climate, the immune system, nervous systems, cells and living things, human beings, human societies, the brain, modern grid energy systems, telecommunication infrastructures, the Internet and many more. The scientific areas that specialize in the interdisciplinary study of complex systems include systems theory, complexity theory, systems ecology and mainly cybernetics.

In the next sections of this chapter we introduce the different areas of complexity, and its mutual intersection, present in Fig. 2.1, namely: computational and algorithmic complexity, fractals, chaos, adaptive behavior and emergence.

I. Zelinka (✉)
Deparment of Computer Science, VSB-TUO, 708 33 Ostrava-Poruba, Czech Republic
e-mail: ivan.zelinka@vsb.cz

J. C. Burguillo
Department of Telematics Engineering, School of Telecommunications Engineering, University of Vigo, 36310 Vigo, Spain
e-mail: J.C.Burguillo@uvigo.es

J. C. Burguillo, *Self-organizing Coalitions for Managing Complexity*, Emergence, Complexity and Computation 29, https://doi.org/10.1007/978-3-319-69898-4_2

2.1 Short Historical Notes

The study of complex systems has its roots in classical economics, starting with the Scottish Enlightenment and continuing with the Austrian School of Economics, along 19th and the beginning of the 20th century; which analyzed market systems as systems that emerge as the result of human actions, but not as the result of human designs [12]. These pioneer approaches open the concept of complexity to the public and to notably lead economists and politicians, and had relevant debates against the dominant Keynesian economics.

Nobel prize economist and philosopher Friedrich Hayek, inspired by Karl Popper's and Warren Weaver's works [3], dedicated much of his work along the 20th century to the study of complex phenomena in fields such as psychology, biology and cybernetics. Gregory Bateson connected anthropology and systems theory; considering that culture interaction work much like ecosystems.

Many other disciplines interacted with Complex Systems Science along the second part of the 20th century, like Systems Theory, Artificial Intelligence or Cybernetics. Many popular scientist have contributed directly or indirectly to the field from that time, including popular ones like John von Neumann, Benoit Mandelbrot, Stuart Kauffmann, John Holland, or Melanie Mitchell among others. They have worked in multiple disciplines ranging from Computer Science to Biology, and the next chapters introduce in more detail some of those disciplines closely connected with Complex Systems, and with the topic of this book.

The Santa Fe Institute was founded in 1984 as the first research institute focused on complex systems, and its initial members included Nobel laureates as Murray Gell-Mann and Philip Anderson in physics or Kenneth Arrow in economics. Nowadays, there are dozens of institutes and research centers focusing on complex systems all over the world.

2.2 Basic Properties

The science of complex systems is a relatively new approach that studies how the relationships among its parts give rise to such collective behaviors, and also how the system interacts with its environment. For our purposes, we classify complex systems into mutually intersected subclasses, that are related as depicted in Fig. 2.1. The aggregate activity in such relations is usually non-linear (even the existence of linear complex systems is also possible) and typically exhibits self-organization and emergence effects under selective pressures. The self-organizing behavior is usually given by the internal system structure, but also by boundary conditions and external forces acting over the observed system.

There are a list of properties associated with the idea of a complex system [17]:

- *Non-linearity*: meaning that the outputs of the system do not satisfy the superposition principle, i.e., the output is not directly proportional to the input or to the sum of inputs.

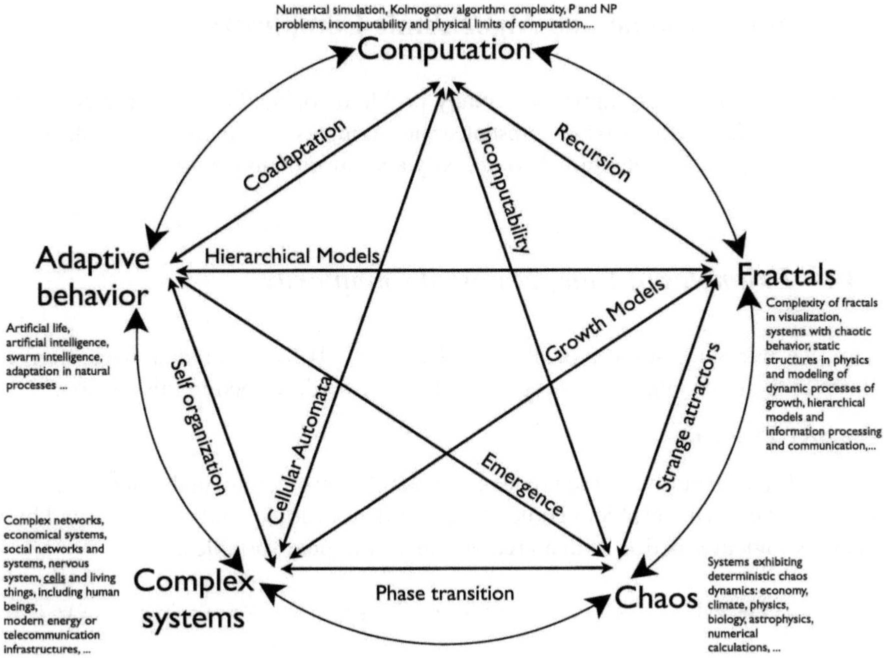

Fig. 2.1 Complexity and mutual relations

- *Feedback*: meaning that the past outputs of the system have influence in processing the new inputs, as part of a chain of cause-and-effect creating a circuit or a loop. This concept is also related with the presence of some memory within those systems.
- *Spontaneous order*: meaning that the order in the system arises from the aggregate of a very large number of uncoordinated interactions among elements.
- *Robustness and lack of central control*: meaning that the system becomes stable under perturbations, but there is no necessarily any central control to take decisions to compensate the perturbations.
- *Emergence*: meaning that larger entities, patterns, and properties appear from the interactions among smaller or simpler entities, that themselves do not exhibit such properties, so they only *emerge* from the system as a whole.
- *Hierarchical organization*: referring to be organized into a variety of structure levels, and each one interacting with an upper and lower one, and presenting various types of symmetry, order and/or periodic behavior.
- *Multiple elements*: in most of the definitions of complex systems, the multiplicity of elements, and the complexity of their interactions are a need in order to describe and understand such systems.

2.3 Computational and Algorithmic Complexity

Algorithmic complexity appears in many problems of mathematics and computer science. In this section, first we present some examples, and then we describe some basic concepts of computational complexity and information theory.

2.3.1 *Examples of Computational Complexity*

Here, we introduce some typical examples (see [24] for a detailed discussion) to illustrate the topic, and the concepts that are later on described in this section.

The SAT Problem

A typical representative is the boolean satisfiability problem (usually denoted as the SAT problem). This problem comes from the field of logic, and is represented by a complex logical function with a great number of logical variables.

$$F(x) = (x_{17} \vee \bar{x}_{37} \vee x_{73}) \wedge (\bar{x}_{11} \vee \bar{x}_{56}) \wedge \ldots \wedge (x_2 \vee x_{43} \vee \bar{x}_{77} \vee \bar{x}_{89} \vee \bar{x}_{97}), \quad (2.1)$$

Take for instance the function presented in Eq. 2.1 that contains 100 variables. The goal is to find the values (true or false) of individual variables for which the resulting value of the function in Eq. 2.1 takes the boolean value *true*. At first sight, this problem looks trivial, but it cannot be solved by classical methods. If we take into account that the expression contains 100 unknown variables, that can assume two values ($true$, $false$), then the number of all possible combinations is 2^{100}, which is approximately 10^{30}. In order to get a better impression on the big size of this number, it is sufficient to imagine how long it would take to evaluate all the combinations; p.e., if 10^{13} of these combinations are evaluated within one second (which is of course impossible on a single processor). The correct answer is 10^9 years, i.e., 1000 million years; which is one order of magnitude less than the age of the Universe.

Another complication related to this problem is the fact that the function defined by Eq. 2.1 does not allow to evaluate the quality of the current solution. This is a substantial drawback, particularly if the evolutionary techniques are used, because there is no possibility how to determine whether the qualities of two subsequently found solutions are close or not. When using evolutionary algorithms, it is of vital importance that the information on the quality of the solution is available for deciding in which direction the optimum solution lays. This is not possible in the SAT problem, because the function does not return how good or bad a given solution is, i.e., only if it is good (true) or bad (false).

The Traveling Salesman Problem (TSP)

As a more practical problem from real life, consider the well known traveling salesman problem. In this problem, a traveling salesman must visit a set of N cities using a shortest path, or in the shortest possible time, or with the smallest fuel consumption, or fulfilling any similar criteria. The condition is that each route must start and end in the same city, and that each city should be visited only once.

The number of all possible combinations of this purely practical problem is $n!$, but if the problem is symmetric (the distance from city A to B is the same as from city B to A), then $2n$ routes are repeated and the final number of all possible combinations is $(n-1)!/2$. As shown in Fig. 2.2, the number of possible combinations grows very quickly with the number of cities. Already for $n > 6$, there are more combinations in the TSP than in the SAT problem as show in Fig. 2.3, which presents a complexity comparison between the SAT problem and the TSP.

The traveling salesman problem has 181, 440 possible solutions for 10 cities. There are 10^{16} possible solutions for 20 cities and 10^{62} for 50 cities. If 60 cities is used, then there is 10^{79} of possible solutions, which is equal to the estimated number of protons in our universe, i.e., if each proton is used as a memory to store one possible TSP combination, then all protons in our universe can store the TSP solution space for a size of 60 cities. It is worth mentioning that at the present time, there are some special types of evolutionary algorithms (ACO, Ant Colony Optimization, [6]) that

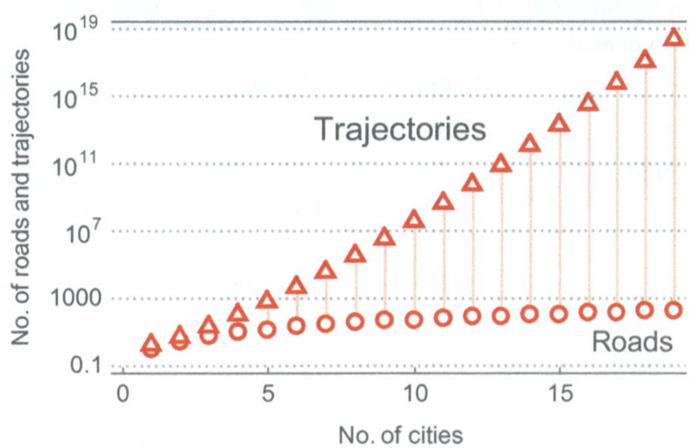

Fig. 2.2 Traveling salesman problem complexity describing the number of roads (circles) and the possible trajectories (triangles) depending on the number of cities

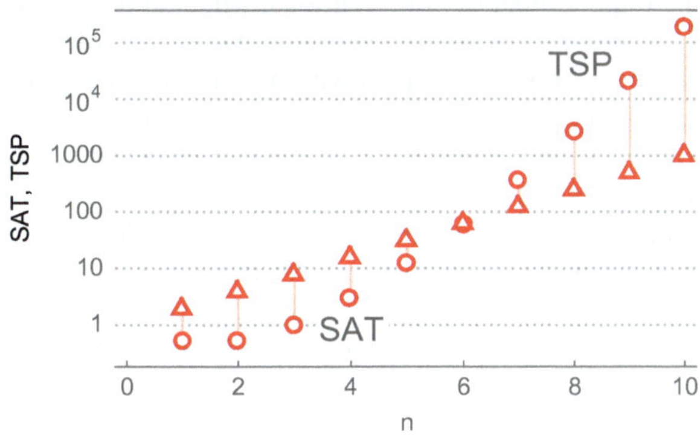

Fig. 2.3 Complexity comparison for SAT (circles) versus TSP (triangles). From 6 cities in TSP (or 6 variables in SAT), the traveling salesman problem is more time consuming

Table 2.1 Estimation the memory needed to store the values of some functions

$n \to$ function	10	50	100	300	1,000
Polynomial					
5n	50	250	500	1,500	5,000
$n\ log_2\ n$	33	282	665	2,469	9,966
n^2	100	2,500	10,000	90,000	1 million (7 digits)
n^3	1,000	125,000	1 million (7 digits)	27 million (8 digits)	1 billion (10 digits)
Exponential					
2^n	1,024	16 digit number	31 digit number	91 digit number	302 digit number
$n!$	3.6 million (7 digits)	65 digit number	161 digit number	623 digit number	giant number
n^n	10 billion (11 digits)	85 digit number	201digit number	744 digit number	giant number

Table 2.2 Estimation of the time needed to calculate $f(n)$ if 1 operation takes 1 μs

$n \to$ function	10	20	50	100	300
Polynomial					
n^2	1/10,000 s	1/2,500 s	1/400 s	1/100 s	9/100 s
n^5	1/10 s	3.2 s	5.2 s	2.8 h	28.1 days
Exponential					
2^n	1/1,000 s	1 s	35.7 years	400 trillion centuries	75 digit # of centuries
n^n	2.8 days	3.3 trillion years	70 digit # of centuries	185 digit # of centuries	728 digit # of centuries

manage up to 10,000 cities satisfactorily. We leave it to the kind reader to calculate the total number of combinations, which appears in this footnote.[1]

Therefore, there are problems whose complexity grows non-linearly, and then we speak about algorithms with polynomial or exponential complexity. Some examples concerning the estimation of the memory needed by some functions appear in Table 2.1, which provides the number of possible solutions for n input parameters. Table 2.2 shows complexity in terms of time, and considers the time needed to search for all possible solutions. It is obvious from these tables that there are many problems that computers cannot solve by brute force, i.e., trying all the possible combinations. Just consider as references for time and space that the number of microseconds from the *Big Bang* has *only* 24 digits, and the number of protons in the visible Universe has approximately 79 digits.

[1]Solution: The total number of combinations is $2.8462596809 \times 10^{35659}$.

2.3.2 Computational Complexity Theory

Computational complexity theory [27] is a branch of computer science and mathematics that focuses on classifying computational problems according to their inherent difficulty, and relating those classes to each other. A computational problem would be a task that may be solved by an algorithm. Under this framework, a computational problem would be inherently difficult if we require significant resources (time, storage, energy, etc.) in order to solve it.

Computational complexity takes its roots in mathematics and the idea of David Hilbert, a relevant german mathematician, that at the beginning of the Twentieth Century tried to completely formalize mathematics. On the one hand, his ambitious attempt was doom to failure; but on the other hand, his endeavor was very successful from the point of view of computing and programming as we will see. Hilbert's unsuccessful attempt hinted mathematician Kurt Gödel, born in nowadays Czech republic, that published his famous and exceptional *Incompleteness Theorem* proving that there is no axiomatic system for mathematics that can be sound and complete at the same time, see [9, 10]. Gödel was able to show it for natural numbers with the operations for addition and multiplication, i.e., every formal system trying to describe the truth, and only the truth, about natural numbers, with addition and multiplication, will be either sound or complete. Thus, there will be sound theorems that can not be proved.

After Gödel's work, and its implications, the basis of mathematical philosophy were really weakened; and then it was the turn of a relevant English mathematician, Alan M. Turing, to formalize Hilbert's ideas in the computing arena, and to discover non-computability. Turing was able to show that it is not possible to create any algorithm able to certify, in advance and for any program, if such program will finish its computation or not. Mathematically, there is no formal axiomatic system to help us in order to decide if a program will stop its computation or not. This result is strongly related with Gödel's incompleteness theorem, and mainly means that any formal axiomatic system for computing cannot be complete. To probe such relevant achievement, Turing designed a machine able to execute algorithms, and nowadays such machine becomes very useful for computational theory and denoted as a Turing machine.

Turing Machines

A Turing machine is an abstract mathematical model that can be used to describe any real computing machine, from the simplest one to the most powerful supercomputer. Basically, it contains an abstract device (the head) able to read and write symbols contained on an infinity strip of tape, that can be shifted left or right, according to a state register and a finite transition table for operations.

The Church–Turing thesis states that if a problem can be solved by an algorithm, then there exists a Turing machine able to solve such problem [4, 33]. Turing machines are easy to analyze mathematically, and that is why they become the most used model in complexity theory.

There are several types of Turing machines, and among them two typical examples are the deterministic and the non-deterministic ones. These types of machines are used to analyze how much time or space are needed to solve a problem. We can use a deterministic Turing machine (DTM), and under this approach the time required by such DTM will be the total number of steps that the machine executes before it halts and outputs the required answer. A Turing machine operates within time $f(n)$ if the upper bound of the time required on each input of length n is at most $f(n)$. Then, a certain problem can be solved in time $f(n)$ if there exists a Turing machine that solves such problem in time $f(n)$. A non-deterministic Turing machine (NDTM) may have a set of rules that prescribes more than one action for a given situation, and such action will be chosen non-deterministically, i.e., the NDTM can clone itself and execute several replicas of the computer and its state, and this process can be repeated in future computation steps creating an execution tree structure, when new replicas are created. These replicas cannot communicate among them, but allow to parallelize execution when needed.

Any problem solvable by a DTM can also be solved by a NDTM, since deterministic algorithms with no clone instructions can be executed in NDTMs. Interestingly, this works also the other way around: any non-deterministic algorithm can be executed in a DTM by sequentially executing, the many (but finite) parallel computation paths. Thus, both types of machines are equally powerful, but NDTMs could be much faster thanks to their arbitrary parallel processing capabilities.

Measures and Classes of Complexity

In order to measure complexity, we need to introduce previously several concepts. One of the basic ones is what we understand by a deterministic algorithm. We say that a deterministic algorithm A solves a problem p, when for any instance n of p, A always produces an answer and such answer is always correct. The time and space needed when computing $f_A(n)$ depends on the hardware where it is run, the selected programming language and how the algorithm is implemented. However, complexity theory will abstract these particular elements to focus on an uniform cost model considering Turing machines to run the algorithms.

Complexity theory is interested in measuring a resource complexity depending on the problem instance size (n), but not for particular samples of the problem itself. For instance, assume that algorithm A orders the values on a list, then we can try to evaluate the time complexity of such problem. Of course, for different lists of n elements the time needed can be very variable, and we can consider different alternatives like:

- *Best-case analysis*: as the shortest time needed for any possible list of n elements.
- *Empirical analysis*: trying to find the most frequent lists of n elements for the problem and the scenario where it must be run.
- *Average-case analysis*: computing the average time needed considering all possible lists of size n.
- *Worst-case analysis*: as the longest time needed for any list of n elements.

Table 2.3 Some relevant complexity classes, considering *time* as the resource

Complexity class	Model of computation	Resource order
DTIME(f(n))	Deterministic TM	Time f(n)
P	Deterministic TM	Time poly(n)
EXPTIME	Deterministic TM	Time 2poly(n)
NTIME(f(n))	Non-deterministic TM	Time f(n)
NP	Non-deterministic TM	Time poly(n)
NEXPTIME	Non-deterministic TM	Time 2poly(n)

The complexity of an algorithm is usually considered under the worst-case conditions, since best-case analysis is too optimistic; empirical test are arbitrary and difficult to obtain, and average-case analysis is usually impossible to compute for high n values. Then, we can define $f_A(n)$ as the worst-case analysis, for algorithm A over a problem of size n, as the upper bound for the function growth rate. However, computing $f_A(n)$ accurately can still be a very difficult task. Luckily complexity theory put the focus on how $f_A(n)$ evolves as n grows, i.e., when $n \to \infty$.

Sometimes, it may happen that different functions have the same growth rate in a certain resource, i.e., having similar complexity. This is mathematically represented by the *big-O notation*, which characterizes functions according to their order of growth. Then, a description of a function in terms of the *big-O notation* usually provides an upper bound to the function growth rate, and represents the worst case input for such function.[2] Some examples, in increasing order of complexity, are: $O(1)$ for a time constant, $O(\log n)$ for logarithmic, $O(n)$ for linear, $O(n^2)$ for quadratic, $O(c^n)$, $c > 1$ for exponential or $O(n!)$ for factorial growing. For example, finding the solution for the *Traveller Salesman Problem* via brute force search presents a factorial growing, as we have seen before.

Table 2.3 presents some relevant complexity classes considering time as the reference resource. Observe that each complexity class is described in the second column in terms of the type of turing machine needed to compute the problem, and in the third column in terms of the order of complexity for *such* turing machine.

Among the complexity classes listed in Table 2.3, the sets P and NP are two fundamental ones. Observe that NP is polynomial in time for a NDTM, but it becomes exponential for a DTM, as it has not the parallel capabilities given by infinity storage that a NDTM has. One of the most intriguing issues in computational complexity is the question of whether the sets P and NP are in fact identical. Assume a DTM (representing a real computer) running the problems, then P problems are identified as the ones that can be solved in polynomial time, while NP problems are those where finding a solution cannot be solved certainly in polynomial time, but they can be verified in polynomial time in such DTM. This is the frequent interpretation of both complexity classes, and it states the intuitive concept that it is simpler to check out if one solution is correct, than to find out the right solution. Looking at Table 2.3,

[2]Another frequent notation is the Ω notation, that provides a lower bound in the function growth rate.

Table 2.4 Some relevant complexity classes, considering *space* as the resource

Complexity class	Model of computation	Resource order
DSPACE(f(n))	Deterministic TM	Space f(n)
L	Deterministic TM	Space O(log n)
PSPACE	Deterministic TM	Space poly(n)
EXPSPACE	Deterministic TM	Space 2poly(n)
NSPACE(f(n))	Non-deterministic TM	Space f(n)
NL	Non-deterministic TM	Space O(log n)
NPSPACE	Non-deterministic TM	Space poly(n)
NEXPSPACE	Non-deterministic TM	Space 2poly(n)

and considering that DTMs are special cases of NDTMs, it is easily observed that each problem in P is also member of the class NP. A classic example to compare these two classes is the traveling salesman problem. As we have seen, finding a solution for the TSP is very difficult for medium size problems (NP time in a DTM), but once a solution is suggested we can check out quickly if it is valid (P time in a DTM).

Table 2.4 denotes similar complexity classes to the ones in Table 2.3, but now considering space as the reference resource.

2.3.3 Information Theory

Information theory is one of the tools most frequently used to analyze and understand complex systems. Information theory describes and quantifies information, and it was originally developed by Claude E. Shannon in his book *A Mathematical Theory of Communication* [32] as a way to understand mathematically the capabilities and limitations of electronic communications. Later on, the number of applications have grown enormously to quantify information in multiple fields of science and engineering.

Shannon entropy is formally defined for a random variable X, with probability distribution $Pr(X)$ and message alphabet $x \in X$, as:

$$H(X) := -\sum_{x \in X} Pr(x).log_2(Pr(x)) \tag{2.2}$$

A very interesting application of information theory is to measure the complexity of a system. Intuitively, a system more complex than another would provide a larger value of a certain complexity measure. A well-known example of such measure is the Kolmogorov's algorithmic complexity, which was formulated independently by Solomonoff, Kolmogorov and Chaitin in the mid-sixties [16]. Algorithmic complexity is defined as the length of the shortest computer program that will generate the system of interest or a complete description of it. If the system is too simple, then a

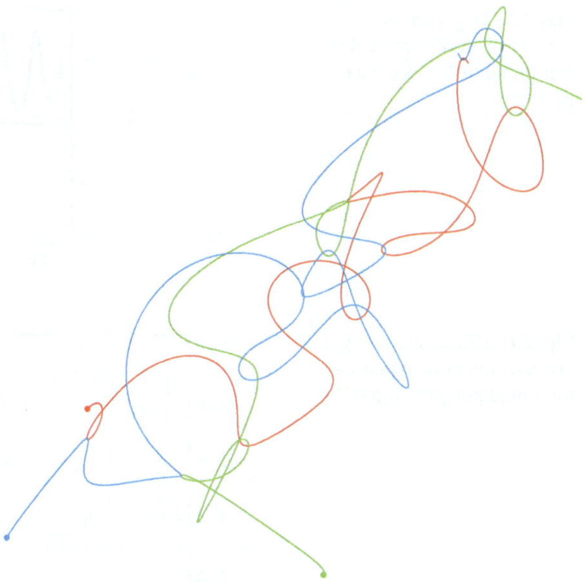

Fig. 2.4 Three body problem - simulation of the three celestial bodies motion

short program will be enough and the algorithmic complexity is low. Therefore, the larger the program, the higher the complexity. Unfortunately, algorithmic complexity is very difficult to calculate, and in many cases impossible.

2.4 Chaos

Broadly used, the term *chaos* covers a rather broad class of phenomena whose behavior may seem erratic at first glance, e.g., the motion of an individual molecule. However, if the word *chaotic* is combined with an attribute such as *stochastic* or *deterministic*, then a specific type of phenomena is involved, having their specific laws, mathematical background and physical interpretation.

As an historical outline, chaos is a discipline which obtained its name only in the 20th century, but whose roots date back to the 18th and 19th centuries, associated with the finding that even simple problems generate very complex and unpredictable behavior. Firstly it was indicated by French mathematician Henri Poincaré in the so called *three body problem* (see Fig. 2.4), that generates unpredictable moves of three bodies linked by gravitation. For historical reasons, these type of systems, denoted as Hamiltonian systems, were the first systems to be studied, and describe problems in celestial mechanics, generally called *billiard problems* (see Fig. 2.7). Many rules, valid for a wide class of Hamiltonian systems, generating chaotic behavior were discovered; and were even found to apply to some dissipative chaotic systems as well.

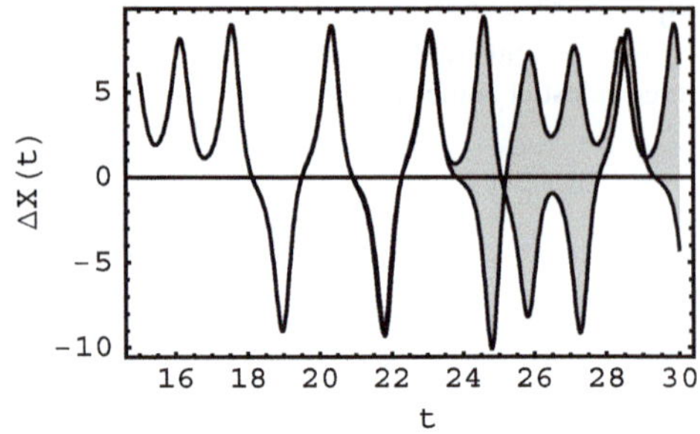

Fig. 2.5 Trajectories quickly diverge despite fact that initial differencc was 10^{-9}

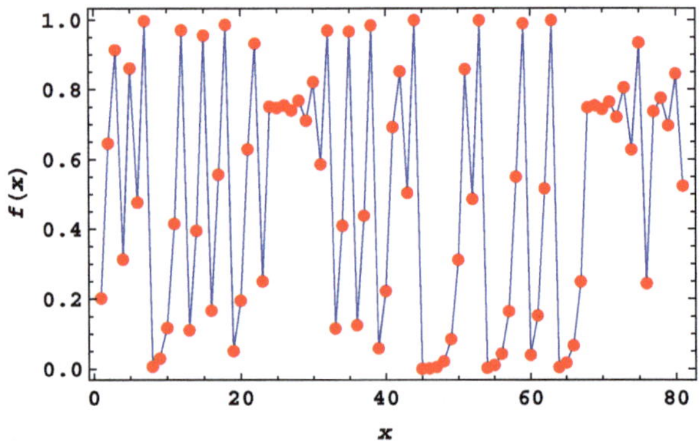

Fig. 2.6 Chaos shown by a discrete algebraic equation for a predator-prey model

Later on, MIT meteorologist E. Lorenz formally discovered chaos performing weather simulations, which resulted extremely sensible to initial conditions (see Fig. 2.5). While Lorenz's weather model [19] was continuous and described by differential equations, a biological system introduced by Robert M. May [22], based on a simple predator-prey model, was described by discrete algebraic equations (see Fig. 2.6).

Considering the class of deterministic chaos systems, as mentioned above, signs of chaotic behavior are usually conditional on the presence of nonlinearities, either in the system itself or in the interaction among linear systems [11]. Usually, such nonlinearities are only visible after describing a mathematical model of the system or analyzing the observed data.

Simple systems exhibiting deterministic chaos include, for instance, a double pendulum, a magnetic pendulum, an electronic circuit or the well-known billiard problem, described before. Figure 2.7 describes a variant of the billiard problem to exemplify the sensitivity to the initial conditions. Although released from the same position, each billiard ball follows a different pathway. This is so because the starting conditions are not absolutely identical; instead, they differ very slightly, even negligibly at first glance. It is because of those differences that the trajectories differ appreciably. In other words, mechanical objects hitting each other do not possess ideally smooth surfaces. Due to this, even the slightest differences in the initial conditions are amplified, ultimately giving rise to different trajectories.

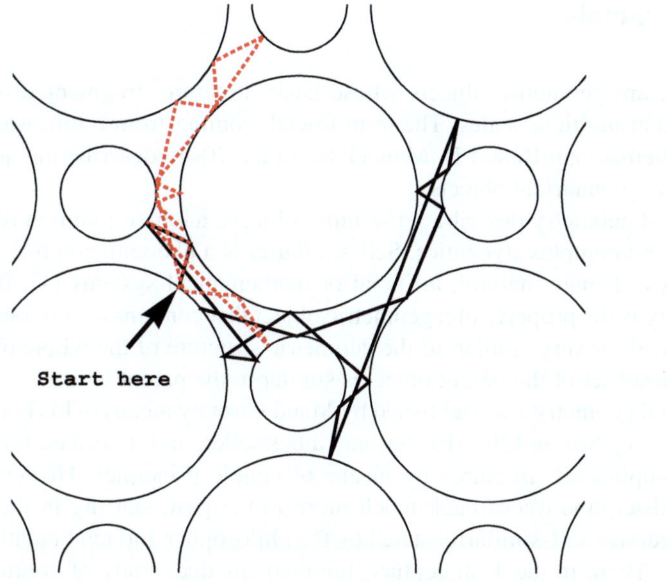

Fig. 2.7 A variant of billiard: trajectories diverge after a few iterations. Starting positions were $(x_1, y_1) = (0.936578, 1.31709)$ and $(x_2, y_2) = (0.936578, 1.3063)$

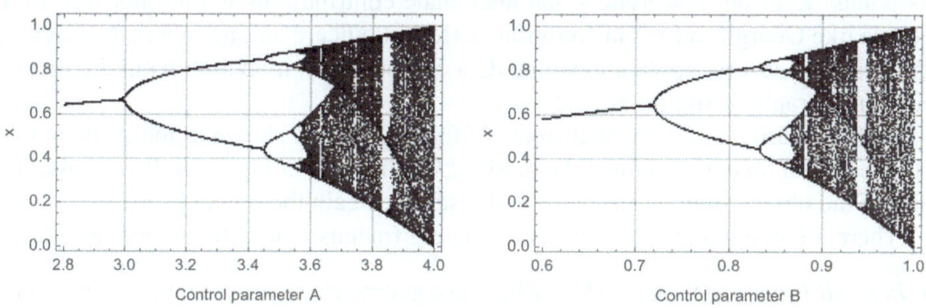

Fig. 2.8 Self-similarity in bifurcation diagrams. Left: diagram for the logistic Eq. (2.3). Right: diagram for Eq. (2.4) containing the trigonometric function

Some graphical visualizations of chaos exhibit the self-similarity property described in Sect. 2.5, when talking about fractals. Self-similarity in chaos may appear within a single visualization, a single bifurcation diagram of a system, or among several bifurcation diagrams[3] of different systems. This is illustrated in Fig. 2.8 for the Eqs. 2.3 and 2.4.

$$x_{n+1} = A.x_n(1 - x_n) \tag{2.3}$$

$$x_{n+1} = B.sin(\pi x_n) \tag{2.4}$$

[3]In dynamical systems, a bifurcation diagram shows the possible long-term values (equilibria/fixed points or periodic orbits) of a system as a function of a bifurcation parameter.

2.5 Fractals

Fractals are geometric objects whose basic structure, fragmented or irregular, is repeated at multiple scales. The term fractal, coming from Latin, was proposed by the mathematician Benoit B. Mandelbrot in the 70s to describe the quality of these *fractured* geometrical objects.

Fractal geometry describe self-similar objects, another common feature of complexity and complex dynamics. Self-similarity is a phenomenon that appears in the dynamics of many natural, artificial or mathematical systems [1]. Basically, self-similarity is the property of a geometric object that contains a component part which is identical, or very similar, to the geometric structure of the whole object. In other words, a subset of the parent object is similar to the parent object.

Fractal geometry was stablished by Mandelbrot by means of his book *The Fractal Geometry of Nature* [20], that became a bestseller, and described how to construct very complicated structures by means of simple principles. However the history of this discipline traces back much more in the past, starting in the 17th century with recursive self-similarity stated by the philosopher and mathematician Gottfried Leibniz. Then, in the 19th century, the mathematical study of continuous, but not differentiable functions, was increasingly rigorous; and it was Karl Weierstrass, who firstly described a function[4] that was continuous, but without derivative at any of its points. Relevant researchers that also made contributions within this area were people like Georg Cantor, Gaston Julia or Henri Poincaré among others. Nowadays, fractals have several applications like data compression, modeling natural patterns, computer graphics, digital art, etc.

Self-similarity, as a typical attribute of fractals, can be shown graphically on two classic fractal objects: a snowflake in Fig. 2.9 and a fern in Fig. 2.10. If you take any part of the object, then its structure will resemble again the object itself.

There are many algorithms for constructing fractals, interesting examples are:

- *Iterated Function System (IFS)*, which use geometric replacement rules, that may be deterministic or stochastic.
- *Escape-time fractals* use a recurrence relation at each point in a certain space.
- *Strange attractors* use the solutions of a system of differential equations that exhibit chaos (see Fig. 2.8).
- *Lindenmayer systems (L-systems)* use string rewriting [18] and a formal grammar to resemble branching patterns present in many biological structures like plants, cells from the immune system, etc.

Among them, we now introduce some examples of the first two well known algorithms, i.e., the Iterated Function System (IFS) and the Escape-time algorithms that present a tight relationship [1].

[4]The function was $f(x) = \sum_{n=1}^{\infty} b^n.cos(x\pi a^n)$, with $a \in \mathbb{Z}^+$ and $0 < b < 1$.

Fig. 2.9 Self-similarity in a snowflake

On the one hand, IFS is based on affine transformations, which are applied over a basic shape iteratively (see. Eq. 2.5).

$$w\begin{pmatrix}x\\y\end{pmatrix}=\begin{pmatrix}a & b\\c & d\end{pmatrix}\begin{pmatrix}x\\y\end{pmatrix}+\begin{pmatrix}e\\f\end{pmatrix}$$

$$or \tag{2.5}$$

$$w\begin{pmatrix}x\\y\end{pmatrix}=\begin{pmatrix}r_1cos(\theta_1) & -r_2sin(\theta_2)\\r_1sin(\theta_1) & r_2cos(\theta_2)\end{pmatrix}\begin{pmatrix}x\\y\end{pmatrix}+\begin{pmatrix}e\\f\end{pmatrix}$$

where r is a scaling factor and sin and cos functions are rotating factors. Parameters e and f produce displacements over x or y directions. Therefore, this transformation allows three different operations: translation, rotation and scaling. See an example in Figs. 2.11 and 2.12. Unifying a sequence of all those affine transformations, like in Eq. 2.6, can produce very complicated shapes (see Fig. 2.13).

$$W=\bigcup_{i=1}^{n} w_i \tag{2.6}$$

An important characteristic of affine transformations is that we just need a few numbers for a perfect description of any object. For instance, for the Koch snowflake in Fig. 2.13 we just need a few iterations. So, each fractal has its own unique description via a few numbers (coefficients of affine transformations, Eq. 2.5), and a slight change over them also changes the shape of the fractal.

Fig. 2.10 Self-similarity in a fern

On the other hand, Escape-time algorithms work in a different way as they are based on the idea of convergent and divergent trajectories over a search space. They take a point from such search space and calculate how many iterations are needed to leave a certain predefined area. The number of iterations is then used to determine the color for such point (see Fig. 2.14).

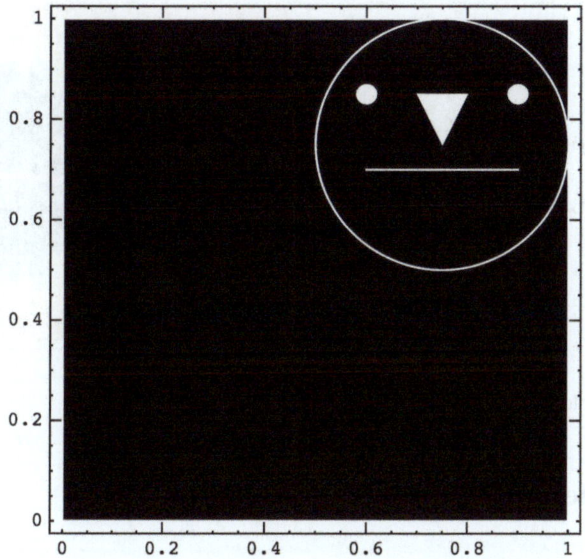

Fig. 2.11 Basic object

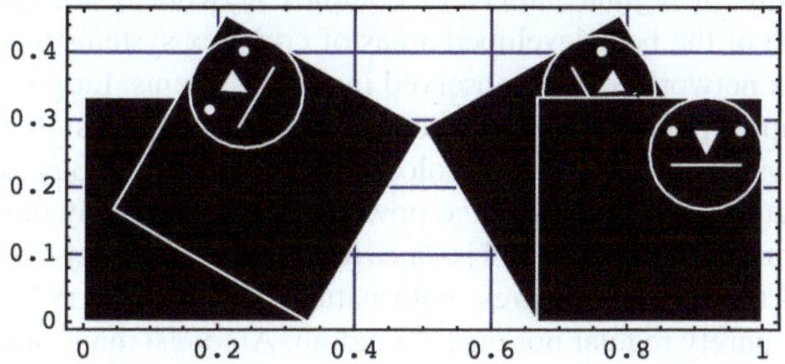

Fig. 2.12 Basic object after applying of 3 affine transformations

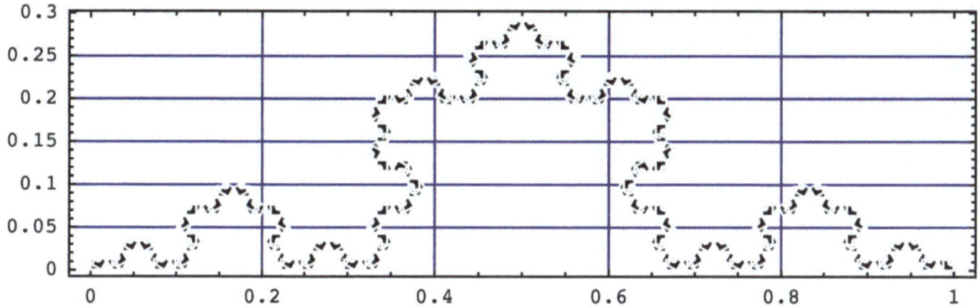

Fig. 2.13 Koch snowflake as a result of operations (Eqs. 2.5 and 2.6) visualized at Figs. 2.11, 2.12

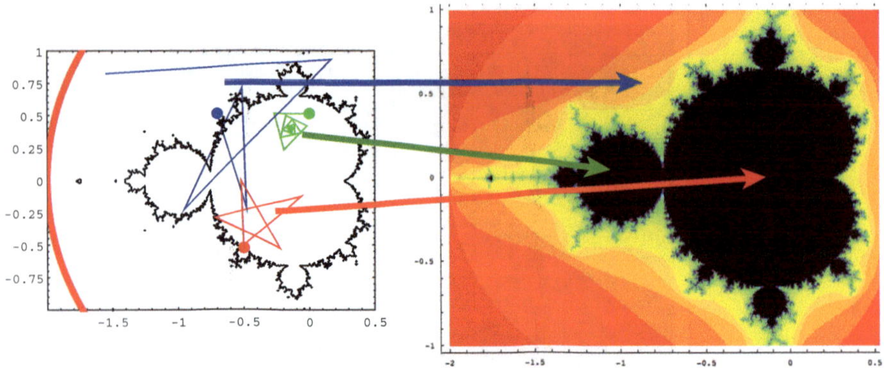

Fig. 2.14 Example of time escape algorithms (TEA)

2.6 Complex Networks

The agents of a complex system can be described by its topology, who interacts with whom, and also by its dynamics concerning such interactions. Topology is usually specified in terms of regular lattices or complex networks [28], which nowadays has become one of the best developed areas of complex systems theory. The structure of complex networks can be observed in many systems. Large-scale networks, exhibiting complex patterns of interaction amongst vertices exist in both nature and in man-made systems like genetics, ecology, economics, sociology, scientific publications, Internet, World Wide Web and power grids among many others.

The name complex networks [2, 7] comes from the fact that they exhibit substantial and non-trivial topological features, with patterns of connection between vertices that are neither purely regular nor purely random. Amongst many studies, two well-known classes of complex networks are the scale-free networks and the small-world networks (see examples in Figs. 2.15 and 2.16). Research in the field of complex

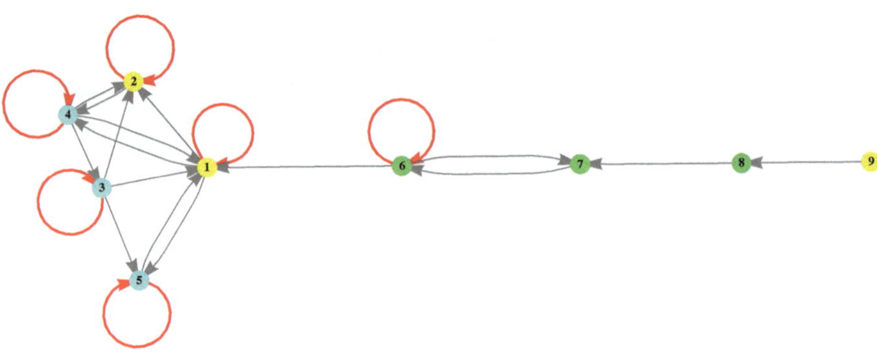

Fig. 2.15 Example of a small network

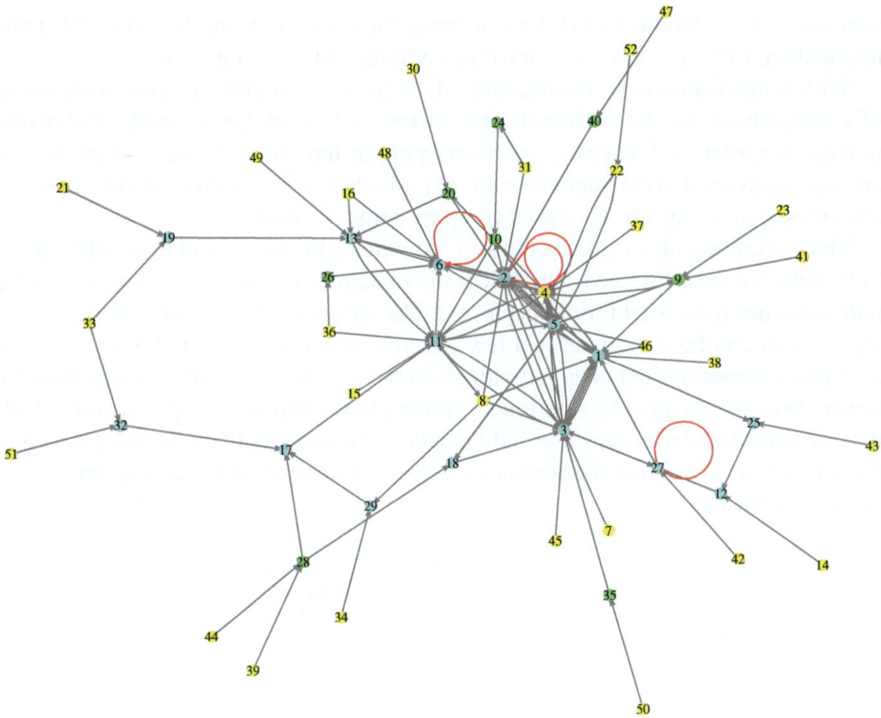

Fig. 2.16 Example of acomplex network with multiple edges and self-loops

networks has joined researchers from many scientific disciplines. Complex networks are very relevant in the scope of this book, so they are introduced in much more detail in Chap. 3.

2.7 Adaptive Behavior and Evolutionary Computation

Even not all, many complex systems show adaptive behaviors, and then are usually named Complex Adaptive Systems (CAS). The adaptation property means that the collective behavior optimizes some set of features along its evolution [29].

From the computational point of view, optimization is the basic aim of evolutionary computation, which is a sub-discipline of computer science. Evolutionary computation belongs to the *bio-inspired* computing area, and it has taken into account the seminal works of Darwin in his theory of evolution based on natural selection [5], and Mendel's theory of heredity [23]. The main ideas of evolutionary computation has been widely disseminated among the scientific community [13, 21], and have produced several well known evolutionary techniques as Genetic Algorithms (GA) introduced by J. Holland [14], Evolutionary Strategies (ES) by Schwefel [31] and Rechenberg [30], and Evolutionary Programming (EP) by Fogel [8].

Within this framework, evolutionary dynamics in complex systems is the result of a merge between two different areas of research: complex networks and evolutionary computation. Basically, evolutionary algorithms need an intensive interaction amongst individual in the population [8, 35], which is, in general, one of the important attributes to consider for the use of complex networks models.

The underlying idea is to represented individuals by nodes, and their relations by links between nodes, in order to reflect the dynamics of the population, p.e., what individual has been used for offspring, etc. The adaptive behavior of such evolutionary systems can be visualized in a few different ways. For example, Figs. 2.17 and 2.18 plot interactions between individuals in the population during the entire evolutionary process. As already mentioned, vertices in the complex graph are individuals that are substituted by other fitter individuals, incrementally from generation to generation. Those interactions between individuals create structures that may evolve into complex networks.

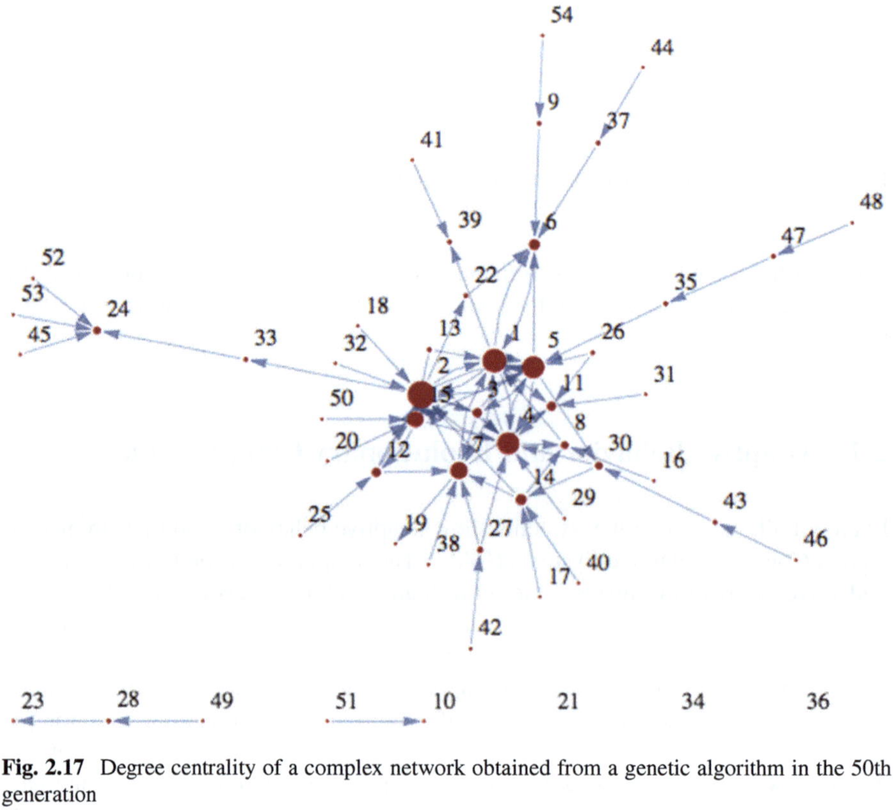

Fig. 2.17 Degree centrality of a complex network obtained from a genetic algorithm in the 50th generation

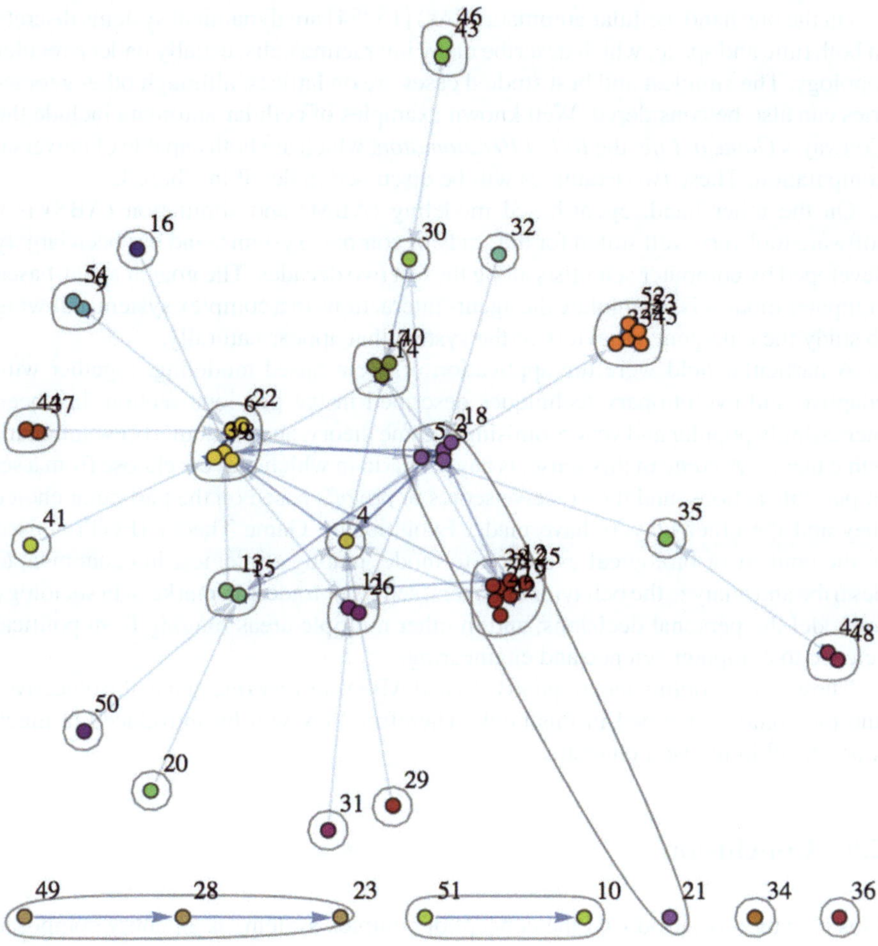

Fig. 2.18 Community graph in the 50th generation

2.8 Modeling and Simulating Complex Systems

In the past sections we have described a set of mathematical tools used to describe and to study complex systems. In this section we shortly describe other approaches to model complex systems by means of more complicated and realistic models, usually involving computer simulations, which are nowadays a powerful tool to analyze many natural and artificial systems. These models describe the interacting parts of a complex system in order to watch and measure the emergence of new global behaviors. The simulation techniques commonly used under this approach include cellular automata simulations or Monte Carlo based methods, usually combined with agent-based simulations.

On the one hand, cellular automata (CAs) [15, 34] are dynamical systems discrete in both time and space, which describe many interacting cells, usually under a regular topology. The simplest and best studied cases are on lattices, although other geometries can also be considered. Well known examples of cellular automata include the Conway's *Game of Life*, the *Rule 110 automaton*, which are both capable of universal computation. These two examples will be discussed in detail in Chap. 4.

On the other hand, agent-based modeling (ABM) and simulation (ABS) is a software tool very well suited for the study of complex systems, and has been largely developed by computer scientists along the last two decades. The goal of agent-based computer models is to simulate the agents interactions in a complex system, allowing to study the emergent behaviors of the system that appear naturally.

A particular field were the application of agent-based modeling, together with adaptive and evolutionary techniques described in the previous section, has been increasingly popular and very promising is game theory, and particularly evolutionary game theory. A *game* in this sense, is any scenario in which players choose from a set of possible actions, and then receive scores or *payoffs* based on the particular choice they and the other players have made. Evolutionary Game Theory (EGT) is used in the context of biological evolution to model mating strategies, in economics to describe and analyze the behavior of buyers, sellers or traders in markets, in sociology to model the personal decisions; and in other multiple areas ranging from political science to computer science and engineering.

These two modeling techniques (CAs and ABS) are very relevant to the objectives and the content included in this book. Therefore they will by introduced in much more detail in the next chapters.

2.9 Conclusion

This chapter has introduced the concept of complex system, as an entity composed of interconnected parts; such that the collective behavior emerging from the interconnection and interaction of those parts can not be derived by the isolated study of the parts.

The contents of the chapter are mainly related with the analysis of Fig. 2.1. First, we introduced the concepts of computational and algorithmic complexity. Then, we presented the concept of mathematical chaos followed by fractal geometry. Then we discussed the basics of complex networks, as a model to describe the interconnection of the parts composing a complex system. Finally, we have presented the concepts of adaptive behavior and evolutionary computation, and its relations with the topics described in the previous sections. Finally, we have explored the most relevant frameworks for modeling and simulating complex systems.

It is clearly obvious that enhancing our knowledge about complex systems, their dynamics, structure, modeling and control is extremely important for todays science and technology, and has become a fundamental part of modern science. In the next chapters we explore in much more detail, some of the mathematical and computing disciplines closely related with the study of complex systems.

2.10 Further Reading

The interesting reader can get a nice introduction written recently by Melanie Mitchell to Complex Systems in the book *Complexity: A Guided Tour* [26], which is targeted to a general audience. More technical descriptions can be found in the books by Miller and Page *Complex Adaptive Systems* [25] and the classical from Mandelbrot *The Fractal Geometry of Nature* [20]. Both describe in detail some of the contents presented on this chapter.

Acknowledgements This work was supported by grant No. GACR P103/15/06700S of the Grant Agency of Czech Republic, and the European Regional Development Fund (ERDF) together with the Galician Regional Government under agreement for funding the Atlantic Research Center for Information and Communication Technologies (AtlantTIC).

References

1. Barnsley, M.: Fractals Everywhere. Academic Press Professional, London (1993)
2. Boccaletti, S., Latora, V., Moreno, Y., Chavez, M., Hwang, D.U.: Complex networks: structure and dynamics. Phys. Rep. **424**, 175–308 (2006)
3. Caldwell, B.J.: Popper and Hayek: who influenced whom? Karl Popper 2002 Centenary Congress (2002)
4. Church, A.: An unsolvable problem of elementary number theory. Am. J. Math. **58**, 345–363 (1936)
5. Darwin, C.: On the Origin of Species by Means of Natural Selection, or the Preservation of Favored Races in the Struggle for Life, 1st edn. John Murray, London (1859)
6. Dorigo, M., Stützle, T.: Ant Colony Optimization. MIT Press, Cambridge (2004)
7. Dorogovtsev, S.N., Mendes, J.F.F.: Evolution of networks. Adv. Phys. **51**, 1079 (2002)
8. Fogel, D.B.: Unearthing a fossil from the history of evolutionary computation. Fundam. Inform. **35**(1–4), 1–16 (1998)
9. Godel, K.: Collected Works, vol. I. Oxford University Press, New York (2001)
10. Godel, K.: Collected Works, vol. II. Oxford University Press, New York (2001)
11. Haken, H.: Synergetics: Introduction and Advanced Topics. Springer, Berlin (2004)
12. Hayek, F.: The Results of Human Action but Not of Human Design in New Studies in Philosophy, Politics, Economics, pp. 96–105. University of Chicago Press, Chicago (1978)
13. Hofbauer, J., Sigmund, K.: Evolutionary Games and Population Dynamics. Cambridge University Press, Cambridge (1998)
14. Holland, J.: Adaptation in Natural and Artificial Systems. University of Michigan Press, Ann Arbor (1975)
15. Ilachinski, A.: Cellular Automata: A Discrete Universe. World Scientific, Singapore (2001)
16. Kolmogorov, A.N.: Three approaches to the quantitive definition of information. Probl. Inf. Transm. **1**, 1–17 (1965)
17. Ladyman, J., Lambert, J., Wiesner, K.: What is a complex system? Eur. J. Philos. Sci. **3**(1), 33–67 (2013)
18. Lindenmayer, A.: Mathematical models for cellular interaction in development. J. Theor. Biol. **18**, 280–315 (1968)
19. Lorenz, E.N.: Deterministic nonperiodic flow. J. Atmos. Sci. **20**(2), 130–141 (1963)
20. Mandelbrot, B.B.: The Fractal Geometry of Nature. Henry Holt and Company, New York (1983)

21. Maynard-Smith, J.: The Theory of Evolution, 3rd edn. Cambridge University Press, Cambridge (1993)
22. May, R.: Simple mathematical models with very complicated dynamics. Nature **261**(5560), 459–467 (1976)
23. Mendel, J.G.: Versuche ber Plflanzenhybriden. Verhandlungen des naturforschenden Vereines in Brunn, Bd. IV fu r das Jahr. Abhandlungen, 3–47 (1865); For the English translation, see: Druery, C.T., Bateson, W.: Experiments in plant hybridization. J. R. Hortic. Soc. **26**, 1–32 (1901)
24. Michalewicz, Z., Fogel, D.B.: How to Solve It: Modern Heuristics. Springer, Berlin (2000)
25. Miller, J.H., Page, S.E.: Complex Adaptive Systems. Princeton University Press, Princeton (2007)
26. Mitchell, M.: Complexity: A Guided Tour. Oxford University Press, Oxford (2009)
27. Moore, C., Mertens, S.: The Nature of Computation. Oxford University Press, Oxford (2011)
28. Newman, M.E.J.: Networks: An Introduction. Oxford University Press, Oxford (2010)
29. Nowak, M.A.: Evolutionary Dynamics: Exploring the Equations of Life. Belknap Press, Cambridge (2006)
30. Rechenberg, I.: Evolutionsstrategie - Optimierung technischer Systeme nach Prinzipien der biologischen Evolution (Ph.D. thesis), Printed in Fromman-Holzboog (1973)
31. Schwefel, H.: Numerische Optimierung von Computer-Modellen, Ph.D. thesis (1974); Reprinted by Birkhauser (1977)
32. Shannon, C.E.: A mathematical theory of communication. Bell Syst. Tech. J. **27**(379423), 623–656 (1948)
33. Turing, A.M.: On computable numbers, with an application to the Entscheidungsproblem. Proc. Lond. Math. Soc. Ser. **2**(42), 230–265 (1936)
34. Wolfram, S.: A New Kind of Science. Wolfram Media, Champaign (2002)
35. Zelinka, I., Davendra, D., Chadli, M., Senkerik, R., Dao, T.T., Skanderova, L.: Evolutionary Dynamics and Complex Networks. In: Zelinka, I., Snasel, V., Ajith, A. (eds.) Handbook of Optimization. Springer, Germany (2012)

Chapter 3
Complex Networks

Juan C. Burguillo

A complex network is a structure made up of nodes connected by one or more specific types of interdependency. Nodes represent individuals, groups, or organizations, while connections (links, edges or ties) represent relations such as friendship, economic deals, internet connections, neuron connections, protein interactions, etc. The resulting graph-based structures are often very complex, being social networks the most popular application, but the analysis of these structures has carried out a great number of research papers in fields like: Economics, Telecommunications, Biology, Artificial Intelligence, Bioinformatics, Anthropology, Information Science, Social Psychology, Sociolinguistics, among others. Network Theory has emerged as a key technique to be applied in order to model, analyze, simulate and understand those complex network topologies; from a static and a dynamic point of view.

One of the most popular applications of complex networks has been for social network analysis to determine the structure of relationships between social entities. These entities are often persons, but may also be groups, organizations, nation states, web sites, scholarly publications, etc. Since the 1970s, the empirical study of networks has played a central role in social science, and many of the mathematical and statistical tools have been first developed in these social sciences.

3.1 Short Historical Notes

Graph Theory has its roots in the 18th century, thanks to the famous mathematician Leonhard Euler, and his work with the famous problem of the bridges of Könisberg [9]. People wondered if the seven bridges connecting different parts of the town could

---

J. C. Burguillo (✉)
Department of Telematics Engineering, School of Telecommunications Engineering, University of Vigo, 36310 Vigo, Spain
e-mail: J.C.Burguillo@uvigo.es

J. C. Burguillo, *Self-organizing Coalitions for Managing Complexity*, Emergence, Complexity and Computation 29, https://doi.org/10.1007/978-3-319-69898-4_3

be traversed without passing any of them twice. Euler found the solution, abstracting the problem by means of nodes and links.

In the 1960s, two mathematicians, Paul Erdös and Alfred Rényi (ER) introduced random graph theory [6], combining graph theory and probability theory, to consider families of graphs rather than individual ones. Random graph theory is to graph theory what statistical mechanics is to Newtonian physics [4], so there was a strong link between both disciplines, and many concepts form statistical physics were reused in the study of networks. Among them, we may cite percolation, scaling, self-similarity, phase transitions and critical exponents.

At the end of the twentieth century, with the apparition of powerful computers and large-scale data, there was a need for mathematical tools to analyze them. The works of Barabási and Albert about the WWW [1], and Faloutsos, Faloutsos and Faloutsos about Internet routers [7], made clear that the Internet; and many other natural networks were not completely random, and could not be described by ER graph theory. These works led to new approaches to graph theory considering also correlations found in other real world networks. This new generalized random graph theory explained previously unsolved puzzles like, p.e., why malware where able to survive in the Internet for very long time.

Nowadays the study of complex networks is a very active area, that have produced thousands of publications in the last two decades.

3.2 Network Models and Applications

In this section we introduce graph theory and network theory as backgrounds for complex networks.

3.2.1 Graph Theory

Graph Theory is a branch of mathematics and computer science that deals with the study of graphs, which are mathematical structures used to model pairwise relations between objects. A graph $G = (V, E)$ is composed by a set of vertices (V) and lines that connect them denoted as edges (E). The order of a graph is $|V|$ (i.e., the number of vertices) and the graph's size is $|E|$ (i.e., the number of edges).

3.2.2 Network Theory

Network theory is a part of graph theory, and also an area of computer and network sciences. It concerns with the study of graphs as a representation of either symmetric relations or, more generally, asymmetric relations between discrete objects.

The attributes of individuals, when using network theory, are less relevant than their relationships with other actors within the network. This approach has turned out to be useful for explaining many real-world phenomena, because many of them rest within the structure of their networks. Network theory has also been used to analyze how organizations interact with each other, characterizing the many informal connections that link their parts, as well as the connections among individuals at different organizations.

3.2.3 *Complex Networks*

In the context of network theory, a complex network is a graph that shows features that do not occur in simple networks, such as lattices or random graphs, but are present those graphs that model real systems. The scientific research on complex networks became very active at this century, inspired by the empirical study of real-world networks from different perspectives. In fact, most social, biological, and technological networks present interesting topological features, with patterns of connection that are neither purely regular nor purely random.

Two well-known and representative classes of complex networks are small-world networks [10] and scale-free networks [1]. Both topologies are characterized by structural features like a high clustering for the small-world, short path lengths in both cases, and power-law degree distributions for the scale-free case. We will review these concepts, and networks topologies along the rest of this chapter.

3.3 Mathematics of Complex Networks

As stated before, a network, referred as a graph in mathematical literature, is a collection of vertices joined by edges, usually denoted as nodes and links in computer science, sites and bonds in physics, or actors and ties in sociology. As this book is mainly related with computer science, we will use the terms nodes and links, and we will denote the number of nodes as n and the number of links as k. We will also work with simple and finite networks.

3.3.1 *Matrix Representation*

While there are different ways to represent a network from a mathematical point of view, one of the simplest ones is to label each node with an unique number as the graph that appears in Fig. 3.1. An extensive representation could be done providing each node and its list of connections. For instance, given the network in Fig. 3.1 we have six nodes ($n = 6$) and the links are $\{(1,2), (1,4), (2,3), (2,4), (3,5), (4,5), (5,6), (6,7)\}$. An alternative representation is the use of an *adjacency matrix* A describing the connections A_{ij} between nodes setting by 1 s in the appropriate rows

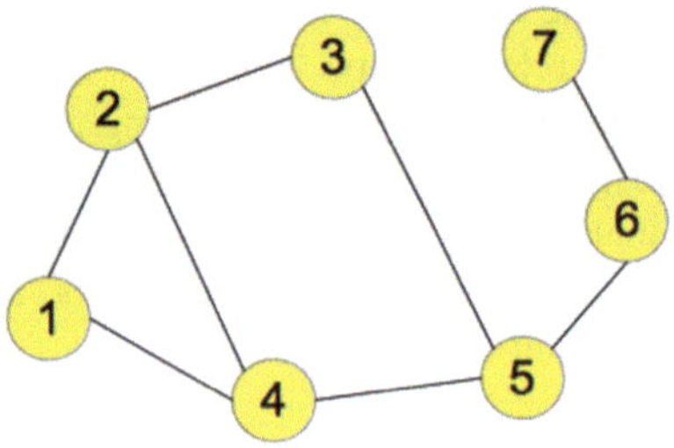

Fig. 3.1 An undirected graph

and columns, i.e., $A_{ij} = 1$ if there is a link between i and j, and 0 otherwise. The adjacency matrix for Fig. 3.1 is

$$A = \begin{bmatrix} 0&1&0&1&0&0&0 \\ 1&0&1&1&0&0&0 \\ 0&1&0&0&1&0&0 \\ 1&1&0&0&1&0&0 \\ 0&0&1&1&0&1&0 \\ 0&0&0&0&1&0&1 \\ 0&0&0&0&0&1&0 \end{bmatrix} \tag{3.1}$$

As there are no self-loops, the diagonal has all its values equal to zero. Besides, as we are using an undirected graph, we have that A is symmetric.

3.3.2 *Directed and Weighted Networks*

We denote by directed graphs those networks where links between nodes (i, j) have a particular direction from node i to node j or vice versa. Directed graphs may produce a non symmetric A matrix.

We can also consider natural values for A_{ij} greater than one to represent multi-edges, i.e., that there is more than one path connecting nodes i and j. Another possibility, when representing networks, is to provide some real values (weights) to the links, in order to represent costs, bandwidth, speed, time, etc. For instance, consider again the simple graph discussed before, but now with some weights in its links (see Fig. 3.2), to represent the delay in seconds associated with communication between nodes. The matrix associated with such figure is now represented as

$$A = \begin{bmatrix} 0&2&0&1&0&0&0 \\ 2&0&1&3&0&0&0 \\ 0&1&0&0&0&0&0 \\ 1&0&0&0&2&0&0 \\ 0&0&4&2&0&2&0 \\ 0&0&0&0&2&1&3 \\ 0&0&1&0&0&3&0 \end{bmatrix} \tag{3.2}$$

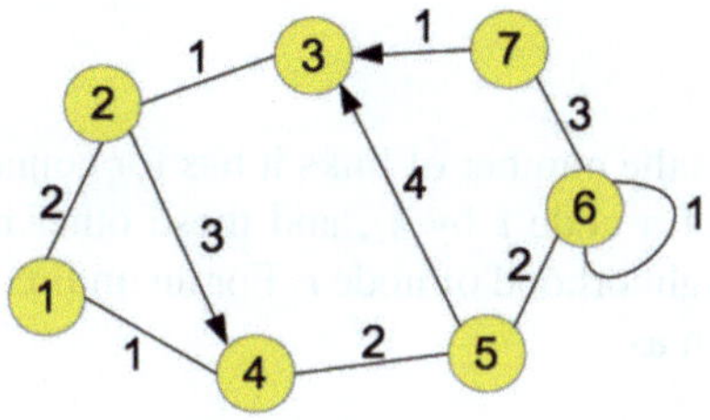

Fig. 3.2 A directed, weighted network with a self-loop

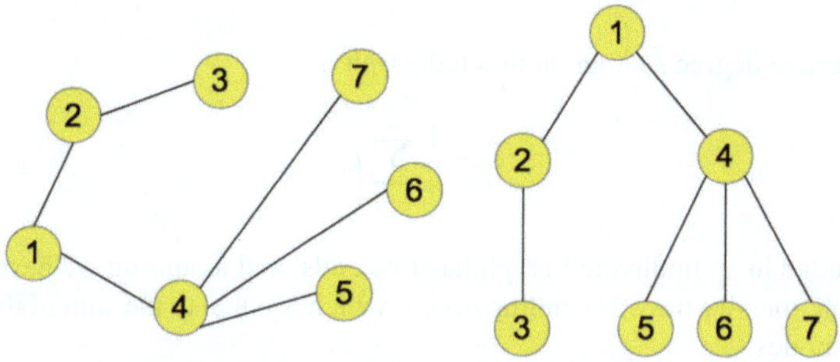

Fig. 3.3 A tree network (left) and a rooted representation (right)

Observe that Fig. 3.2 also contains some direct links between nodes $\{(2, 4), (5, 3), (7, 3)\}$ and a self-loop for node (6). This causes that the resulting matrix is not symmetric, and it also has values in its diagonal.

3.3.3 Tree

A tree is a connected network without closed loops, where *connected* means that every node in the network is reachable from any other via some path, and *closed loops* means no self-loops neither loops passing through several nodes. An example appears in the left side of Fig. 3.3. If our network is composed by several disconnected networks, each one being a tree, the complete network is denoted as a *forest*.

Trees are usually represented as *rooted graphs*, as shown in the right side of Fig. 3.3, starting with a top root node and a set of branches going down. A node connected only with one node is denoted as a *leaf*. From a topological point of view, a tree has no particular root and can be represented from any node as the root. Nevertheless, in some applications, and in directed graphs, the use of certain nodes as roots may be necessary. As trees have no closed loops, there is exactly one path joining any pair of nodes.

3.3.4 Degree

The degree of a node i is the number of links it has for connecting with other nodes. We denote the degree of a node i by k_i, and these other nodes linked to i as the neighbors of i or the neighborhood of node i. For an undirected graph with n nodes, the degree can be written as

$$k_i = \sum_{j=1}^{n} A_{ij} \tag{3.3}$$

The average degree $\overline{k}$ of an undirected graph is

$$\overline{k} = \frac{1}{n}\sum_{i=1}^{n} k_i \tag{3.4}$$

Every edge in an undirected graph has two ends, and assuming we have m edges in the graph, then we have $2m$ ending nodes, which is equal to the sum of the degrees for all the nodes

$$\sum_{i=1}^{n} k_i = 2m \tag{3.5}$$

so now joining the last two equations we have

$$\overline{k} = \frac{2m}{n} \tag{3.6}$$

which means that the average degree in a graph equals to two times the number of edges divided by the number of nodes.

3.3.5 Graph Density

The density ρ of a simple graph with n nodes is the fraction of edges that are actually present (m), in relation to the potential number of edges for such graph $\binom{n}{2} = \frac{n(n-1)}{2}$. Note that here we have simple, undirected links and no self-loops. Therefore, using Eq. 3.6 we have

$$\rho = \frac{m}{\binom{n}{2}} = \frac{2m}{n(n-1)} = \frac{\overline{k}}{n-1} \tag{3.7}$$

Given its definition, density is in the range $0 \leq \rho \leq 1$. A network keeping a constant density ρ as $n \to \infty$ is referred as *dense*, and this means that the average degree $\overline{k}$ is also growing. On the contrary, a network where $\rho \to 0$ as $n \to \infty$ is referred as *sparse*.

3.3.6 Paths

A path in a network is a sequence of nodes that are connected by links, i.e., it is a route across the network by using the actual links that connect its nodes. The *length* of the path is the number of links traversed by the path, which is different to the number of nodes (note that a path with length one contains one link and connect two nodes). In general paths can pass two times through the same node, or use a link more than once. Paths that do not intersect themselves are called *self-avoiding paths*.

A *geodesic path*, also called *shortest path*, is a path connecting two nodes such that no shorter path exists in such network. The length of such geodesic path is the shortest distance connecting those two nodes. Of course, a shortest path does not intersect itself, i.e., it has no loops. However, there can be more than one shortest path in a network for connecting two particular nodes, or even there can be no path connecting them. The *diameter of a network* is the length of the longest geodesic path connecting a pair of nodes in the network.

An Eulerian path in a network traverses each link in the network exactly once, while a Hamiltonian path visits each node exactly once. The number of Eulerian or Hamiltonian paths in a particular network can range from zero to several ones. These type of paths have several practical applications in computer science, from job sequencing, to garbage collection, or parallel programming [5].

3.3.7 Random Walks

A random walk is a path connecting a pair of nodes in a network, where the links have been selected randomly. Random walks may include self-loops, loops or may be selected avoiding them, and then called *self-avoiding random walks*.

3.3.8 Distances

The distance between two nodes of the network is the shortest path one has to travel to go from one node to the other. The average path-length is the average of all these distances between all pairs of nodes, and indicates the capacities of long distance information transmission.

Sometimes we want to know the number of paths in a network with a given length r. In order to calculate this we know that element A_{ij} in the adjacency matrix is 1 if there is a link between nodes i and j. If we take the product $A_{ik}A_{kj}$, we will have again 1, if there is a path connecting i and j via k with length 2, as we have used two links. If there is no path connecting i and j we have a zero in the product, and any of the terms is zero. Then, the total number $N_{ij}^{(2)}$ of paths in the network from i to j with length two, i.e., via other node k, is

$$N_{ij}^{(2)} = \sum_{k=1}^{n} A_{ik}A_{kj} = [A^2]_{ij} \tag{3.8}$$

which is the ij element of the matrix A multiplied by itself. If we consider paths with length three between i and j then we need to visit two intermediate nodes kl and the product $A_{ik}A_{kl}A_{lj}$ has a value 1 if there is such path or zero otherwise, so the total number of paths with length three is

$$N_{ij}^{(3)} = \sum_{k=1}^{n}\sum_{l=1}^{n} A_{ik}A_{kl}A_{lj} = [A^3]_{ij} \tag{3.9}$$

Generalizing to paths with length r, we have

$$N_{ij}^{(r)} = [A^r]_{ij} \tag{3.10}$$

Therefore, the shortest distance L_{ij} between nodes i and j is the smallest value of r such that $[A^r]_{ij} > 0$.

3.3.9 Components

When there is a path connecting any two nodes in a network we say that such network is *connected*. Otherwise we refer to such network as *disconnected*, meaning that it is broken in several subnetworks that are its *components*. Therefore, a component is a subset of nodes in the network, where there is at least one path from every node to any other of the subset, and where we can not add another node from the whole network and still preserve this property.

Intuitively, and from a graphical point of view, this means that we have the whole network divided into several graphs, without connections among them. The forest graph described before is a particular type of network composed by tree graphs, which are its components.

3.4 Metrics in Complex Networks

This section presents several metrics to measure the properties of a given network. They will help us to understand and recognize different types of complex network topologies presented in the next section.

3.4.1 Degree Centrality

The most basic measure of centrality in a network is the degree of a node, i.e., the number of links it has, and it is sometimes referred as *degree centrality*. This measure gives a rough indication of the social power of a node, based on how well it *connects* the network, and helps to determine what are the most significant nodes. Intuitively, it means that nodes with a higher number of links can have a bigger influence to propagate information to the rest of the network, or to get updated information from the other nodes.

3.4.2 Eigenvector Centrality

This is a natural extension of the degree centrality in order to determine the importance of a node in a network. It assigns relative scores to a node i in the network, based on the principle that having links to other high connected nodes contribute more to the score of the node i itself. Basically *eigenvector centrality* gives each node a score proportional to the sum of the scores of its neighbors. Intuitively, a person in a social network can be relevant because she knows many people, or because she knows a few people very well connected.

Mathematically, the eigenvector centrality x_i of a node i is proportional to the sum of centralities of its neighbors:

$$x_i = \frac{1}{\lambda} \sum_{j=1}^{n} A_{ij} x_j \tag{3.11}$$

where λ is the is the highest eigenvalue of matrix A. We could use the matrix notation to describe the equation $Ax = \lambda x$ referring to all the nodes n in the network. Google's PageRank is a variation of eigenvector centrality for directed networks.

3.4.3 Closeness Centrality

This measure determines how close is a node from the rest of the nodes in a network, and it reflects the ability to quickly spread information to other network members. Hence, the *closeness centrality* for a node i can be defined as the inverse of the sum of the shortest distances from i to the rest of the nodes j in the network. If we denote by d_{ij} the shortest distance from node i to node j, and we define $\overline{d}_i$ as this distance averaged over all the nodes j in the network, then we have

$$\overline{d}_i = \frac{1}{n}\sum_{j} d_{ij} \tag{3.12}$$

now we define the closeness centrality C_i^c for node i as the inverse of $\overline{d}_i$

$$C_i^c = \frac{n}{\sum_j d_{ij}} \tag{3.13}$$

This quantity takes lower values for nodes worse connected than others, while nodes well connected have higher C_i^c values; representing their ability to quickly propagate information. The value of C_i^c ranks in the interval $[1, log(n)]$. One of the drawbacks of closeness centrality is that small fluctuations in the network structure can change its values substantially, and make it difficult to distinguish between central and less central nodes in dynamic networks. Other centrality measures like degree or eigenvalue centralities do not suffer for this problem (see [8] for a detailed discussion).

3.4.4 Betweenness Centrality

The *betweenness centrality* measure gives higher values for nodes that are in between many shortest paths, reflecting how good is a node for connecting the network. Nodes with a high betweenness centrality have a relevant influence as information passes through them frequently, and if we remove them, the information-flow in the network could be compromised.

The betweenness for a node i can be computed as:

1. For a pair of nodes j, k find all the shortest paths σ_{jk} between them (there may be more than one) and determine its number of elements $g_{jk} = |\sigma_{jk}|$, that refers to the number of geodesic paths.
2. For that set σ_{jk} select σ^i_{jk} as the set of paths that pass through node i and let $g^i_{jk} = |\sigma^i_{jk}|$ be its number of elements.
3. From both sets, determine the fraction of paths passing through i as g^i_{jk}/g_{jk}.
4. Sum these fractions over all pair of nodes in the network.

This can be expressed mathematically as

$$C_i^b = \sum_{jk}^{n} \frac{g^i_{jk}}{g_{jk}} \tag{3.14}$$

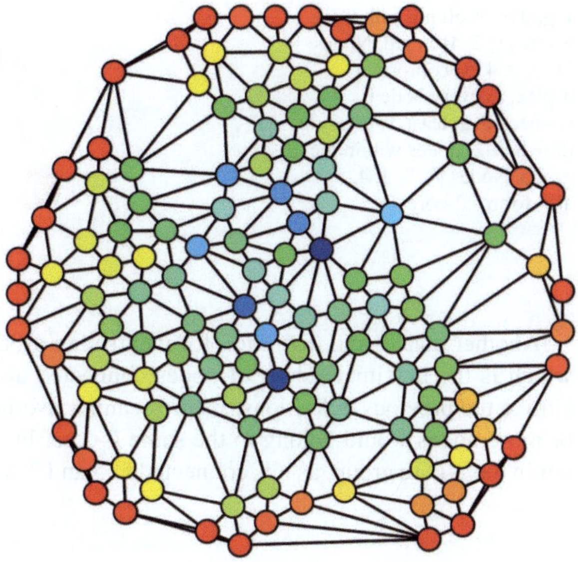

Fig. 3.4 Representation of node betweenness from minimal in red to maximal in blue (source Wikimedia Commons [13])

Sometimes this value is normalized in some network analysis tools dividing by the square of the number of nodes

$$C_i^b = \frac{1}{n^2} \sum_{jk}^{n} \frac{g_{jk}^i}{g_{jk}} \tag{3.15}$$

where the values of C_i^b lie strictly between zero and one, meaning low and high connectivity, respectively (see Fig. 3.4).

3.4.5 Groups: Cliques, Plexes and Cores

A clique is a maximal subset of nodes in an undirected network such that every pair of nodes in the clique are connected by a link. Defining this subset as maximal we mean that we cannot add any other node preserving the property that all the nodes keep directly connected. A clique indicates the occurrence of a cohesive group within a network (see Fig. 3.5).

The requirement that every pair of nodes in a clique must be connected can be very strict, and it can be relaxed using *plexes*. A *k-plex* with size m is a group of m nodes in a network connected to $m - k$ nodes in the group (see an example in Fig. 3.5). With $k = 1$ we have a clique, but increasing the value of k we relax the number of connections. Cliques and plexus can overlap, i.e., a node can be part of two or more groups at the same time.

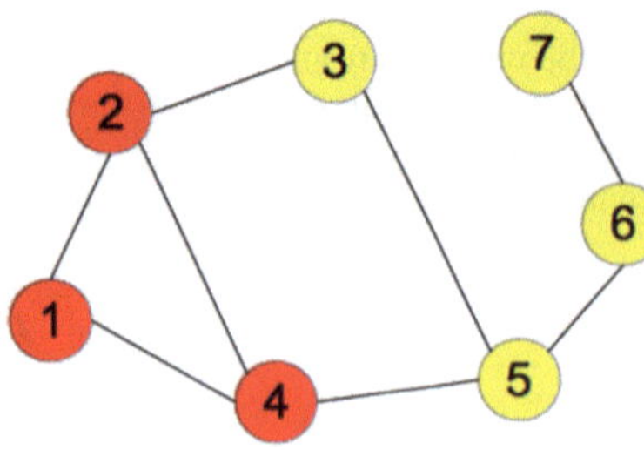

Fig. 3.5 A clique with three nodes {1, 2, 4} in red. Nodes {1, 2, 3, 4, 5} conform a 3-plex, as every node is connected, at least, to $(5 - 3 = 2)$ nodes within the group. Nodes {1, 2, 3, 4, 5} also form a 2-core

Another conceptual definition for identifying groups in a network is the *k-core*, which is the maximal subset of nodes connected at least to other k in the group. Unlike the previous definitions, k-cores cannot overlap, as two k-cores sharing one or more nodes would belong to the same k-core. In Fig. 3.5 observe that node 6 is not in the 2-core group, as it is connected to 5 and 7, and the last one is not in the set.

3.4.6 *Transitivity*

The *transitivity* property in complex networks refers that if an element a is related $(\circ)$ to another element b denoted as $(a \circ b)$ and b is related to c as $(b \circ c)$ then we have $(a \circ c)$. We can find examples of such property in relations like equality, ordering, etc. In networks we can also model this relation using nodes and the links that connect them, so if a node i is connected with a node j and j is connected with k, then i must also connected to k. If a network holds this property for all its nodes, then it is called a *transitive network*.

3.4.7 *Clustering Coefficient*

Cliques are a type of networks where we can find perfect transitivity, but we can also consider *partial transitivity*. If we have that a node i is linked to node j and j is linked to node k, then if i is linked with k, then we have a cycle among the three nodes denoted as a *closed triad*, i.e., the transitivity property fulfilled by a path of length two that is closed. We define the *clustering coefficient* as the fraction of closed paths of length two P_2^c divided by the total number of paths with length two P_2:

$$cc = \frac{P_2^c}{P_2} \tag{3.16}$$

where $cc \in [0, 1]$. If $cc = 1$ then we have perfect transitivity, but if $cc = 0$ there is no closed triads (p.e., a tree graph has no loops).

There is a local version for the clustering coefficient, which is equivalent to the cc definition, but centered around the paths where a node i acts as the origin cc_i. The global and the local version of the cc reflect the network or node capacities, respectively, for local information transmission. In general, a higher clustering coefficient indicates a higher *cliquishness*.

3.4.8 Degree Distributions

We have previously defined the degree of a node as the number of links with its neighbors. We now define p_k as the fraction of nodes in a network with degree k. The different values that we obtain for p_k as a function of the degree k compose the *degree distribution* of the network. Therefore, this value p_k can also be seen as the probability of randomly choosing a node i with a certain degree k. This can be used also to determine the number of nodes in a network with a certain degree as

$$n_k = n.p_k, \tag{3.17}$$

where n is the whole number of nodes in the network.

It is usual to represent graphically the degree distribution of a network as a function of k (see Fig. 3.6). In real systems, these plots usually present many nodes of the network with a low connection degree, while there is a very long tail, where some nodes get a very high degree. This high degree nodes show a very good potential for connecting the network, and are called *hubs*. Most of real-world networks have degree distributions with very long tails and high-degree hubs.

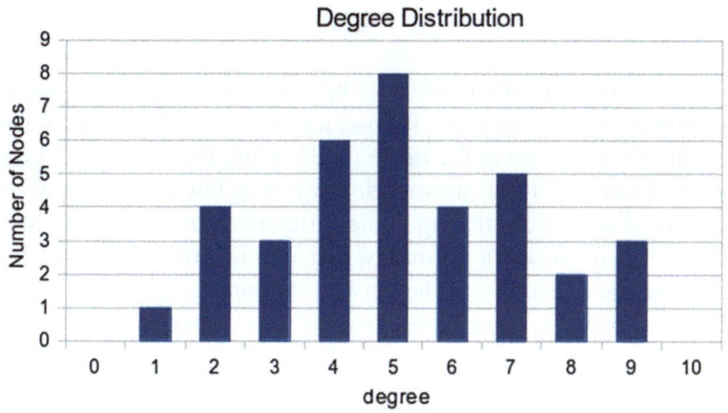

Fig. 3.6 A possible degree distribution

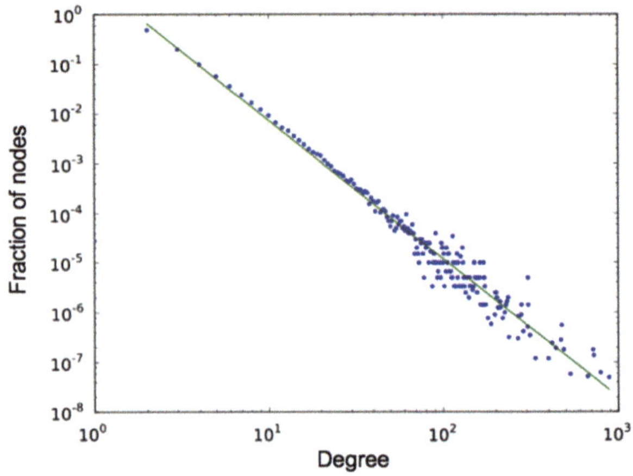

Fig. 3.7 A power law degree distribution, using logarithms in both axis (source Wikimedia Commons [12])

3.4.9 *Power Laws*

When plotting some real degree distributions, e.g., Internet connections, using logarithmic scales for both axis (the degree and the number of nodes), the distribution gets closer to a straight line (see Fig. 3.7). This distribution can be approximated using the logarithm of the degree distribution p_k as a linear function of the degree k as

$$ln(p_k) = -\gamma.ln(k) + c \tag{3.18}$$

where γ and c are positive constants, and the minus sign represents the negative slope. Taking exponentials in both sides we can write the equation as

$$p_k = C.k^{-\gamma} \tag{3.19}$$

where $C = e^c$ is another constant. When we have a network that follows a distribution varying as a power of k we say that it follows a *power law*. The most common values for γ in real distributions are in the range [2, 3], while the constant C is commonly used for normalization. In real networks, the power law is usually approximated in the tail of the distribution, having some differences in the initial values for the degree k. Networks following a power law distribution are usually called scale-free networks to refer the capability of the network to appropriately scale its properties (p.e., the Internet) because power laws have the same functional form at all scales.

3.5 Relevant Topologies in Complex Networks

In this section we present some network topologies that we will use along this book.

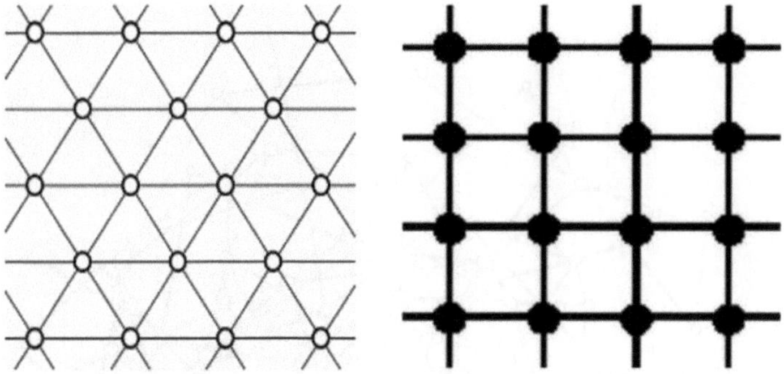

Fig. 3.8 Trigonal (left) and square (right) lattices

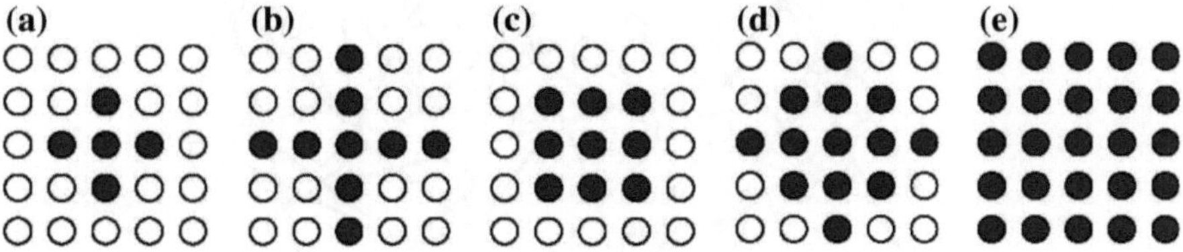

Fig. 3.9 Five possible spatial neighborhoods. Among them, von Neumann's neighborhood in (**a**) and Moore's neighborhood in (**c**) are the most popular

3.5.1 Regular Networks

A *regular network* has all its nodes with the same degree k, and it is sometimes denoted as *k-regular*. For instance, a triangular lattice has degree 3, so it is *3-regular*, while a square lattice is *4-regular* (see Fig. 3.8). Observe that a triangular lattice has perfect transitivity (i.e., its clustering coefficient $cc = 1$), while this does not happen with a square lattice ($cc = 0$).

Figure 3.9 shows some possible neighborhoods for a spatial two-dimensional network. In this spatial networks there is some concept of locality, as nodes are linked with nearer ones, considering a geometrical distance.[1] Besides, in finite lattices, the nodes in the external frontier may have a special neighborhood, depending on its position. In this book, we will consider that they are fully connected creating a torus.

3.5.2 Random Networks

Random Networks (RN) are randomly generated by placing a fixed number of links between nodes with a certain probability (see Fig. 3.10). The simplest and most

[1] In this book we use the Cartesian distance to determinate how far is a node from another in a spatial network. However, the Manhattan distance is also popular in square lattices, and counts the number of horizontal and vertical segments between two nodes.

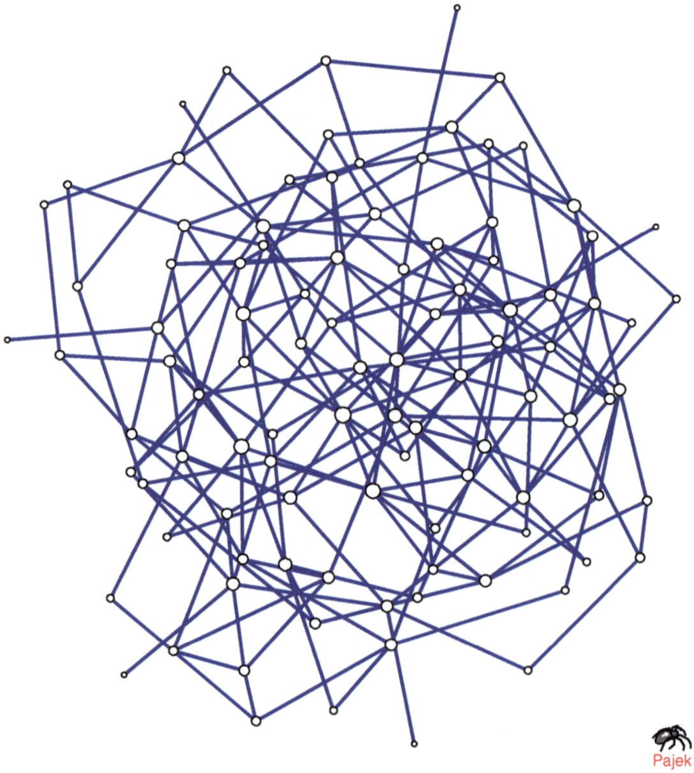

Fig. 3.10 Random network with 100 nodes

common reference for a random network model is the Erdös-Rényi random network (ER) [6]. The resulting random networks from this model exhibits an average shortest path length L that grows proportionally to the logarithm of the number of nodes n in the network ($L \propto \log n$). This means that the average maximum distance between any two nodes in the graph is short.

The simplest ER generative model is the $RN(p, n)$ (usually denoted as R_n^p), where n is the number of nodes, and each link is included in the graph with probability p, independent from every other link of the network. The probability p can be considered as a weighting function; and increasing p from 0 to 1 we obtain networks more and more connected.

Percolation theory studies the robustness of random networks. Given a random network with n nodes and an average degree k, if we remove randomly a fraction of $(1 - p)$ nodes, then there exists a critical percolation threshold $p_c = \frac{1}{k}$, where for lower values the network becomes fragmented in several subnetworks, while for above values the network still remains connected.

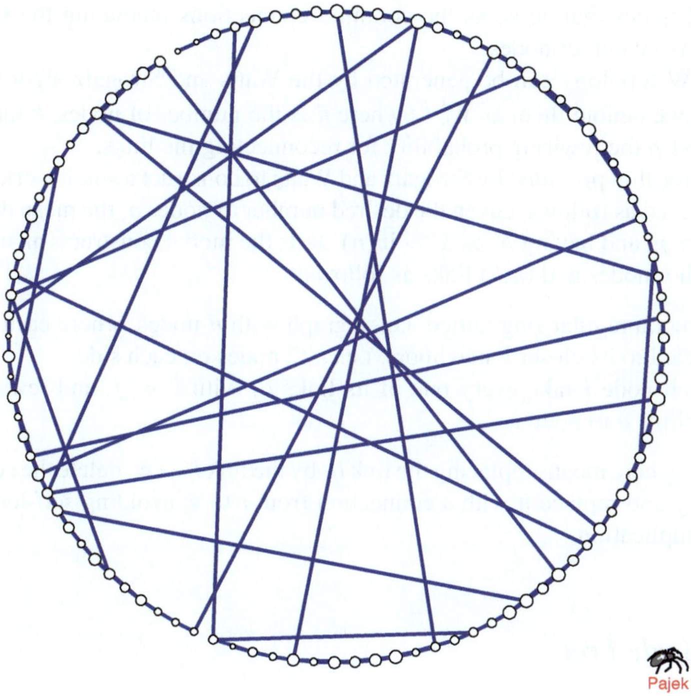

Fig. 3.11 Small-world network with 100 nodes

3.5.3 Small World

In a *small-world* (SW) network all nodes are not direct neighbors, but most of them can be reached from every other by a small number of links (see Fig. 3.11). These networks present the small-world phenomenon, in which nodes have small neighborhoods, but yet it is possible to reach any other node in a small number of hops.[2] This type of networks are also highly-clustered (i.e., have a high clustering coefficient). In 1998, Duncan J. Watts and Steven Strogatz published the first small-world network model [10]. They show that adding a small number of long-range links to a regular graph, where the diameter is proportional to the size of the network, we can obtain a small world network in which the average shortest path length L grows proportionally to the logarithm of the number of nodes n in the network ($L \propto \log n$), while the clustering coefficient stays large.

A wide variety of real world networks such as the World Wide Web, gene networks, social networks, electric power grids, brain neuron connections, and the metabolic network also exhibit these properties. Typically there is an over-abundance of well

[2]The small-world phenomenon is popularly known in as the *six degrees of separation*, which is based on the idea that the average diameter of a whole social network is shorter than six.

connected nodes that serve as the common connections mediating the short path lengths between other nodes.

This SW topology can be generated by the Watts and Strogatz algorithm [10]. Formally, we denote them as $W_n^{k;p}$, where n is the number of nodes, k the average degree, and p the rewiring probability for reconnecting the links.

The algorithm provided by Strogatz and Watts to construct a small world network can be defined as follows. Given the desired number of nodes n, the mean degree k, a probability p, and having $N \gg k \gg \ln(n) \gg 1$; the model constructs an undirected graph with n nodes and $(n.k)$ links as follows:

1. Construct a regular ring lattice, i.e., a graph with n nodes, where each node i is connected to its closer k neighbors, i.e., $k/2$ nodes on each side.
2. For each node i take every one of its links l_{ij} with $i < j$, and rewire it with probability p to node v.

Rewiring here means replacing the link l_{ij} by the link l_{iv}, i.e., delete the connection from i to j and replace it with a connection from i to v, avoiding self-loops $v \neq i$ and link duplication.

3.5.4 Scale Free

Scale-free (SF) networks follow a power law concerning degree distribution, at least asymptotically. These networks are characterized by having a few nodes acting as highly-connected hubs, while the rest of them have a low connectivity degree (see Fig. 3.12). Scale-free networks are low-clustered, as a fundamental difference with small-world ones. Formally we denote them as $S_n^{k;-\gamma}$, where n is again the number of nodes, k the average degree, and the probability p_k that a node in the network connects with k other nodes is roughly proportional to $k^{-\gamma}$. Scale-free networks show characteristics present in many real world networks, like the presence of hubs connecting almost disconnected sub-networks. Their degree distributions are heavy tailed, and follow a power law (see Sect. 3.4.9). Among these type of networks we may cite the Internet, the World Wide Web, email network, protein interactions, etc. Cohen and Havlin [3] showed analytically that scale-free networks are ultra-small worlds; and, due to hubs, the shortest paths become significantly smaller and scale as [2]:

$$L \propto \frac{\log(n)}{\log(\log(n))} \tag{3.20}$$

Derek de Solla Price showed in 1965 that the number of papers citations have a heavy-tailed distribution following a power law, but he did not use the term scale-free. Later, in 1976, Price also proposed a mechanism to explain the occurrence of power laws in such citation networks, which he called *cumulative advantage*. Recent interest in scale-free networks started in 1999 with work by Albert-László Barabási

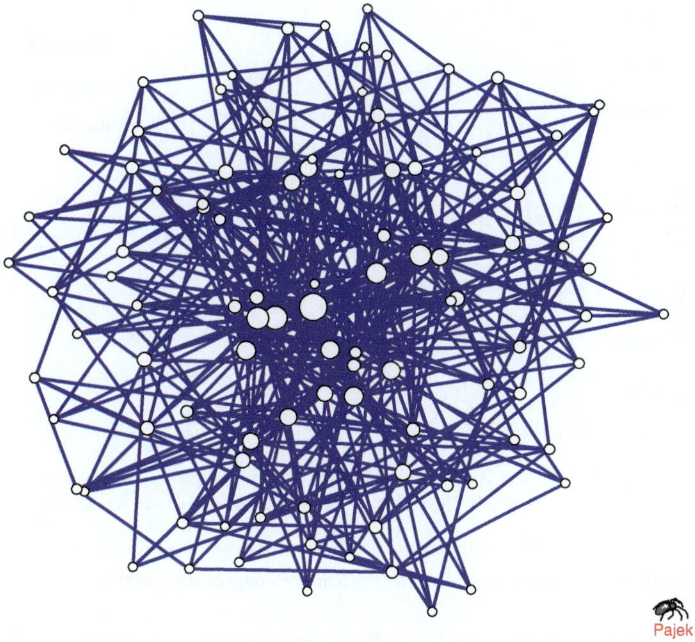

Fig. 3.12 Scale-free network with 100 nodes and some hubs

and collaborators, who mapped the topology of a portion of the World Wide Web [1], finding that some nodes, which they called *hubs*, had many more connections than others, and that the network as a whole had a power-law distribution on the number of links.

After finding that a few other networks, including some social and biological networks, also had heavy-tailed degree distributions; Barabási and collaborators coined the term "scale-free network" to describe the class of networks that exhibit such power-law degree distribution. Barabási and Albert proposed a generative mechanism to explain the appearance of power-law distributions, which they called *preferential attachment*, which corresponds to the cumulative advantage proposed by Price, and reflects the dynamical aspect of those networks. The preferential attachment method works as follows:

1. The network begins with an initial connected network of n_0 nodes.
2. New nodes v are added to the network one at a time, and each new node is connected to $n \leq n_0$ existing nodes with a probability proportional to the number of links that the already existing nodes have.

Mathematically, the probability p_i that the new node v is connected to a pre-existing node i is

$$p_i = \frac{k_i}{\sum_j k_j} \tag{3.21}$$

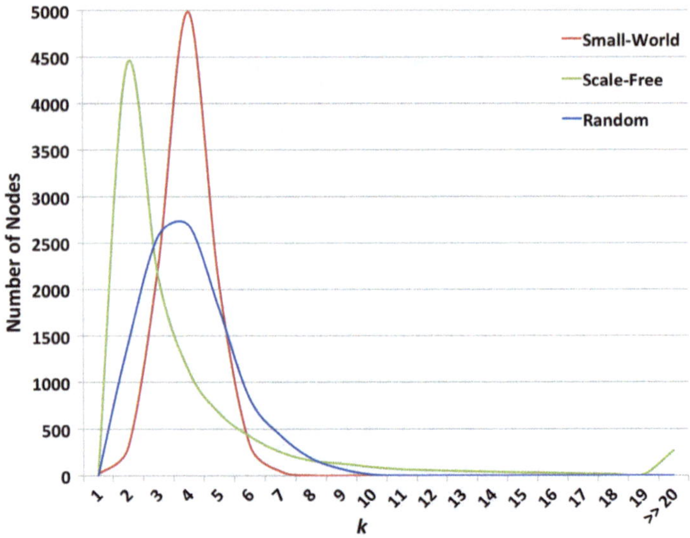

Fig. 3.13 Small-world versus Scale-free versus Random degree distributions

where k_i is the degree of node i and the sum is made over all nodes j. With this method, hubs tend to collect more links than less connected nodes, i.e., new network nodes have a *preference to attach* to the already heavily linked nodes.

The scale-free property has a strong correlation with the network's robustness to failure studied by Percolation (also mentioned above for random networks). Given the heavy-tailed distributions of node degrees, we obtain a heterogeneity of node connections in the network, that allows for a fault tolerant behavior. In general, scale-free networks keep their properties for information transmission if we randomly remove a few number of links, or even completely disconnect some nodes. Only if a few major hubs are disconnected, then the network is turned into a set of rather isolated graphs. Thus, in scale free networks hubs are both a strength, as they produce the interesting properties showed by these networks, and a weakness, as they are critical points.

Figure 3.13 provides a comparison of the degree distribution among a small world, a scale free and a random network. In the horizontal axis appear the degree connectivity for the nodes (note that the last point of the axis refers to sum of all nodes where $k > 20$). In the vertical axis we have the number of nodes with a certain degree connectivity. The figure has been generated from three networks having 10.000 nodes each, and an average connectivity degree of 4 links per node. Observe how most of the nodes in the scale free network have a very low connectivity $k \in [2, 4]$, but still there are a significative number of hubs with very high connectivity $k > 20$. The small-world network has a higher frequency of nodes around $k = 4$ (i.e., the average degree connectivity), but most of the nodes are in the interval $k \in [2, 7]$. Finally, the node degree of the random network is more dispersed than the small-world one, but it has not such a long tail that appears in the scale-free network.

3.6 Conclusion

Nowadays approaches to complex systems usually consider a collection of entities interacting in some specified way. To model such system interactions, we need to specify its topology (who interacts with whom) and the system dynamics (how the individual entities interact). Most complex systems have complicated non-regular topologies, that require a complex network framework for their representation.

This chapter has introduced complex networks as a model to describe complex system topologies. First, we have introduced basic concepts from graph and network theories. Then we have described what a complex network is, and how it can be used to perform social analysis. Next, we have proceeded to introduce some mathematics to analyze complex networks, and they allow us to calculate some basic metrics to describe relevant network properties. Finally, we took a look on some popular complex networks that can be found in multiple complex systems, among them small world and scale free networks; which have attracted much interests along last years, and will be used extensively along this book.

3.7 Further Reading

Several books have appeared along the last decade to introduce complex networks. Among them, the book *Networks: An Introduction* [8] from Newman is a very readable and complete introduction. The book *Six Degrees: The Science of a Connected Age* [11] from Watts is oriented for a more general audience. Finally, the book *Complex Networks: Structure, Stability and Function* [4] from Cohen and Havlin is shorter and more selective.

References

1. Barabási, A.L., Albert, R.: Emergence of scaling in random networks. Science **286**(5439), 509–512 (1999)
2. Chen, F., Chen, Z., Wang, X., Yuan, Z.: The average path length of scale free networks. Commun. Nonlinear Sci. Numer. Simul. **13**(7), 1405–1410 (2008)
3. Cohen, R., Havlin, S.: Scale-free networks are ultrasmall. Phys. Rev. Lett. **90**(5), 058701 (2003)
4. Cohen, R., Havlin, S.: Complex Networks: Structure, Stability and Function. Cambridge University Press, Cambridge (2010)
5. Cormen, T.H., Leiserson, C.E., Rivest, R.L., Stein, C.: Introduction to Algorithms, 2nd edn. MIT Press, Cambridge (2001)
6. Erdös, P., Rényi, A.: On random graphs. Publ. Math. **6**, 290–297 (1959)
7. Faloutsos, M., Faloutsos, P., Faloutsos, C.: On power-law relationships of the Internet topology. Comput. Commun. Rev. **29**, 251 (1999)
8. Newman, M.E.J.: Networks: An Introduction. Oxford University Press, Oxford (2010)
9. Shields, R.: Cultural topology: the seven bridges of Konigsburg, 1736. Theory Cult. Soc. **29**(4–5), 43–57 (2012)

10. Watts, D.J., Strogatz, S.H.: Collective dynamics of small-world networks. Nature **393**(6684), 440–442 (1998)
11. Watts, D.J.: Six Degrees: The Science of a Connected Age. Norton, New York (2003)
12. Wikimedia Commons by Arpad Horvath - Own work, CC BY-SA 3.0. https://commons.wikimedia.org/wiki/File:Barabasi-albert_model_degree_distribution.svg
13. Wikimedia Commons by Claudio Rocchini - Own work, CC BY-SA 2.5. https://commons.wikimedia.org/wiki/File:Graph_betweenness.svg

Chapter 4
Cellular Automata

Juan C. Burguillo

A basic Cellular Automata (CA) is a regular grid of cells (a lattice), each one having a finite number of states [11] (i.e., a finite state machine). Every cell, also denoted as *cellular automaton*, has a defined neighborhood to interact with. Time is discrete, and in every iteration any cell interacts with its neighborhood to find its new state depending on its own state and its neighbors' state. CAs are simulated by a finite grid, which can be a line in one dimension, a rectangle in 2D or a cube in 3D. In the cellular space, cells go from their current state to a next state according to their *local rule*, that describes how every state machine evolve depending on its own state and its neighbors' state. All the cells use the same local rule, and they change synchronously from their current state to the next one.

Scientist have found that simple rules allow complex behaviors to emerge; and that is what makes cellular automata so interesting. Due to that, CAs have been used as models for describing physical and computational phenomena [6] with many applications in physics (like particle simulations), computational theory (like image processing or pattern classification), mathematics (like random number generation, or cryptography), complexity theory, urban traffic, biology, bioinformatics, etc.

4.1 Short Historical Notes

The CA model was originally introduced in the 1940 s by Stanislaw Ulam and John von Neumann, both working at Los Alamos National Laboratory. Ulam was studying

---

J. C. Burguillo (✉)
Department of Telematics Engineering, School of Telecommunications Engineering, University of Vigo, 36310 Vigo, Spain
e-mail: J.C.Burguillo@uvigo.es

J. C. Burguillo, *Self-organizing Coalitions for Managing Complexity*, Emergence, Complexity and Computation 29, https://doi.org/10.1007/978-3-319-69898-4_4

the growth of crystals using lattice networks, while, at the same time, von Neumann was interested in self-replicating systems.

Von Neumann's initial ideas considered 3D factories described by partial differential equations. Soon, he discarded that model and started to consider a kinematic model [10], with a set of robots, each one building another like a toy set. But, along the design process, von Neumann realized about the big challenge it was to build a self-replicating robot.

Then, his colleague Ulam suggested to use a 2D discrete system, like the CA, for creating a simpler model of self-replication. Von Neumann considered the use of 2D cellular automata, similar to Ulam's lattice networks, with algorithmic self-replication. The result was a universal copier and constructor machine within a cellular automaton, with a small neighborhood composed by the four nearest orthogonal cells (von Neumann's neighborhood), and with 29 states per cell [14]. Von Neumann also designed a 200,000 cell initial configuration, and provided an existence proof to show that such particular pattern would be able to self reproduce within the given cellular universe.

In the 1960s, cellular automata started to be studied as a particular type of dynamical systems, and in 1968 Codd summarizes in [3] his research into the requirements for universal computation and construction, showing that there is no cellular space with 2 states and a local rule, using the von Neumann neighborhood, that exhibits universality. He then proceeds to develop a cellular space with simpler rules than von Neumann's space, and he showed how universal computation and construction can be performed in his simpler space. In 1969, computer scientist Konrad Zuse published his book Calculating Space [16] (*Rechnender Raum* in german), proposing that the physical laws of the universe are discrete by nature, and that the entire universe is the output of a deterministic computation on a single cellular automaton. Such theory became the foundation of digital physics.

In 1970 Conway released the most popular CA, the *Game of Life*, which was published by Martin Gardner at Scientific American [5]. Life is a 2D Cellular Automata game, where the neighborhood of a cell is composed by the cell itself, and the 8 ones surrounding it (known as Moore's neighborhood, see Fig. 4.2). Cells, in Life, can be dead or alive (2 states) and evolve using 3 very simple rules:

1. *Birth*: a dead cell gets back to live if it has 3 neighbors.
2. *Death*: an alive cell dies if it has less than 2 cells around (isolation) or more than 3 (overpopulation).
3. *Survive*: alive cells stay alive with 2 or 3 cells around.

Life is one of the simplest CA to exhibit universal computation and universal construction, meaning that with an initial appropriate distribution of cells, Life becomes a Turing Machine [1, 2, 8]. The rules for Conway's Game of Life only consider the number of neighbors that are in a certain state, but their positions are not taken into account. While in von Neumann's and Codd's spaces many rules consider only neighbors in certain positions and in a certain state, i.e., they have a directional perspective; in the case of Conway's rules they are completely symmetric. An this is an

Fig. 4.1 *Conus textile* exhibits an emergent cellular automaton pattern on its shell (source Wikimedia Commons [12])

important property meaning that Conway's rules are much simpler than Codd's and von Neumann's.

Moving forward in time, in the 80s Stephen Wolfram systematically investigates a very basic, but essentially forgotten class of cellular automata, which he denoted as elementary cellular automata [15]. Giving the emerging computer power at that time, he considered to do an exhaustive survey over cellular automata. But, given that there are too many 2D cellular spaces, he looked at the simplest possible one: the 1D automata with only two states and a three cell neighborhood. From his experiments, he defined four classes where csellular automata can be divided depending on their behavior. The unexpected complexity in the behavior of these simple models led Wolfram to consider if complexity in Nature may be due to similar mechanisms. Figure 4.1 presents a cone shell exhibiting an emergent cellular automaton pattern.

Around the time that Wolfram started to work on his elementary cellular automata, Christopher Langton, building over Codd's work, designed another self-replicating structure, the *Langton Loop*. Langton realized that such new self-replicating structure had several properties of living organisms (p.e., like Autopoiesis[1]), and he coined the term *Artificial Life* to describe such structure. Langton considered that there was a closer relation between his work and Wolfram's classification, and he also postulated that the most interesting rule-sets would be found on the boundaries between Wolfram classes. This reflection about the space between those classes, after a series

[1]From Greek *auto*, meaning self, and *poiesis*, meaning creation or production, refers to a system capable of reproducing and maintaining by itself.

of experiments performed by Langton and Packard, became popular as the *Edge of Chaos*. Melanie Mitchell [7], who worked extensively on the evolution of rule sets for cellular spaces, argue that the interpretation of the original experiments performed by Langton and Packard was not correct, and that the edge of chaos concept would need a more rigorous and useful definition.

In 2002 Wolfram published the book titled *A New Kind of Science* [14], which extensively argues that the discoveries about cellular automata are not only mathematical or isolated facts, but are robust and have significance for all disciplines of Science.

4.2 Basic Notation

Some basic notations popular for cellular automata are introduced in this section, in order to clarify some of the concepts that will be presented later.

- *Cell*: is a single element of a cellular space, i.e., the smallest unit of such space.
- *Cellular Space*: is lattice space made up of cells.
- *Cellular Automaton*: is a structure built in a cellular space, i.e., an automaton composed by its cells.
- *Cellular Automata*: is the plural of Cellular Automaton, and nowadays commonly used instead of the singular form.
- *Local Rule*: is the rule governing the transition between states in the cell's finite state machine. The term local refers to the fact that it only uses its local neighborhood as input.
- *Neighborhood*: refers to a set of cells surrounding a particular one, that is used as input to determine its next state. There are multiple choices of neighborhood alternatives, and the one selected clearly influences the behavior of the automata.
- *Configuration*: is a snapshot of the cellular automata at a particular point in time. The initial configuration of the cellular automata is very relevant, as it is the starting point of a simulation, and it usually has influence over the final results.
- *Generation*: a generation is every synchronous step in the evolution of a cellular automata, and it can be described by an intermediate configuration. Therefore, time in the CA is discrete, and evolves as a sequence of generations.
- *Universal Computation*: this concept describes if the cellular space configurations are able to compute any possible function. This property has been usually proved by showing that a cellular automata local rule, and its initial configuration, is able to build a Turing Machine.
- *Universal Construction*: this concept describes that the CA is able to create any type of construction in the cellular space, for instance, the machine built by von Neumann was able to construct copies of itself, i.e., to self-replicate.

4.3 Basic Cellular Automata Definition

A d-dimensional cellular automata (or d-CA), A, is a 4-tuplet $(\mathbb{Z}^d, S, N, f)$, where:

- $\mathbb{Z}^d$ is the finite or infinite d-dimensional lattice,
- S is a finite set of cell states for A,
- N is a finite cell subset of $\mathbb{Z}^d$, $N = \{c_j \mid c_j = (x_{1j}, \ldots, x_{dj}), \quad j \in \{1, \ldots, n\}\}$, and denotes the neighborhood of A,
- $f : (S, N) \rightarrow S$ is the local transition function, or local rule, of A.

4.4 Types of Neighborhoods

Cells in 2D are usually represented by squares, but triangles and hexagons can also be used. In these domains neighborhoods are usually based on the Euclidean distance (see Fig. 4.2).

The designer of the CA model has to define the behavior of the cells in the lattice frontier and, to solve this, several alternatives have been considered:

- *Open frontier*: the cells outside the lattice have a constant state.
- *Periodic frontier*: the frontiers of the lattice are in contact, i.e., a circumference in 1D, a toroidal shape in 2D, etc.
- *Reflection frontier*: outside cells take the value of their internal mirror cells.
- *No frontier*: the frontier expands when needed.

As said before, usually 1D (as in the case of Wolfram's elementary cellular automata) and 2D (as in the case of Conway's Game of Life) constructions have been used in practice, but there can be neighborhoods and local rules to model the interaction on 3D or further dimensions. Nevertheless, it is not strictly necessary to restrict to lattices and Euclidean distances; as other conventions can be considered as stochastic distance functions, complex networks, etc. In this book we shall explore interactions among cells from this open perspective.

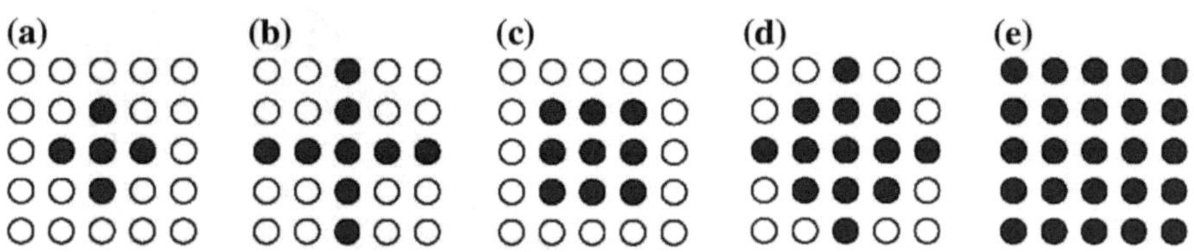

Fig. 4.2 Five possible neighborhoods used in CA. Among them, we can see von Neumann's neighborhood in (**a**) and Moore's neighborhood in (**c**)

4.5 Cellular Automata Classification

Wolfram defined four classes into which cellular automata, and several other simple computational models, can be divided depending on their behavior. These four classes are:

1. *Class 1 (Stable)*: patterns evolve into homogeneous stable final configurations.
2. *Class 2 (Oscillating)*: patterns evolve into oscillating structures.
3. *Class 3 (Chaos)*: patterns evolve in a pseudo-random or chaotic manner.
4. *Class 4 (Complexity)*: patterns evolve into stable and repetitive structures, that interact in very complex ways, and do not belong to any of the previous three categories.

As an example of an elementary cellular automata investigated by Wolfram, consider a one-dimensional CA, with two possible states per cell, and two adjacent cells behaving as neighbors. Then, a cell and its two neighbors form a neighborhood of 3 cells, so there are $2^3 = 8$ possible patterns for a neighborhood (see table in Fig. 4.3). In this model, a rule consists of deciding, for each pattern, whether the center cell will be a 1 or a 0 in the next generation. Thus, as can be clearly seen in the figure, there are then $2^8 = 256$ possible rules. These 256 CAs are generally denoted by their Wolfram code, a naming convention proposed by Wolfram that gives each rule a number from 0 to 255. Among them, rule 110 is particularly interesting. Figure 4.3 shows the time evolution of rule 110, when the starting configuration of the one-dimensional CA consists of a 1 (at the top of the image) surrounded by 0's. In the figure, each row of pixels represents a generation in the history of the automaton, being $t = 0$ the top row, and each pixel is colored white for 0 and black for 1. Rule 110 exhibits class 4

current pattern	111	110	101	100	011	010	001	000
new middle state	0	1	1	0	1	1	1	0

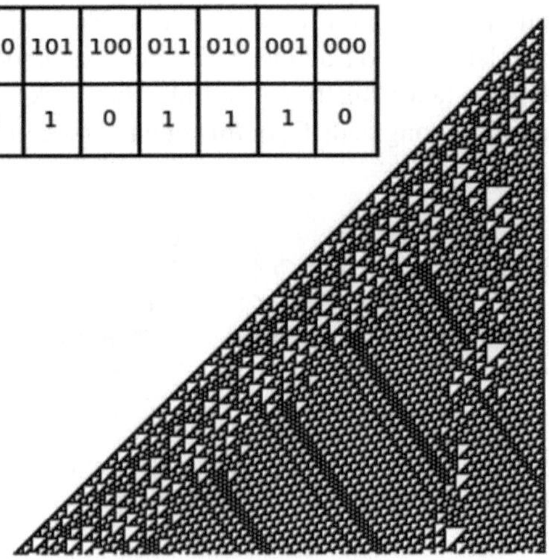

Fig. 4.3 One-dimensional CA evolution using rule 110

current pattern	111	110	101	100	011	010	001	000
new middle state	0	0	0	1	1	1	1	0

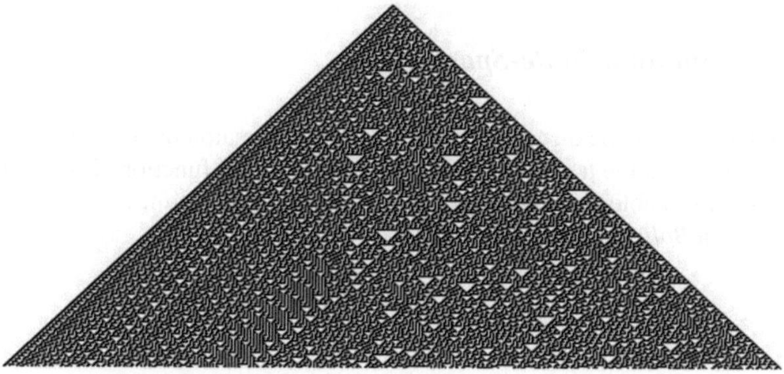

Fig. 4.4 One-dimensional CA evolution using rule 30

complex behavior, and Wolfram has conjectured that many, if not all, class 4 cellular automata are capable of universal computation. This has been proven for Conway's Game of Life, and for rule 110 by Matthew Cook [4], a Wolfram's research assistant.

Figure 4.4 shows the time evolution of rule 30, when the starting configuration of the one-dimensional CA consists of a 1 (at the top of the image) surrounded by 0's.

4.6 Extended Cellular Automata

There have been several proposals to modify the initial model, that lead to different types of automata, and here are some of them.

4.6.1 Asynchronous

In very large spaces, specially in hardware implementations, the synchronous operation of the classical CA model can limit the size of spaces that can be simulated efficiently. One approach considered to solve this problem was the use of a fully asynchronous model with a set of equivalent rules to be used under synchronous or asynchronous updating. In general, most local rules behave differently when the updating strategy is changed, and emergent behavior cannot appear if cells are updated asynchronously [9]. Another variation of this approach is the use of asynchronous block operation, where the whole cellular space is divided into sub-spaces, and cells within the same block are updated synchronously, but communication among blocks is asynchronous. This solution potentially allows much larger cellular spaces, but keeping self-organization and emergent properties at a certain level.

4.6.2 Continuous State-Space

This model replaces the discrete states within the automaton by a continuous space representation, and the transition rule table by a transition function. The result is an interesting model able to simulate gases and fluids, and sometimes denoted as a *Gas Lattice* or as a *Boltzmann Lattice* [13].

4.6.3 Non-homogeneous

In this case, the set of rules is not homogeneous for the whole set of cells, and there are several types of hybrid cellular automata depending if there are different rules for every cell, for different regions (p.e., cells at the boundaries of the cellular space could have different rules), or if the rules are determined depending on inputs external to the cellular space.

4.6.4 Stochastic

Researchers have used non-deterministic local rules when the cellular automata models entities with a non completely deterministic behavior. In this case, the next state depends not only on the neighborhood, but also on some probability distribution.

4.6.5 Memory-Based

Up to now we have only considered that the present state of the CA influences the next state of the cell, but it could be considered that the past has also some influence over the next cell state.

4.6.6 *Mobile*

We could consider that the lattice could be occupied only at some positions by cells, while the rest of them are empty, and available to be occupied by new cells. Under this framework, cells can move from one place to another along the lattice.

4.6.7 *Dynamic Lattices*

In the classical CA, the underlying lattice is a static and passive background, where cells evolve. But we also can consider the possibility to provide a more active role to the lattice, evolving depending on cell states and interacting with them.

4.6.8 *Nested and Hierarchical*

This concept refers to the multiple scale properties inherent to complex systems. As complexity can be found at multiple levels in a given entity depending on the scale one is looking at. For instance, a society is a complex system composed by individuals, which at the same time are composed by complex system: organs, cellular interactions, and so on. Therefore depending on the scale we look at, we can find complex interactions following different rules, and some kind of hierarchical systems, where entities have several organization levels, and are usually specialized.

4.7 Conclusion

Cellular automata are a class of systems that mainly posses the next characteristics: a discrete lattice of homogeneous cells, taken a finite number of discrete states, interacting with its local neighborhoods and using discrete dynamics to update each cell state taking into account also the states of its neighbors. Basically, they have been traditionally used for [6]: parallel computation, to simulate discrete dynamical systems, to study pattern formation, to model fundamental physics and, of course, to study complexity.

This chapter has briefly introduced the basic concepts of cellular automata. First, we introduced some basic notation in order to define precisely what a cellular automata is, and how cells interact with their local neighborhood. Then, we have considered a popular classification for cellular automata, that depends on the results obtained after a system is run. Finally, we have introduced other types of cellular automata that extend and go beyond the approach considered for the traditional model.

As it happened with other areas studied by John von Neumann during the 1940s, his original concerns about self-replication systems derived into a new and relatively successful field of theoretical research. There are a huge amount of papers and interesting books about cellular automata, from computer science to philosophy. In this book we shall explore some extended models rooted on cellular automata ideas.

4.8 Further Reading

Two books can be considered good and extensive introductions to cellular automata. First, the book *Cellular Automata: a Discrete Universe* from Andrew Ilachinski [6] is a very complete introduction, and relates cellular automata with many other scientific disciplines. The book *A New Kind of Science* from Stephen Wolfram [14] has a first part that describe the basics of the field, while the second part of the book is mainly oriented to discuss the author's research and perspective about how other disciplines may take advantage of the potential of cellular automata for modeling and describing the world around us.

References

1. Adamatzky, A. (ed.): Game of life cellular automata, vol. 1. Springer, Berlin (2010)
2. Berlekamp, C, Conway, J.H., Guy, R.K.: Winning Ways for Your Mathematical Plays, vol. 2. Academic Press.(1982)
3. Codd, E.F.: Cellular Automata. ACM Monograph Series. Academic Press Inc., New York and London (1968)
4. Cook, M.: Universality in elementary cellular automata. Complex Syst. **15**, 1–40 (2004)
5. Gardner, M.: Mathematical Games: The Fantastic Combinations of John Conway's New Solitaire Game Life. Scientific American, (1970)
6. Ilachinski, A.: Cellular Automata: A Discrete Universe, World Scientific Publishing (2001)
7. Mitchell, M., Crutchfield, J.P., Peter T. Hraber, V.: Dynamics, Computation, and the 'edge of chaos': A re-examination. In: Cowan, G.A., Pines, D., Meltzer, D., (eds.) Complexity: Metaphors, Models, and Reality. Santa Fe Institute Studies in the Sciences of Complexity, vol. 19 pp. 497–513. Addison-Wesley, (1994)
8. Rendell, P.: Collision-based computing. Turing universality of the game of life, pp. 513–539. Springer, Berlin (2002)
9. Sipper, M., Tomassini, M., Capcarrere, M.S.: Evolving asynchronous and scalable non-uniform cellular automata. In: Smith, G.D., Steele, N.C., Albrecht, R.F., (eds.) Proceedings of International Conference on Artificial Neural Networks and Genetic Algorithms (ICANNGA97) (1997)
10. von Neumann, J.: The general and logical theory of automata. In: Jeffress, L.A. (ed.) Cerebral Mechanisms in Behavior The Hixon Symposium, pp. 1–31. John Wiley & Sons, New York (1951)
11. von Neumann, J.: The Theory of Self-reproducing Automata. University of Illinois Press, Urbana, IL (1966)
12. Wikimedia Commons by Richard Ling - Own work, CC BY-SA 3.0. https://commons.wikimedia.org/wiki/Conus_textile#/media/File:Textile_cone.JPG

13. Wolf-Gladrow, D. A.: Lattice-gas Cellular Automata And Lattice Boltzmann models: An Introduction. Springer Science & Business Media (2000)
14. Wolfram, S.: A New Kind of Science (2002)
15. Wolfram, S.: Theory and Application of Cellular Automata. World Scientific, Singapore (1986)
16. Zuse, K.: Rechnender Raum. Friedrich Vieweg & Sohn, Braunschweig (1969)

Chapter 5
Multi-agent Systems

Juan C. Burguillo

The term agent derives from the latin participle *agens*, coming from the verb *agere* that means *to do* and denotes the capability of an entity to do or to act. But in practice the term refers to multiple meanings in different contexts, like a human agent, a hardware agent, a chemical agent, etc. In this book we are interested in the concept of autonomous agents, and for that purpose one of the classical and simple definitions comes from [21] and states that: *an agent is any entity that can perceive its environment through sensors and change it by means of actuators*. Under this agent conceptual image we still could find many simple hardware or software objects that, being rather deterministic, would fit within the agent philosophy. A more general discussion and description about autonomous agents appears in [28]. According to it, the term agent refers to a hardware or (more usually) software-based computer system that usually enjoys the following properties:

- *Autonomy*: agents operate without the direct intervention of humans, and have some kind of control over their actions and internal state.
- *Social ability*: agents interact with other agents (and possibly humans) via some kind of agent-communication language.
- *Reactivity*: agents perceive their environment, and respond in a timely fashion to its changes.
- *Pro-activeness*: agents do not simply act in response to their environment, they are able to exhibit goal-directed behavior by taking the initiative.

There are some other attributes that can be present, but usually they are not considered as a requirement: mobility, veracity, benevolence, rationality and adaptability (or learning) (see [28] for a more detailed discussion).

Agents coexist and interact with other agents in different ways. A system consisting of an interacting group of agents is called a Multi-agent System (MAS). Within

J. C. Burguillo (✉)
Department of Telematics Engineering, School of Telecommunications Engineering,
University of Vigo, 36310 Vigo, Spain
e-mail: J.C.Burguillo@uvigo.es

J. C. Burguillo, *Self-organizing Coalitions for Managing Complexity*, Emergence, Complexity and Computation 29, https://doi.org/10.1007/978-3-319-69898-4_5

the context of this book, a complex software system can be treated as a collection of many agents, each one with its own local functionality and properties. Some of the benefits of MAS technology in large scale software systems are [24]:

- Speedup and efficiency, due to asynchronous and parallel computation.
- Robustness and reliability, in the sense that the whole system can undergo a *graceful degradation* when one or more agents fail.
- Scalability and flexibility, since it is easy to add new agents to the system.
- Cost, assuming that an agent is cheap compared to the whole system.
- Development and reusability, since it is easier to develop and maintain modular than monolithic software.

So far, multi-agent systems have been applied in different domains. Some examples are:

- Social sciences, where MAS technology is used for simulating interactivity and other social phenomena.
- Distributed data mining and information retrieval.
- Virtual reality and computer games use agents to implement intelligent behavior.
- In robotics, a common application is to manage a group of robots, so that they can locate themselves to navigate through their environment.
- Internet auctions and e-commerce.
- Management of telecommunication networks.
- Artificial life and biological simulations.

These are just some examples, since MAS have also been applied to control, scheduling and planning of manufacturing, air traffic management, medicine, e-Learning, etc. The interested reader can find examples of MAS applications in these domains in [27].

5.1 Short Historical Notes

Historically, the origins of autonomous agents and multi-agent systems trace back to the 70s with the works done by Victor Lesser [7, 15] that state the cooperation of multiple entities in order to solve complex problems. Such cooperation help to jointly solve tasks that could not be solved by individual entities, or at least to provide a higher solution efficiency. These works somehow complement the classical centralized approach followed by many AI researchers on those days, more centered on expert systems or knowledge-based systems that derived in Knowledge Engineering. Some of the problems faced on those days were the limited context of application, the difficulties to model the context to reason with intelligence, and the difficulties to exchange knowledge with similar entities, i.e., to cooperate. Then, distributed approaches appear to enhance and extend the capabilities of classical isolated and centralized systems. The design and development of such cooperating systems face several challenges, like how to model knowledge to be exchanged among several entities, how to coordinate effectively the distributed entities, and how to obtain a higher efficiency than centralized systems.

Such working area within the AI community was called Distributed Artificial Intelligence (DAI) and considered several approaches to model, and to solve complex problems in AI involving reasoning, planning, learning, etc. The conditions for applying DAI techniques show characteristics like large data samples, distributed systems and/or loosely coupled autonomous processing. Distributed Problem Solving (DPS) and Multi-agent systems (MAS) are the two main approaches resulting from DAI.

- On the one hand, distributed problem solving techniques divide the tasks among a set of nodes, and the knowledge is shared. Its main concerns are task decomposition, knowledge synthesis and how to provide a global effective solution.
- On the other hand, in multi-agent systems, agents provide a more flexible and autonomous approach to coordinate their knowledge and activities, in order to model the problem to be solved, and this may involve cooperation or competition among different groups of agents.

Multi-agent systems become an excellent approach to model and simulate complex systems as they can apply the traditional top-down approach of AI as well as a bottom-up approach, becoming a good vehicle for the emergence of new system properties, which are one of the basic characteristics of complex systems. Along the next sections we introduce in a more detailed perspective the main characteristics of multi-agent systems.

5.2 Intelligent and Autonomous Agents

An autonomous agent is usually modeled by a certain system architecture. In this section we first introduce the three basic agent architectures, and from them we consider concepts related with the architecture of the whole multi-agent system.

5.2.1 Deliberative Agents

Deliberative architectures use a symbolic knowledge representation, where agents start from an initial state, and then generate plans in order to achieve their goals. Therefore a deliberative agent has a symbolic model of its environment, and decisions are taken considering logic reasoning mechanisms and symbolic manipulation.

The main difficulty when using deliberative architectures is to provide and adequate symbolic representation of the problem, because; given the complexity of symbolic manipulation algorithms, it is possible that the agent cannot face the real time requirements inherent to most of real world environments.

The BDI (Beliefs, Desires and Intentions) architecture [20] has been the most popular deliberative architecture in the past years, and there are a relevant number of implementations. This agent architecture considers a philosophic model based on Michael Bratman's theory of human practical reasoning. The basic concepts related with the BDI architecture are:

1. *Beliefs*: represent the knowledge that the agent has about its environment. This knowledge can be modeled as variables in a database, symbolic expressions from predicate calculus, or inference rules to create new beliefs.
2. *Desires*: represent what the agent wants to accomplish, i.e., some of them can become the next goals or the final objectives that the agent will pursuit.
3. *Intentions*: describe what the agent has chosen to do, and they usually include a set of plans oriented to achieve the agent desires. These intentions should be changed appropriately in dynamic environments if the agent perceive that the context has evolved and one or more plans are not valid anymore.

Besides these three main BDI elements, the system usually works based on *events* that may update beliefs, trigger plans or modify the agent goals, p.e., when some of the desires becomes unfeasible. Algorithm 1 describes the typical cycle for the BDI interpreter as described in [20]: "At the beginning of every cycle, the option generator reads the event queue and returns a list of options. Next, the deliberator selects a subset of options to be adopted and adds these to the intention structure. If there is an intention to perform an atomic action at this point in time the agent then executes it. Any external events that have occurred during the interpreter cycle are then added to the event queue. Internal events are added as they occur. Next, the agent modifies the intention and desire structures by dropping all successful desires and satisfied intentions as well as impossible desires and unfeasible intentions."

Procedure 1: The BDI interpreter cycle

```
initialize-state();
repeat
    options:= option-generator (event-queue);
    selected-options:= deliberate (options);
    update-intentions (selected-options);
    execute();
    get-new-external-events();
    drop-unsuccessful-attitudes();
    drop-impossible-attitudes();
end repeat
```

Among the critiques faced by BDI, and other deliberative architectures, are the difficulties to obtain solutions fast enough to face dynamic and complex environments that sometimes can only be managed by choosing suboptimal actions due to real-time restrictions.

5.2.2 *Reactive Agents*

The inherent complexity of symbolic representations in deliberative agents, and the difficulties to meet real time constraints have conducted researchers to analyze other agent architectural alternatives based in simpler models and bottom-up approaches.

Among them, the most popular has been the subsumption architecture [3] by Rodney Brooks, based on the hypothesis that intelligence is an emergent property on certain complex systems, and that there is no need to include symbolic models to obtain it. Subsumption architectures usually are organized in hierarchical layers (see Fig. 5.1), from lower to higher abstraction levels [3], that decompose the complete behavior of the agent into a set of sub-behaviors. Within this architecture, higher levels subsume lower ones in order to obtain a feasible practical behavior.

The major application of reactive agents, and subsumption architecture, has been in robotics. The reason is that robots need to model the complex inputs coming from the outside environment, and they must usually act fulfilling real time constraints, which become difficult for deliberative architectures and need more agile approaches. Reactive architectures only consider a set of basic actions, sometimes denoted as instincts or behaviors, and the problem is to decide which one should be executed next.

Typically the problems with these architectures appear when we face complex problems with a huge search space, and becomes difficult to the inherent short time view of the agent behavior. Besides, learning from the environment and from past actions are not naturally considered within these models.

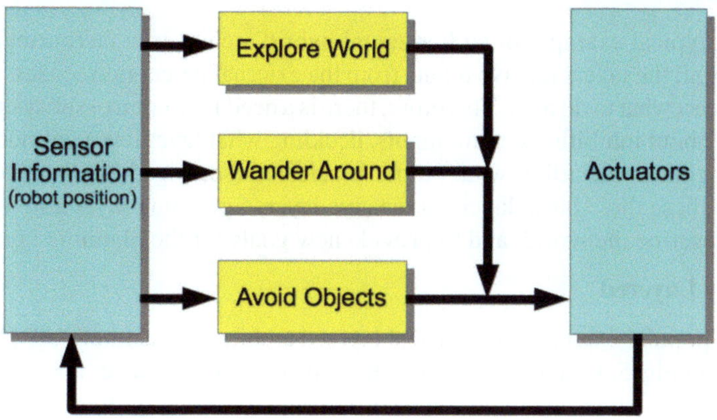

Fig. 5.1 An implementation of the subsumption architecture for robot navigation

5.2.3 *Hybrid Agents*

Due to the limitations, faced by deliberative and reactive architectures; a new type of architectures was considered to combine the advantages of both models, but trying to avoid their inherent limitations. These hybrid architectures typically consider a set of vertical or horizontal layers. Some of these layers work under a reactive model, for simple or urgent events that do not need complex reasoning, i.e., imitating animal instincts. Other layers work using a deliberative architecture, providing a more abstract model to generate medium and long term plans.

In a similar way to the Brooks' subsumption architecture described before, the layers are organized hierarchically over several abstraction levels, which in most of the models are three defining:

1. A reactive level, at the lower layer, to react to inputs in real time and that it is usually implemented using a subsumption architecture.
2. An intermediate knowledge level, describing the knowledge that the agent has about the problem and its environment, usually modeled with a deliberative architecture.
3. An upper social level, that manages the interactions with other agents in order to cooperate, compete or interact in a general term.

The two basic hybrid architecture models are horizontally layered, where all the layers have access to external devices; and vertical layered, where there is only a layer that has access to external sensor and actuator devices.

Horizontally Layered

The most typical example of an horizontal hybrid architecture is TouringMachines [9], where all the layers receive inputs from the external perception sensors; and each layer suggest what to do next. Therefore, there is a need for a control subsystem to take decisions about inhibiting certain inputs, deciding what layer has a temporal control over the agent, or deciding what output is selected (see Fig. 5.2). This architecture considers these three basic layers: a reactive layer, a planning layer and a modeling layer to describe the world, and to provide new goals for the planning layer.

Vertically Layered

The most popular example of a vertical layered architecture is InteRRaP [16], that is described in Fig. 5.3 and contains a reactive layer, a planning layer and a social layer to interact with other agents in the environment. As shown in the figure, the main difference is the interaction between the layers and the external systems, which in this case implies that all sensor inputs are managed vertically by each layer before feeding the upper one. The same happens with the action outputs, that are submitted by upper layers to lower ones to reach the external interface. Therefore, the information here first flows down-top, and then top-down when interacting with the environment.

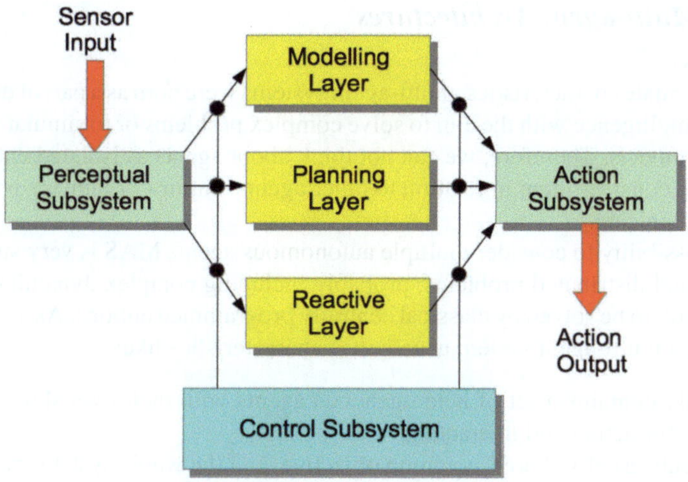

Fig. 5.2 TouringMachines: a horizontally layered agent architecture

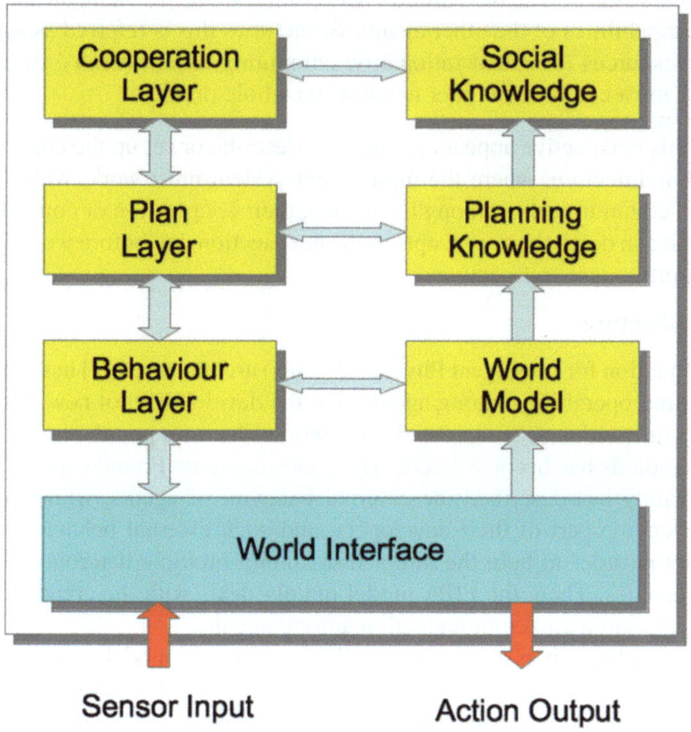

Fig. 5.3 InteRRap: a vertically layered agent architecture

5.2.4 Multi-agent Architectures

Due to its innate characteristics, multi-agent systems were born as a part of distributed artificial intelligence with the aim to solve complex problems or to simulate physical or virtual models. Therefore, we can not think about agents as isolated entities, and we have to consider them in relation to other agents, entities or humans with whom they communicate.

The possibility to consider multiple autonomous agents MAS is very suitable for complex and distributed problems, probably including complex dynamics, that are very difficult to be solved by classical centrally programmed entities. As a distributed approach, a multi-agent system usually has characteristics like:

- It usually contains a set of heterogeneous agents with their own skills, data and abilities for acting and interacting.
- The agents usually share a common objective, and the whole system must be able to divide the tasks to be performed by the agents, taking into account their abilities and processing capabilities.
- Each agent in the system has a limited view of the whole problem, the environment and the capabilities of the other agents. Sometimes this is referred as agents have limited resources (bounded rationality, computing time, memory storage, etc.), that they must combine in order to solve the whole problem.

Under this perspective appears the need to describe or set up the conditions, and the global architecture, where the multi-agent system must work, which normally implies the communication among the agents, their cooperation or competition. We will introduce in detail these concepts in the next section, but before we will describe the FIPA multi-agent architecture.

FIPA Architecture

FIPA (Foundation for Intelligent Physical Agents) architecture [12] has mainly work in solving interoperability among agents, and the development of new extensions to the architecture within a standardized framework. The most popular implementation of those standards has become JADE[1] (Java Development Framework).

FIPA mainly has described the external behavior of agents, while the internal decisions become part of their developers, and such external behavior follows an open model in order to help the interaction among multiple heterogeneous agents or agent societies. Then, the FIPA model mainly deals with the creation, deletion, registering, location and communication among agents.

The agent platform is the core of FIPA's reference model, and provides the infrastructure for the agent development and their use, including: the operating system, communications, middleware and agent management. The FIPA standard defines the services that must provide any agent platform like:

[1] See http://jade.tilab.com.

- *Agent Management System (AMS)*: the AMS is the main managing element concerning the platform and the agents it contains. It offers services for creation, deletion, state control, registration, mobility, sharing resources and communication. Besides it also provides a domain name service (Agent Name Service, ANS), also known as white pages. The AMS allows to control the agents' life cycle, and every agent may be: started, active, suspended or in stand-by.
- *Directory Facilitator (DF)*: the DF is a complement to the ANS, and in this case it provides a yellow pages service; allowing to search agents with certain skills or characteristics.
- *Message Transport System (MTS)*: all FIPA agents have access to the MTS, that supports communications between agents in the local platform or between the local and remote platforms. Agents communicate among them using the Agent Communication Language (ACL), which is another FIPA standard. The message communication model is asynchronous, and use queues for sending and receiving messages, with their own management policies.

Security is a relevant concept in any distributed system, and even more using multi-agent systems, given their autonomy and the complexity that the system may achieve. FIPA has considered several policies to maintain high levels of security within such open platforms. FIPA also has recommended existing standards whenever possible, like X.509 for public key infrastructures, or SSL (Secure Sockets Layer) among others.

Mobility is another service considered in the FIPA platform, and the standard tries to provide interoperability solutions for mobile agents traveling between different platforms. This mobility can consider cloning and running the local agent in a remote platform, which is the simplest case; or migrating the agent among different platforms, which is a more complex execution model. In the migration case, the agent is suspended in the local platform and sent over the network, with its code and data, to continue its execution on the remote platform.

The FIPA platform also provides an Ontology service to support communication among agents using ACL messages. This is helpful to allow agents to understand each other. To use this service, when an agent registers at the DF it informs about the ontologies it can manage, so other agents may know in advance how to communicate with him. The main functions of the ontology service are:

- Keeps a set of ontologies for public use, available for the agents.
- Translate expressions between different ontologies.
- Answers queries about terms of the managed ontologies.
- Helps to identify and to use shared ontologies among agents.
- Discovers new ontologies and make them available for the agents.

5.2.5 Mobile Agents

Mobile agents are a particular, and time ago popular, type of agents with the capability to decide when and where to move themselves (their code, state and data) between several devices connected through a communication network, in order to resume their previous execution at every new device. Therefore, besides the classical properties any agent may enjoy, mobile agents add the mobility feature to their profile.

The idea behind any mobile agent technology is, basically, to substitute remote procedure calls, between a client and a server, and execute the code directly at the server side gaining efficiency and reducing bandwidth consumption. Remote procedure calls invoke a procedure (or method in the OO terminology) on another process, which usually is remotely located. Normally, remote procedure calls block the client, while waiting for the answer of the server, i.e., they usually follow a synchronous communication model. An alternative, in the mobile agent paradigm, is to send the agent, i.e., the client process itself, to perform the task(s) directly on the server side and return back the results. Potentially, this could reduce significantly the bandwidth used in the classical client-server remote procedure call paradigm, and obtain an overall improvement of network and computing resources. Consider, for example, a classical search process done by any web spider along the Internet day after day. Classical web spiders request and index web pages, causing a big redundancy in the data sent by the web servers to the web search machines. It would be much more efficient to just obtain what has been modified since the last search. So, instead of requesting all the information and analyzing it at the web search machines, it would be faster to send an agent to execute its code in the web servers, and to detect what has been changed since the last visit.

There are a number of challenges to support mobile agent technology like: how to serialize all the agent's code and data to continue the execution on the remote device, how to execute the code if running on a different hardware or operating system, or how to protect both, the host and the agent from malicious attacks. There have been a number of attempts to try to minimize these difficulties. For the most cases, even not all of them, the use of the Java language has simplified (but not definitively solved) these difficulties, and has been adopted by a number of mobile agent platforms and solutions. The main reasons are that Java includes the serialization process within its own features; uses a Java Virtual Machine (JVM), so the code potentially can run in any device capable of running a JVM; and, finally, security has been one of the basis from where Java was developed. Nevertheless, a huge problem deals with security. The idea of allowing the execution of external code in the network devices seems potentially insecure, even if you trust in the network agents and nodes. Besides, it is very difficult to protect the spreading of malware, once the agent is attacked by a malicious host; which could have been infected previously.

Along the 90s, a set of mobile agent platforms were developed, being Aglets[2] the most popular mobile agent framework. Aglets takes advantage of the Java technology to overcome some of the challenges described above.

[2]http://aglets.sourceforge.net/.

5.3 Ontologies

What is the aim of communication and how it can influence communicating entities and the external environment? These questions have been addressed along the second half of the XX century by analytical philosophers like Austin [1] and Searle [22], which drive to the study and classification of *speech acts*. Their reflections have been reused by the agent community when trying to communicate different agents.

The problem of communicating two different entities make necessary the establishment of some agreement between both parts about the terminology they will use to describe the domain in order to effectively exchange knowledge. In order to solve such problem, an *ontology* specifies a set of terms that must provide a basis for understanding each other on a certain domain. Ontologies have arisen and become popular in the last fifteen years due to the emergence of the *Semantic Web*. The idea was to enrich web pages with semantic information in order to simplify how the information is processed by computers.

There have been a number of languages used for creating ontologies. Now we enumerate a few of them, which became popular:

- *RDF*: the Resource Definition Framework is not an ontology language, but very closely connected with ontologies and the semantic web. RDF was developed with the idea of providing a framework to represent knowledge in web pages or resources. RDF is very simple, and this makes things easier in general, but sometimes limited in expressive power. RDF basically represents subject-predicate-object triples describing knowledge about its elements, denoted as resources, and is based in XML. Another language related with RDF is RDF-Schema (RDFS), which is a simplified set of resources and standardized properties to create RDF vocabularies, and, for instance, to check out if a set of RDF triples is valid (p.e., a clock has not surnames) and the linked values and their types are also valid (p.e., the price of a product is not "sunny"). As XMLS models syntactic relations among data, RDFS models semantic relations, i.e., knowledge. But RDFS only allows to define very simple ontologies, so more complex and powerful languages have been defined for such purposes.
- *OWL*: the Web Ontology Language is presently the most relevant language for writing ontologies. There are three main levels of OWL: *OWL Lite* (which is the simplest and less expressive version, but easier to manage computationally and to understand by humans), *OWL DL* (which extends OWL Lite to include *description logic*) and, finally, *OWL Full* (which is a very expressive and powerful framework, but much more difficult for checking out consistency properties).
- *KIF*: the Knowledge Interchange Format is closely related with first-order logic, that emerged in the agents area, as a difference with the three previous languages mentioned before. KIF allows agents to express properties of objects, relationships between objects and general properties in a domain. KIF uses first-order logic elements, such as the Boolean connectors (*and, or, not*) and the universal and existential quantifiers *for all* and *exists*. KIF provides a basic vocabulary of objects (numbers, characters, and strings) and some standard functions and rela-

tions among objects (e.g., *less than* or *addition* for numbers). KIF uses a LISP-like notation for handling list of objects.

A set of software tools have been developed in the past years to design and refine ontologies. Among them, the most popular probably are Ontolingua server [8] and Protégé [19]. Ontolingua is a web-based service intended to provide a common platform to share and unify ontologies among different groups. The main component is a library of ontologies, expressed in the Ontolingua ontology definition language (based on KIF). Protégé has been a very popular tool for ontology development, and uses a platform-independent editor for creating class hierarchies, properties and instances that can be imported or exported in several formats, including OWL.

5.4 Communication

The theory of speech acts is based in the seminal work of the philosopher John Austin [1], as he reflected that a certain class natural language expressions (*utterances*) could influence the world in a similar way of physical actions do. He denoted such utterances as *speech acts* and examples could be the word *yes* when accepting a marriage or a confession in front of a jury. Then, Austin denoted the act of speech as a set of *performatives*, which refer to a set of contents that could be true or false, but also could express intentions, decisions, promises, etc.; which are not necessarily true or false. Examples of such performative verbs are: request, inform and promise. Austin also distinguishes three different aspects of speech acts: locution (the propositional content, e.g., 'please, give me the key'), illocution (the action that the performative does, e.g., 'she requested me to give her the key'), and perlocution (the effect really caused on the receiver, e.g., 'I gave her the key').

John Searle extended Austin's work in the book *Speech Acts* [22], where he identified several properties necessary for a speech act to succeed. These conditions are:

- *Normal I/O conditions*: state the speaker and the hearer are under normal circumstances, p.e., the speaker is not dumb, the hearer is not deaf, and the environment in not so much noisy.
- *Preparatory conditions*: state that the hearer must be able to perform the action potentially requested by the speaker.
- *Sincerity conditions*: that describe if the speaker really wants the action to be performed.

Searle also identified five types of speech acts:

- *Assertives*: commits a speaker to the truth of the expressed proposition (p.e., informs).
- *Directives*: cause the hearer to take a particular action (p.e., request, commands, etc.).

- *Commissives*: commits the speaker to do some future action (p.e., promises).
- *Expressives*: express the speaker emotions towards the proposition (p.e., congratulations, excuses, thanks, etc.).
- *Declaratives*: change the reality in accordance with the proposition (p.e., baptisms, guilty sentence, marriage acceptance, etc.).

The technical group *Knowledge Sharing Effort* from DARPA (Defense Advanced Research Projects Agency) worked in standardizing a communicating language among agents in the early nineties. The main outcome was KQML (Knowledge Query and Manipulation Language)[10], a message-based language for agent communication. KQML is both a message format and a message-handling protocol to support run-time knowledge sharing among agents. KQML can be used to interact with an intelligent system, or by two or more intelligent systems to share knowledge among them. KQML contains an extensible set of performatives, which comprise a substrate on which to develop higher-level models of agent interaction such as contract nets and negotiation, and can be divided in three main classes: discourse, intervention, and network communication. Several critics has been done to this language, especially concerning its interoperability problems among different implementations, the lack or rigorous semantics, the absence of performatives of type commissive and a performative set too large [27].

Nevertheless, KQML was a relevant pillar for FIPA to develop its own Agent Communicative Language (ACL), which defines the structure that must have a message exchanged among agents in a network, and was defined having in mind all the critics done to KQML. ACL, which looks very similar to KQML, but with a reduced set of performatives, and it was defined considering a comprehensive formal semantics. The specification consists of a set of message types, and the description of their pragmatics, that is the effects on the mental attitudes of the sender and receiver agents. Every communicative act is described with both a narrative form and a formal semantics based on modal logic. The specification also provides the normative description of a set of high-level interaction protocols and several kinds of auctions. Find next an intuitive example taken from [11]:

```
(ask-if
 :sender I
 :receiver j
 :content (= (weather England (summer 1997)) wet)
 :ontology meteorology
 :reply-with query-17)

(inform
 :sender j
 :receiver I
 :content true
 :in-reply-to query-17)
```

Basically agent I ask to agent j if the weather was wet in England during the summer of 1997. Agent j answers that the answer is *true*, i.e., it was wet indeed. FIPA ACL has been considered as part of FIPA-OS.

5.5 Coordination

As we have seen in previous sections Multi-agent Systems are usually composed by multiple computing entities interacting in an autonomous and sometimes intelligent way. In order to manage appropriately the resulting complex system, the capability to coordinate with other agents, implicitly or explicitly, is a key element of any agent.

Given a particular problem, or a global task to be solved, once agents have agreed with respect to a common goal, it is needed to specify the details of the collaboration. For instance, providing the tasks that must be performed by every one, when they must be finished, what to do if problems appear, when one or several particular tasks have failed, etc.

One relevant tool to coordinate the agents is the use of Distributed Problem Solving (DPS) [5], that usually assume that agents are benevolent and cooperative, i.e., not necessarily self-interested. This means that they do not try to maximize their own utility function, but the whole MAS maximizes the global one. Usually this condition is accepted assuming that the whole MAS is built by the same designer or, more realistically, that self-interested agents have reached a previous agreement. The use of DPS has a set of advantages:

- It takes advantage of specialized agents, and allows to distribute the tasks depending on the characteristics, load, and skills of the agents in the system.
- This distribution allows to take decisions locally, avoiding to overload the networks with inefficient managing information.
- Besides it allows agents to be physically distributed, taking advance of several computing infrastructures; and to use parallel computing, that can enhance the performance depending on the system characteristics and the tasks to be done.

Under this approach, when an agent must perform a certain set of tasks it can search other agents to ask them for help. Usually, task decomposition has a number of steps like: dividing a big task into subtasks, finding the appropriate agents to perform those subtasks, and gathering the results from the subtasks to get the original task solved. Note that this problem decomposition can be performed recursively by any of the agents participating in the process.

If the systems is homogeneous, then task decomposition becomes trivial among the agents, and only will take into account their availability, usually depending on their overload at a certain moment. The problem becomes more difficult, but interesting, when the agents are heterogeneous and have different skills that allow them to perform the different tasks more or less efficiently.

One classical approach to assign subtasks in a MAS is the Contract Net Protocol (CNP) [23]. In CNP there is an agent that behaves as the manager, and it is the responsible for dividing a task into a set of subtasks, and for finding the appropriate agents to solve the global tasks under a required deadline. Usually the manager broadcast the subtasks to the set of available agents, consider their bids to solve the subtasks; and, depending on the offers, perform the best possible assignment.

Sometimes, when accomplishing a certain task, it is useful to distribute the results among several agents in order to improve the global performance. This usually happens when several agents face the same problem, and they exchange their approach and results in a certain agent neighborhood, so they can increase their confidence in their own solutions. At the same time, an agent can use other agents' solutions to refine its particular approach. This can be done, distributing the information among the interested agents ad-hoc, or creating some kind of structures to adequately organize and store the relevant information by a coordinator or by repositories. Nevertheless, the sharing of results should be done if the agents know how to adequately integrate other agents context and results. Besides, the communication level among the agents must be kept under a certain threshold, avoiding to exchange big amounts of information that can use too much network bandwidth.

In a MAS system, where agents collaborate in a distributed problem solving framework, it is usually needed to conceive a global plan, that considers the actions for every agent, taking into account other agents' actions too. This is known as distribute planning, and it can be designed from three points of view:

1. The system can do a central planning, that afterwards is distributed among the agents, and then they execute their own sub-plan distributively.
2. The different agents in the system can create distributed and partial sub-plans, that afterwards are composed and organized, by a set of organization agents, to create a global plan for the whole system.
3. We also can consider the creation of distributed sub-plans, that must be somehow revised to avoid conflicts, and then will be implemented in a distributed way.

One classical planning framework is the Partial Global Planning [6], which mainly works as follows:

- An agent analyzed and decompose its own tasks to create a local plan with its main objectives.
- The local plan is abstracted, to its main elements, and send to other related agents to check its common goals and conflicting elements.
- Local plans can be integrated into a global plan, trying to improve coordination and avoiding conflicts or redundancies.
- Then, it is decided how to communicate the agents to provide the results of their own tasks.
- Afterwards, the local plans are provided to the local agents to be performed.
- During the execution of the plan, the results and unexpected problems are considered, in order to decide when to do a re-planning. The idea is to avoid that minor changes cause too many plan changes or even a complete reorganization.
- Also, if some agents become overloaded, they can try to get help from idle agents, changing a bit the organization to load distribution.

5.6 Methodologies

As commonly happen with Software Engineering technologies, there is an increasing effort to support the development of new systems using them. Developments in the framework of Multi-agent Systems have not been an exception. Methodologies for the analysis and design of agent-based systems fall into two main groups: the ones that take their models from object-oriented development, adapting OO methodologies to the purposes of Agent-oriented Software Engineering (AOSE); and the rest of them that adapt knowledge engineering or alternative techniques.

In general, the most common approach for developing methodologies belong to the first category, probably because it is highly common to have a previous background in object-oriented development. Nevertheless, there are some remarkable differences between agents and objects [27], for instance proactivity in agents systems, or the cooperation or negotiation among self-interested agents. Several approaches have been extended in order to improve those deficiencies, but the development of multi-agent systems using formal methodologies has not became yet mainstream.

Some of the most popular methodologies include:

- The Object Management Group [25] and FIPA collaborated to support the development of notations from object oriented design for modeling agent-based systems. Users with experience with UML (Unified Modeling Language) and object oriented techniques, can find in MaSE (Multi-agent Systems Engineering, [4]) a possibility that greatly reduces the learning curve.
- The AAII methodology used the PRS-based Belief-Desire-Intention technology, and the distributed multiagent reasoning system (DMARS) [13]. This methodology is also based on an object-oriented background, enhanced with some agent-based concepts.
- Gaia [29] takes its roots from object-oriented analysis and design, but it also provides a set of agent concepts to support agent-oriented development. Mainly, Gaia encourages the developer to construct agent-based systems as a continuous process of *organizational design*, where each refining step gets systematically closer to the final implementation.
- Tropos [2] provides a conceptual framework for modeling agent-oriented systems based on the development of an *actor model* and a *dependency model*. The former models the key stakeholders and their goals, while the latter models the dependencies among actors and resources. Then a *goal model* is developed to detail those particular goals, and finally a *plan model* to describe how to achieve them.
- The Prometheus methodology [17] consists on three main stages: (i) a *system specification* to identify the goals, the basic functionalities and the system interfaces; (ii) an *architectural design* to identify the agent types, the system structure and the interaction among the agents; and, finally, (iii) a *detailed design* to refine agents capabilities and detail the system processes.
- MAS-CommonKADS [14], is another popular methodology especially for those designers familiarized with knowledge engineering systems.

- INGENIAS [18] became a great alternative methodology for modeling properties closer to the agent world paradigm, like intentional properties or agent sociability.

5.7 Modeling and Simulating Complexity

Multi-agent systems and its associated models for processing and communication have become excellent tools to model and simulate complexity in Science. For instance, they have been used to analyze emergent properties, such as big structures emerging from simple local interactions, which lead to wonder how do natural systems self-organize by themselves to reach such high levels of complexity. At the same time, selfish agents can cooperate with others in order to create larger coalitions that drive to a higher common wealth. All these ideas together with the aggregation (composition) and specialization (inheritance) properties of multi-agent systems are very interesting for analyzing and understanding natural systems from a scientific point of view, and for creating new efficient ones, from an engineering perspective.

5.8 Conclusion

Agent-based simulations have become an interesting framework to model and explore complex systems, as they support the idea that the global behavior of a multi-agent system derives from low-level interactions among its composing agents. The fundamental element of many models to describe complex adaptive systems is the adaptive agent, which is an entity that uses its experience in a changing environment to continually improve its abilities and to reach its goals.

This chapter has introduced the concept of agents, as the interacting entities that may compose a complex system. We have seem several types of agent architectures, from deliberative to reactive ones, and their composition as hybrid agents. From such agent definitions, the concept of multi-agent system has been derived. Then, we have introduced the use of ontologies to effectively support the exchange of knowledge among the agents. We also have seen several approaches to coordinate and plan the actions in a multi-agent system. Then, several methodologies for agent-oriented engineering have been described. Finally, the potential of the multi-agent model to simulate complex systems has been addressed.

The next chapters extend the possibilities of multi-agent systems to describe frameworks where agents play a fundamental role. Chapter 6 describe scenarios, where a multi-agent system behaves as a self-organized global entity, with possible emergent properties and behaviors, not explicitly present in the individual agents. Chapter 7 is the last one of this background, and introduces Game Theory as a framework where agents collaborate or compete depending on their own interest and the type of problem or game they have to face.

5.9 Further Reading

The interesting reader can get a nice and overall introduction to the topic at the book *Introduction to Multiagent Systems* [27] by Mike Woorldridge. A more technical book is *Multiagent Systems* [26], written by several authors leading their respective fields, and edited by Gerhard Weiss.

References

1. Austin, J.L.: How to Do Things with Words. Harvard University Press, Cambridge (Massachusetts) (1962)
2. Bresciani, P., Perini, A., Giorgini, P., Giunchiglia, F., Mylopoulos, J.: Tropos: an agent-oriented software development methodology. Auton. agents multi-agent Syst. **8**(3), 203–236 (2004)
3. Brooks, R.A.: Intelligence without representation. Artif. Intell. **47**, 139–159 (1991)
4. Deloach, S.A., Wood, M.F., Sparkman, C.H.: Multiagent systems engineering. Int. J. Softw. Eng. Knowl. Eng. **11**, 231 (2001)
5. Distributed problem solving and planning. Chapter in the book Multi-agent Systems and Applications. Lecture notes in computer science, pp. 118–149 (2001)
6. Durfee, E.H., Lesser, V.R.: Partial global planning: a coordination framework for distributed hypothesis formation. Syst. Man Cybern. IEEE Trans. **21**(5), 1167–1183 (1991)
7. Erman, L.D., Lesser, V.R.: A multi-level organization for problem solving using many diverse cooperating sources of knowledge. In: Advanced Papers of the Fourth International Joint Conference on Artificial Intelligence, Tiblisi, Georgia (URSS) (1975)
8. Farquhar, A., Fikes, R., Rice, J.: The Ontolingua server. Int. J. Hum. Comput. Stud. **46**(6), 707–727 (1997)
9. Ferguson, I.A.: TouringMachines: an architecture for dynamic, rational, mobile agents. Doctoral dissertation, University of Cambridge (1992)
10. Finin, T., Fritzson, R., McKay, D., McEntire, R.: KQML as an agent communication language. In: Proceedings of the third International Conference on Information and Knowledge Management, ACM (1994)
11. FIPA ACL Message Structure Specification. http://www.fipa.org
12. Foundation for Intelligent Physical Agents. IEEE Computer Society standards organization. http://www.fipa.org
13. Georgeff, M.P.: Distributed multi-agent reasoning systems (dmars). Technical report, Australian AI Institute, Level 6, 171 La Trobe Street, Melbourne, Australia (1994)
14. Iglesias, C.A., Garijo, M., Gonzlez, J.C., Velasco, J.R.: Analysis and design of multiagent systems using MAS-CommonKADS. Intelligent Agents IV Agent Theories, Architectures, and Languages. Lecture Notes in Computer Science, vol. 1365, pp. 313–327 (1998)
15. Lesser, V.R., Corkill, D.D.: Functionally accurate, cooperative distributed systems. IEEE Trans. Syst. Man Cybern. SMC **11**(1), 81–96 (1981)
16. Müller, J.P.: The Design of Intelligent Agents: A Layered Approach. Springer Science & Business Media, New York (1996)
17. Padgham, L., Winikoff, M.: Prometheus: a methodology for developing intelligent agents. Agent-oriented Software Engineering, pp. 174–185. Springer, Berlin (2002)
18. Pavón, J., Gómez-Sanz, J.: Agent oriented software engineering with INGENIAS. Multi-agent Systems and Applications III. Lecture Notes in Computer Science, vol. 2691, pp. 394-403. Springer, Berlin (2003)
19. Protégé Website. http://protege.stanford.edu/
20. Rao, A.S., Georgeff, M.P.: BDI Agents from Theory to Practice. In: Proceedings of the First International Conference on Multi-agent Systems (ICMAS-95) (1995)

21. Russell, S.J., Norvig, P.: Artificial Intelligence: A Modern Approach, 3rd edn. Prentice Hall, New Jersey (2014)
22. Searle, J.R.: Speech Acts: An Essay in the Philosophy of Language. Cambridge University Press, Cambridge (1969)
23. Smith, R.G.: The Contract net protocol: high-level communication and control in a distributed problem solver. IEEE Trans. Comput. **29**(12), 1104–1113 (1980)
24. Sycara, K.: Multiagent systems. AI Mag. **10**(2), 79–93 (1998)
25. The Object Management Group. http://www.omg.org
26. Weiss, G.: Multiagent Systems, 2nd edn. MIT Press, Cambridge (2013)
27. Wooldridge, M.: Introduction to Multiagent Systems, 2nd edn. Wiley, New York, NY, USA (2009)
28. Wooldridge, M., Jennings, N.R.: Intelligent agents: theory and practice. Knowl. Eng. Rev. **10**(2), 115–152 (1995)
29. Zambonelli, F., Jennings, N.R., Wooldridge, M.: Developing multiagent systems: the Gaia methodology. ACM Trans. Softw. Eng. Methodol. (TOSEM) **12**(3), 317–370 (2003)

Chapter 6
Self-organization

Juan C. Burguillo

Many natural systems work based on local interactions among their component entities, and some of them are described as self-organized (selforg). This term usually refers to a system that is able to change its internal organization in order to adapt to internal or external changes without the need for an explicit external control [8]. These collective systems are particularly robust, because of the inherent redundancy, provided by their multiple components; that allow them to adapt to changes in order to ensure their own survivability.

Strongly linked with self-organization, we can find the concept of emergence [7], describing that even individual entities may perform very simple behaviors, the whole system can carry out very complex tasks, that emerge from the local interactions of those individuals.

Examples for natural self-organizing systems include social insects, such as ants and termites, where communication among individuals occurs through a stigmergy process, by the means of pheromones deposited within their environment. Other animals have evolved self-organizing behaviors such as flocks of birds and schools of fish, usually oriented to collectively avoid predators. Human societies also present self-organizing behaviors, when they work by combining local and network interactions to give rise to emergent complex global phenomena. But, perhaps, the human brain is the most complex and interesting system showing self-organization and emergence [21].

Autonomous agents and multi-agent systems are natural candidates to evolve in societies with self-organizing capabilities. Usually agents and multi-agent systems have been used for simulating self-organizing systems, in order to better understand them or to engineer new models. An open arena is the development of distributed

J. C. Burguillo (✉)
Department of Telematics Engineering, School of Telecommunications Engineering,
University of Vigo, 36310 Vigo, Spain
e-mail: J.C.Burguillo@uvigo.es

J. C. Burguillo, *Self-organizing Coalitions for Managing Complexity*, Emergence, Complexity and Computation 29, https://doi.org/10.1007/978-3-319-69898-4_6

systems, where components self-organize and work in a decentralized manner towards the achievement of a given (global) possibly emergent functionality. The use of self-organization techniques in artificial multi-agent systems follows a bottom-up approach, and ranges from imitating naturally-inspired self-organizing models to designing new self-organization models for artificial systems. At the same time, most prototypes of self-organizing systems have been developed using MAS architectures. This does not happen by chance, presently, the MAS paradigm has all the necessary means for the design, development, implementation and simulation of self-organizing systems.

Some applications of this biology inspired computation already have been very successful at computational and mathematical domains like swarm techniques applied in optimization, when the search space is too big and/or involves non linear problems. Besides, the concept of self-organization is present in almost all sciences related with complexity, including: Physics, Biology, Chemistry, Economics and Computer Science among others.

6.1 Short Historical Notes

The term "self-organizing" was introduced to contemporary science in 1947 by the psychiatrist and cybernetics pioneer W. Ross Ashby [1]. In 1959, Grassé [15] proposed the theory of stigmergy, that shows that direct interactions are not necessary to coordinate a group, and indirect communications through the environment can be enough. Stigmergy describes how coordination is performed by means of information deposited over the environment, that allows the indirect local interaction among the entities without the need for any external or central control.

In 1967 Koestler [18] defined the concepts of holons and holarchies, where holons are functional systems, but at the same time they are also parts of larger systems. The holonic hierarchy model involves structural patterns that form nested hierarchies of self-replicating structures named holarchies. The elements of holonic systems are denoted with the term holon, which is a combination of the Greek word *holos*, meaning whole?, with the suffix *on* meaning part, as it is used in the words proton or neuron.

Also in the 1970s, Francisco Varela [23] established the notion of *autopoiesis* (meaning self-production) as the process where a system is able to self-reproduce. Autopoiesis applies to closed systems made of autonomous components whose interactions self-maintain the system itself, such as living organisms or body cells.

After Ilya Prigogine's got his Nobel Prize in 1977, scientific researchers started to migrate from the cybernetic view to the thermodynamic concept of self-organization, which describes how open systems decrease their entropy (increase ordering) when an external energy is applied onto the system.

During the last 20 years, research in artificial systems has been oriented towards introducing self-organization mechanisms specifically for software applications. These works have taken diverse inspirations from stigmergy to autopoiesis, or to the holon concept itself. Recently, a relevant effort to analyze the state of the art, to structure the research effort and to define a roadmap in the selforg area was made in the European AgentLink project [10], where a working group for studying self-organizing MAS was created.

Recent research efforts have been oriented towards introducing self-organization mechanisms specifically for software applications. A detailed and extensive introduction to self-organizing models together with a review of applications can be found in [13, 14].

6.2 Concepts of Self-organizing Systems

Self-organization generally refers to the internal driving force that leads to an increasing level of organization, i.e., improving the structure and the interactions among the distributed components. Self-organization is basically an adaptive process, meaning that when the environment changes the system evolves to a new stable equilibrium. We can consider two different types of self-organization systems [7]:

- *Strong* self-organizing systems are those ones where there is no explicit central control either internal or external. For instance, we can consider here an ideal economic market that evolves according to liberal economic rules.
- *Weak* self-organizing systems are those ones where there is an internal (central) control or planning. In this case we can consider a termite society where there is a queen that has a higher influence, p.e., than termite soldiers, over nest behavior (see Fig. 6.1).

Therefore, the absence of explicit external control is a mandatory property for these types of systems, stating their autonomy. If the system also works under decentralized control, then there is no internal central authority or centralized information flow. In this type of systems information spreading, among the agents, is mainly done by local interactions.

As a working definition, we consider *self-organization as a dynamical and adaptive process, where systems reorganize and/or maintain inner structures without the need of external control.*

Self-organization is strongly coupled with the concept of emergent behavior, and this concept refers to the possibility of having some kind of emergent phenomena arising from the local interactions. Emergence is reviewed in more detail in the next section. Here, we consider a set of other relevant characteristics that self-organized systems sometimes may exhibit [6, 7]:

- *Endogenous global order*: that allows the system to reach some (stable) global state produced by the system itself.
- *Simple local rules*: the overall complex system behavior observed is usually based on simple individual behavioral rules. Therefore, local interactions determine the global behavior, but not codify the pattern itself.

Fig. 6.1 A termite "cathedral" produced by a termite colony

- *Dissipation*: this concept refers to some possible stable states reached by the system, where some emergent properties can be observed, and those states imply some kind of energy dissipation.
- *Instability*: is a property exhibit by many chaotic systems usually characterized by nonlinear dynamics, where minor perturbations in the environmental conditions produce significant variations in the overall system behavior. Such chaotic systems exhibit sensitivity to initial çonditions and parameter values, making much more difficult to understand present states and future system evolution.
- *Multiple equilibria*: Multiple equilibria are observed when the systems presents several possible attractors for stable states.
- *Criticality*: describing the presence of threshold effects or phase changes.
- *Redundancy*: this property is present when the elements in the system have similar replicas, so if a *few* of them suffer any damage, then the system can continue working without special difficulties.
- *Self-maintenance*: describing the capacity to self-repair its damaged elements, components or modules.
- *Adaptivity*: referring to the capability of adaption to environment changes.
- *Dynamic*: the whole system is a process in continuous change.

- *Autonomy*: the different components of the system exhibit a high degree of freedom in their behavior.
- *Hierarchies*: are present in a system when multiple nested self-organized levels can be observed.
- *Complexity*: this characteristic usually is present when observing the global patterns, emerging from the interaction among local behaviors.

Measuring the presence of self-organization mechanisms is not an easy task. A self-organizing system must be studied from local or global points of view, and even considering multiple perspectives; if the self-organization process spans along multiple nested hierarchical levels. Measurements can modify the system structure (re-organization), the process (system dynamics) and the function (purpose) for each level.

6.3 Emergence

The roots of the emergence concept go back to ancient Greeks, and more recently to the English philosopher G.H. Lewes who used the term. Nowadays, this concept has been revised in multiple domains such as philosophy, mathematics, physics, thermodynamics, systemics and complex systems [12]. The concept of emergence relates that larger entities, patterns, and properties arise through interactions among smaller or simpler ones, that themselves do not exhibit such properties that *emerge* from the system as a whole. Therefore, such phenomena are observed at a global level, but they cannot be necessarily foreseen by looking only at the individual behavior. The reason is that such emergent phenomena usually arise from local interactions, occurring among the individual components. In this case we can apply the popular sentence describing that *the whole is more that the sum up of its parts*. A natural example can be the flocking of birds or fish schools (see Fig. 6.2), where the whole swarm structure emerges from very simple rules.

The emergence concept can be analyzed from two perspectives: one concerning what *observed properties* are sufficient to identify emergence, and another focusing on the *system characteristics* that enable the system to produce emergence. Besides, to show emergence a system must be perceived, at least, at two interdependent levels: a macro and a micro-level. Even there are multiple definitions of emergence [6], here we define it as: *A system exhibits emergence when there are coherent features at the macro-level (properties, behavior, structure, patterns, etc.) that dynamically arise from the interactions between the parts at the micro-level.*

The outcome of emergence is often called an *emergent phenomenon* [16]. For instance, the shortest path between an ant nest and a food source, is perceived by the external observer that monitors the density of individuals and pheromones in the trails of the ants; but it is not a concept considered by the ants themselves.

Fig. 6.2 Fish schools emerge from simple rules: keep distance, stay aligned and avoid predators (source Wikimedia Commons [24])

Since emergent phenomena are externally observable over a certain period of time, some dynamic equilibria are needed [12], when emergence may appear between conditions that are either too ordered or too disordered. This has been vaguely described as the *edge of chaos* [17, 19]. Usually, the appearance of emergence, around these equilibria, implies that the system also exhibits some kind of self-organization.

In order to identify emergent phenomena, some authors establish some key properties to characterize them [6, 7]:

1. *Observability*: describing that the phenomenon needs to be observable at least at a macro-level to perceive the emergent phenomena.
2. *Novelty*: referring that global macro-level properties are novel or different from individual properties at micro-level and cannot be predicted from them.
3. *Irreducibility*: Churchland [5] describes the irreducibility properties of complex systems as those ones that are not present in their individual parts.
4. *Interdependency*: this property describes the strong dependency between the dynamics observed at both macro and micro-levels [19]. The micro-level causes the emergent phenomena, and the macro-level contains the micro-level entities, and constrains their behavior.
5. *Nonlinearity*: emergent phenomena originate from nonlinear activities at the micro-level, usually in the form of positive and negative feedback loops.

6.4 Self-organization Versus Emergence

The main similarity in the concepts of emergence and self-organization is that both are dynamic processes, where local interactions at microlevel manifest at macrolevel. However, albeit strongly related, the concepts of emergence and self-organization do not always appear together [6]. For instance, an ant nest shows a self-organizing process with emergence (e.g., the shortest path when foraging food), that can be observed visually and emerges as a result of local rules and stigmergy. On the other hand, many physical phenomena only show emergence without self-organization, for instance, tornados or hurricanes where atmospheric conditions determine the creation of such macro-level structures from the micro-level interaction of gas particles.

Nevertheless, the practice has shown that the coexistence of self-organization and emergence concepts in complex systems is quite natural. Self-organization enables a complex system to adapt itself to a dynamic environment, based only on the local interactions of its elements. The interactions work in such a way as to generate emergence properties of the system as a whole.

6.5 Mechanisms for Self-organizing Multi-agent Systems

One of the main issues when engineering multi-agent systems showing self-organizing and/or emergent properties is that there must be a correlation between the design of the local behaviors, and the global goal that the designer wants to accomplish. Therefore, the central question becomes how to program single agents that self-organize when working as a whole. This is usually difficult to perform, as many times the global goals, or some emergent properties, are not directly predictable from the local behaviors (an inherent characteristic of self-organizing systems). Thus, simulation and testing become main tools in order to validate, up to a certain level, the performance of the system [8].

There is no general neither easy classification of self-organization mechanisms in natural or artificial systems, as they can be analyzed from different points of view, and many mechanisms combine several basic principles [13]. Along this section we explore different perspectives to classify selforg mechanisms.

6.5.1 Information-Based Perspectives

One possible analysis of a system may consider the information flow among the system components. Therefore, there could be three concepts to be considered [13]:

1. What information is exchanged among the agents?
   We can consider two alternatives: markers or sematectonic information.

   (a) In interactions based on markers, agents explicitly exchange symbols or signals having a special and identical meaning for every receptor. For instance hazard audio signals, chemical substances like pheromones, etc.
   (b) When using sematectonic information, agents tend to provide implicit information without a specific shared meaning. For instance, letting a pile of objects in a specific place could influence other agents to continue doing it.

2. How does the information flow?
   For instance, the communication can be point-to-point, locally restricted or generally disseminated. In the first case an agent communicates with another agent in its neighborhood, in the second case, the agent can communicate with all its direct neighbors, and in the third case, the agent may disseminate a broadcast message accessible by all the members of the multi-agent system.
3. How do the agents use the information received?
   There are two possible uses of such information: *event-driven* or *follow-through*.

   (a) In the *event-driven* or *trigger-based* alternative, the reception of certain messages triggers specific agent activities. For instance, a hazard audio message may trigger the birds to escape flying assuming the presence of any threat.
   (b) The *follow-through* alternative implies a sequence of actions for the recipient agent. For example, an agent leaves a trail that other recipient agents follow.

6.5.2 Interaction-Based Perspectives

A natural classification of self-organization mechanisms was proposed in [8] and considers the use of direct or indirect interactions among the agents in the system.

Direct Interactions

In this approach, an organization or agent structure of the agents may already exist or it may emerge from the interactions among the agents. This model can be considered when a certain organization must be created and maintained on a set of autonomous agents (see [25]). These approaches use a few basic principles, such as localization and broadcasting, coupled with local interactions and computations done by agents in order to provide a desired global state. The final goal of these mechanisms is to converge and to provide robustness to maintain such global stable state, even under perturbations happening at the micro-level or coming from the environment. Therefore, these mechanisms focus on the structural aspects of the agent organization, such as spatial placement of agents and/or agent communication topology.

These mechanisms have the significant advantage that they enable the design of specific self-organized behaviors with the required outcomes and robust properties.

Indirect Interactions

Under this subset we can consider self-organization mechanisms based on the stigmergy concept. Here, the indirect interactions happen due to changes made by the agents over the environment, that are perceived by other agents that modify their behavior and, eventually, lead the system towards the desired global state.

In these approaches, the design phase involves selecting an appropriate self-organizing model and the evaluation of its correctness is usually performed usually by means of simulation and prototyping [9], where the model is calibrated via iterative refinements based on the experimentation results.

On the one hand, these mechanisms have reused the strengths and robustness of known self-organization mechanisms taken from biology to build self-organizing software. Besides, the agent behaviors are usually very simple and easy to implement, resulting in fast prototyping and low development cost.

On the other hand, due to the non-linearity, and the complexity of the phenomena involved; it is not possible to have direct control on the system behavior; neither it can be proven that the desired behavior will be always achieved, and multiple solutions can be reached.

6.5.3 Other Self-organizing Mechanisms

In [8, 13] we can find other self-organizing mechanisms that share features with the ones just described in the previous classifications. Here we mention some of them.

Reinforcing Behaviors

In some approaches, self-organization is based on the capabilities of the agents to adapt their behavior according to some kind of individual utility function, that determines a reward that they try to maximize modifying their behavior along time. The multi-agent system as a whole also has a global utility function that is not necessarily known by its agents. These approaches are usually based on distributed reinforcement learning techniques, and are also studied under the multi-agent reinforcement learning domain [3].

Cooperation

Cooperation is the *leitmotiv* used in the AMAS (Adaptive Multi-Agent Systems) theory [11], where the desired collective behavior must always occur as the result of cooperation among the agents in the system. In this framework, self-organization is founded on the capability that agents possess to behave cooperatively with other agents. These cooperation capabilities do not imply that agents are always helpful or altruistic, but that they are able to recognize cooperation failures called Non Cooperative Situations (NCS) and handle them. The local handling of NCS maximizes the flexibility and adaptability of the system to unexpected situations, which can happen due to the inner dynamism of the agent interactions and the environment itself.

Generic Architectures and Holonic Systems

A particular class of self-organization mechanisms is based on generic reference architectures or meta-models of agents organizations, which are dynamically modified as needed according to the requirements of the particular application. The common technique for representing such architectural meta-models is as a typed, directed complex graph.

Two classical examples of such reference architectures are the Mediator [20] and the PROSA [2] architectures, that are both based on the holonic hierarchy model and applied in manufacturing systems. When agents are organized according to the holonic metaphor, they participate in holons forming holarchies [18], and self-organization then refers to altering and adapting such holarchies following the perturbations coming from other agents or from the environment [22].

6.6 Conclusion

In this chapter we have reviewed the concepts of self-organization and emergence that have attracted attention within the multi-agent system community. This interest is mainly explained because multi-agent systems are well adapted to simulate and implement self-organizing models due to the inherent characteristics of selforg systems. Self-organizing systems consist of a huge number of autonomous entities distributed over space, and connected locally or using a network topology, but with a limited communication range. These entities usually have selfish goals, that not necessarily coincide with the goals of the global system. Altogether, these characteristics have made multi-agent systems a reference model for engineering selforg systems by means of design and simulation.

Hardware and software engineering nowadays manage large-scale distributed systems, usually composed by a large number of elements. It is unrealistic to be able to achieve distributed optimal control of such systems, and even more from a centralized point of view. This is not feasible because of the huge size of those systems, the unpredictability of their dynamic organization, their interactions with the environment, and the diversity of the goals pursued by the different devices. As the building blocks are autonomous entities inherently dynamic, that work distributed and decentralized in a loosely coupled model over a continuously changing environment. The result is a dynamic system in terms of composition and topology.

On the one hand, the problem is how to design individual agent behavior that collectively produce emergent properties that fit with the overall goal to be addressed. In general, the possibility to engineer and to predict the outcomes of emergent properties is feasible only in trivial problems, and that is why most of the artificial selforg systems have been inspired by natural systems. On the other hand, classical and object-oriented software engineering techniques do not fit well within the self-organizing context, as usually they follow a top-down approach, or they just define the global behavior as a direct function of its components or modules. Besides, as we have seen, the environment plays an important role as a source for self-regulating the agents' behavior in selforg systems.

Nowadays, the growing complexity of the ICT ecosystem and the appearance of concepts like sensor networks, traffic management, autonomic computing, ubiquitous computing, ambient intelligence, Internet of Things, etc.; needs new solutions to support the design and the analysis for autonomous, adaptive and robust complex distributed systems. Selforg models are potentially good candidates to understand such complex behavior, where emergent phenomena may appear from their numerous interacting components, and where self-organization can be a powerful tool to manage complexity.

6.7 Further Reading

The book *The Origin of Order: Self-Organization and Selection in Evolution* [17] by Stuart Kaufman is a classic introduction to the topic from the theoretical biology perspective. An alternative book is *Self-organization in biological systems* [4] by Scott Camazine et al., which is mainly oriented to the study of natural systems. Finally, the book *Swarm intelligence: from natural to artificial systems* by Bonabeau, Dorigo and Theraulaz provides an approach more related with the concepts described in this chapter in order to engineer artificial systems.

References

1. Ashby, W.R.: Principles of the self-organizing dynamic system. J. Gen. Psychol. **37**, 125–128 (1947)
2. Bongaerts, L.: Integration of scheduling and control in holonic manufacturing systems. Ph.D. Thesis, Katholieke Universiteit Leuven (1998)
3. Busoniu, L., Babuska, R., De Schutter, B.: A comprehensive survey of multiagent reinforcement learning. IEEE Trans. Syst. Man Cybern. **38**(2), 156–172 (2008)
4. Camazine, S., Deneubourg, J.L., Franks, N.R., Sneyd, J., Theraulaz, G., Bonabeau, E.: Self-Organization in Biological Systems. Princeton University Press, Princeton (2003)
5. Churchland, P.: Matter and Consciousness. MIT Press, Cambridge (1984)
6. De Wolf, T., Holvoet, T.: Emergence versus self-organization: different concepts but promising when combined. Engineering Self-Organising Systems. Lecture Notes in Artificial Intelligence, pp. 1–15. Springer, Berlin (2005)
7. Di Marzo-Serugendo, G., Gleizes, M.P., Karageorgos, A.: Self-organization in multi-agent systems. Knowl. Eng. Rev. **20**(02), 165–189 (2005)
8. Di Marzo-Serugendo, G., Gleizes, M.P., Karageorgos, A.: Self-organization and emergence in MAS: an overview. Informatica **30**, 45–54 (2006)
9. Edmonds, B.: Using the experimental method to produce reliable self-organised systems. In: Brueckner, S. (ed.) Engineering Self-Organising Systems. Lecture Notes in Artificial Intelligence, pp. 84–99. (2005)
10. European co-ordination action for agent-based computing. http://www.agentlink.org/index.php
11. Gleizes, M.P., Camps, V., Glize, P.: A theory of emergent computation based on cooperative self-organisation for adaptive artificial systems. Fourth European Congress of Systems Science, Valencia (1999)
12. Goldstein, J.: Emergence as a construct: history and issues. Emergence **1**(1), 49–72 (1999)
13. Gorodetskii, V.I.: Self-organization and multiagent systems I: models of multiagent self-organization. J. Comput. Syst. Sci. Int. **51**(2), 256–281 (2012)

14. Gorodetskii, V.I.: Self-organization and multiagent systems II: applications and the development technology. J. Comput. Syst. Sci. Int. **51**(3), 391–409 (2012)
15. Grassé, P.: La reconstruction du nid et les interactions inter-individuelles chez les bellicositermes natalenis et cubitermes sp. La theorie de la stigmergie: essai d'interpretation des termites constructeurs. Insectes Sociaux **6**, 41–83 (1959)
16. Holland, J.H.: Emergence: From Chaos to Order. Addison-Wesley, Reading (1998)
17. Kaufman, S.: The Origin of Order: Self-Organization and Selection in Evolution. Oxford University Press, New York (1993)
18. Koestler, A.: The Ghost in the Machine, Reprint edition 1990. Penguin, East Rutherford (1967)
19. Langton, C.G.: Computation at the edge of chaos: phase transitions and emergent computation. Physica D **42**(13), 12–37 (1990)
20. Maturana, F., Norrie, D.H.: Multi-agent mediator architecture for distributed manufacturing. J. Intell. Manuf. **7**, 257–270 (1996)
21. Searle, J.R.: The Rediscovery of the Mind. MIT Press, Cambridge (1992)
22. Valckenaers, P., Van Brussel, H., Holvoet, T.: Fundamentals of holonic systems and their implications for self-adaptive and self-organizing systems. In: Proceedings of 2nd IEEE International Conference on Self-Adaptive and Self-Organizing Systems, pp. 168–173. Italy (2008)
23. Varela, F.: Principles of Biological Autonomy. Elsevier, New York (1979)
24. Wikimedia Commons by Avoini - Own Work, CC BY-SA 3.0. https://commons.wikimedia.org/wiki/File:Jack_fish_and_reef_sharks.jpg
25. Zambonelli, F., Gleizes, M.P., Mamei, M., Tolksdorf, R.: Spray computers: frontiers of self-organisation for pervasive computing. Second International Workshop on Theory and Practice of Open Computational Systems (TAPOCS 2004). In: 13th IEEE International Workshops on Enabling Technologies: Infrastructure for Collaborative Enterprises (WETICE'04), pp. 397–402. Los Alamitos, CA, USA (2004)

Chapter 7
Game Theory

Juan C. Burguillo

Game Theory (GT) is the formal study of conflict and cooperation among several agents, denoted as players, representing individuals, animals, computers, groups, firms, etc. The concepts of game theory provide a mathematical framework to formulate, structure, analyze, and understand such game scenarios, i.e., it provides useful mathematical models and tools to understand the possible strategies that agents may follow when competing or collaborating in games. The list of games to apply game theory is almost endless: entertaining games, political scenarios, competitions among firms, geopolitical issues between countries, and so on. This branch of applied mathematics is used nowadays in disciplines like economics, social sciences, biology, political science, international relations, computer science and philosophy among others.

We can consider two main parts in game theory represented by noncooperative games and cooperative ones. On the one hand, noncooperative (or competitive) games assume that each participant acts independently, without collaborating with the others, and chooses her strategy for improving her own benefit. On the other hand, cooperative game theory studies the behavior of players when they cooperate. Within cooperative games, we find coalition games, in which a set of players seek to form cooperative groups to improve their performance in a competitive game, and to enable players to succeed in reaching objectives that they may not accomplish independently.

Coalitions usually emerge as a natural way to achieve better conditions for defending its members against the outside players. Game theory provides a natural framework to analyze the partitions that can be formed in a multiplayer game, and the relative power or influence they can achieve over the whole community. Unfortunately, many times coalition formation can become intractable since all possible coalition combinations in a game depend exponentially on the number of players.

J. C. Burguillo (✉)
Department of Telematics Engineering, School of Telecommunications Engineering,
University of Vigo, 36310 Vigo, Spain
e-mail: J.C.Burguillo@uvigo.es

J. C. Burguillo, *Self-organizing Coalitions for Managing Complexity*, Emergence, Complexity and Computation 29, https://doi.org/10.1007/978-3-319-69898-4_7

Therefore, finding the optimal partition by checking the whole space may be too expensive from a computational point of view.

In this chapter, we are going to review the basic concepts of game theory, and relate them with the use of coalitions as a way to obtain cooperation in Complex Systems, with self-interested agents.

7.1 Short Historical Notes

The roots of game theory trace back to the twentieth century, when in 1921 the mathematician Emile Borel suggested a formal theory of games, which was taken further by the mathematician John von Neumann in 1928 in a *Theory of Parlor Games*. But this discipline really becomes popular since 1944, when John von Neumann and Oskar Morgenstern publish the book *Theory of Games and Economic Behavior*, where they analyze competitions in which one individual does better at another's expense, i.e., zero sum games [36]. From that moment, traditional applications of game theory attempted to find equilibria in these games, where each player adopts a strategy that is unlikely to change.

In 1950 John Nash demonstrated that finite games have always an equilibrium point [24], where all players choose the best actions for them given their opponents' choices. This result was lately denoted as the Nash equilibrium, and it made game theory became very active along the 50s and the 60s, when it was broadened theoretically and applied to war and political models. The scientific community developed new concepts such as the core, the extensive form game, fictitious play, repeated games, and the Shapley value. Besides, the theory became to spread to philosophical, political and social sciences. Among them, the Nash equilibrium became a central concept of noncooperative game theory, and a focal point of analysis mainly in economic theory.

Another approach which has been very popular and interesting is the interdisciplinary combination of evolutionary models from Biology with game theory, which gave birth to Evolutionary Game Theory (EGT) [23]. EGT models the application of interaction dependent strategies in populations along generations, and differs from classical game theory by focusing on the dynamics of strategy change more than the properties of strategy equilibrium. In evolutionary games participants do not possess unfailing Bayesian rationality. Instead they play with limited resources. The only requirement is that players learn by trial and error, incorporate what they learnt into their future behavior, and disappear or somehow change if they do not. Along those years, an interesting set of spatial and social evolutionary games, based on the iterative version of the Prisoner's Dilemma [2] had been suggested and deeply analyzed by Nowak and other authors [18, 27] trying to understand the role of local or social interactions in the maintenance of cooperation.

Since the end of the 90s game theory started to consider mechanism design, for instance to design auction mechanisms for efficient resource assignment concerning electromagnetic spectrum bandwidth by the mobile telecommunications industry.

7.2 Representation of the Games

Assume that there is a number of players, represented by agents that take decision in a game, which will be denoted by n; which are labelled by the integers 1 to n, and we denote the set of players by $N = 1, 2, \ldots, n$. We will study mostly two person games, $n = 2$, where the concepts are clearer. There are also one-player games, and in this case the theory is simply called Decision Theory. There are even zero-person games, such as the Conway's game of life [13], where an automaton gets in motion without any person making decisions. In several games, and in macroeconomic models, the number of players can be very large; and sometimes those games are mathematically modeled with an infinite number of players.

We assume that, depending on their game actions, players receive a certain payoff; as a number that reflects the desirability that a player has about a certain outcome, i.e., its utility.

The concept of rationality is a central assumption in many variants of game theory. A rational player always plays to maximize his own payoff, assuming certain actions that the other players will do. The goal of game-theoretic analysis, under a rational approach, is to give advice on how to play the game against other rational opponents. This rationality assumption can be relaxed or limited, and the resulting models have been more recently applied to the analysis of observed behavior.

Next we introduce the main representations used to describe a game: the extensive form, the strategic form and the coalitional form.

7.2.1 Strategic Form

The strategic (or normal form) of a game is usually represented by a matrix which shows the players, the strategies and the outcomes of the game. The outcomes are represented by payoffs, which are real numbers (also called utilities) that measure how much each player likes an outcome. This strategic form representation is normally used to describe non-cooperative games. In this chapter we will usually refer to players as one and two along the text, and player I and II, respectively, in figures.

In Fig. 7.1 we find a two players game, where player one can choose a strategy between its two row strategies (A or B), and player two will choose between its two column strategies (C or D). The payoffs are directly provided in the matrix cells, with the first value corresponding to player one, and the second value to player two. For example, if player one plays B and player two plays C, then player one gets 5 units, while player two receives 0 units.

In a game in strategic form, a strategy is one of the given possible alternatives of a player. Here, it is presumed that each player acts simultaneously or, at least, without knowing the actions of the other. We must differentiate between strategy and action, being the first a plan to play a set of actions along the game, and perhaps achieve a certain goal; while the action would be the particular choice made by a

Fig. 7.1 Game described in a strategic (or normal) form

Player I \ Player II	C	D
A	3,3	0,5
B	5,0	1,1

player at a concrete game iteration. Sometimes, in simple one-shot games, like the one in Fig. 7.1, both concepts coincide and are prone to confusion.

7.2.2 *Extensive Form*

The extensive form is used to formalize games as graphs (usually trees) that describe a time sequencing of moves (see Fig. 7.2) and the information each player has at each node. Graphs nodes represent a point of choice for a player, and the links between nodes represent a possible action for the player. Final payoffs are specified at the bottom of the graph. This extensive form representation is more detailed than the strategic form of a game, as it describes how the game is played over time, including the order in which players take actions, the information that players have at the time they must take those actions, and the times at which any uncertainty in the situation appears.

In the game depicted at Fig. 7.2 there are two players. Player one moves first and chooses either A or B. Player two sees player one's move and then chooses between C or D. In the terminal node we have the payoffs for every player, being the first value to player one and the second to player two. For instance, if player one chooses A, and player two chooses D, then player one gets 0 and player two gets 5.

In an extensive form game, a strategy is a complete plan of choices, one for each decision point of the player. Games where players have information about choices

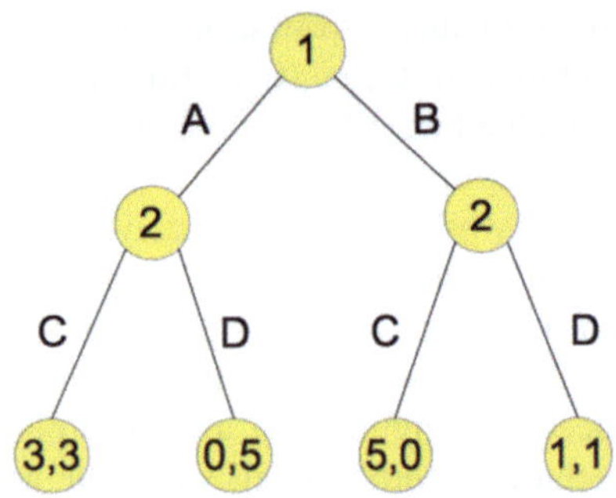

Fig. 7.2 Game described in an extensive form

of the other players are usually presented in extensive form. Every extensive form game has an equivalent strategic form representation, but such transformation may result inadequate due to the exponential growth of strategies for each player, making it computationally unfeasible.

7.2.3 Coalitional Form

In many-player games, there is a tendency for the players to form coalitions to favor common interests. In the coalitional form of a game the notion of a strategy disappears, and the main elements are coalitions, and the value or the worth a coalition has. It is assumed that each coalition can guarantee its members a certain amount, called the value of the coalition.

The coalitional form of a game is a part of cooperative game theory with transferable utility. Under these circumstances it is natural to assume that a grand coalition, consisting of all the players, can appear in the game; and then the question is how to share the payoff received among all its players.

7.3 Types of Games

In this section we introduce several types of games according to their characteristics.

7.3.1 Cooperative, Competitive and Hybrid Games

A game is considered cooperative if the players collaborate establishing binding commitments, while in noncooperative (or competitive) games they just compete among themselves. Often it is assumed that communication among players is allowed in cooperative games.

Hybrid games contain elements from cooperative and non-cooperative games, usually creating coalitions of players that cooperate among themselves, but play in a competitive style with the rest of the players or coalitions. As an example, a football championship is a pure competition among teams, i.e., coalitions with a set of players playing cooperatively against the other teams.

7.3.2 Symmetric Versus Asymmetric Games

Symmetric games model situations where both players have the same opportunities to play and payoffs, and therefore, the strategic form of the game is represented by a symmetric payoff matrix. Asymmetric games usually model different player roles that provide asymmetric payoffs.

Rock-paper-scissors is a symmetric game, as all the players may choose any of the strategies and they have the same opportunities. Chess or checkers are asymmetric games, as one of the players play first.

7.3.3 Zero-Sum Versus Non-zero-Sum Games

Zero-sum games are those ones where the global payoff is divided among the players, therefore choices done by the players can neither increase nor decrease the amount of available resources. Hence, in zero-sum games a player gets a benefit at the expense of the others. For instance, poker is a classical example of zero-sum game.

Non-zero-sum games are those ones where the gains obtained by one player does not necessarily correspond with losses for the rest of the players. In these type of games, usually a cooperation among players produce higher payoffs than a pure selfish playing.

7.3.4 Simultaneous Versus Sequential Games

Simultaneous games are the ones where all the players play their actions at the same time. Even, if the movement is not effectively simultaneous, then the players cannot know the others movement at the same round. Rock-paper-scissors is an example of a simultaneous game.

In sequential games, there is a sequence of movements, and each player has some information about the previous action done by the rest of the players. There is no need of perfect information about the previous moves, but a little knowledge is required. Chess or checkers are sequential games.

Usually, simultaneous games are represented by the normal or strategic form, while sequential games are represented by the extensive form.

7.3.5 Perfect, Imperfect and Complete Information Games

In sequential games, a game has perfect information if each player, when making any decision, is perfectly informed about all the events that have previously occurred, i.e., the player knows all the actions that have been made previously by the rest of the players, and the obtained payoffs. For instance, chess or checkers can be games with perfect information if players have access to previous moves. As imperfect information games we can refer to many card games, where a player does not know the previous actions done by the other players.

Perfect information must not be confused with complete information. In a game with complete information, a player knows the strategies and payoffs available to the other players at a certain round, but not necessarily the past events or moves performed by them. Examples include Poker, tic-tac-toe, Battleship, etc. Games with incomplete information are called pseudo-games.

Another typical concept used is *common knowledge*. A fact is considered as common knowledge when all players know it, and know that the others know it, and so on. It is often assumed that players' rationality is also a common knowledge.

Finally, games in which players remember all past information they once knew, and all past moves they made are called games of perfect recall.

7.3.6 Combinatorial Games

Combinatorial games are those games with perfect information, no random moves and with a win-or-lose outcome. Such a game is determined by a set of positions, including an initial position, and the player whose turn it is to move. Playing the game means to move from one position to another, with the players usually alternating moves, until a terminal position is reached where no more moves are possible. Then one of the players is declared the winner. Chess is a typical example of a combinatorial game.

A combinatorial game where the players have the same set of legal moves from each position is denoted as impartial, otherwise it is denoted as partizan. Chess or checkers are examples of partizan games, since one player moves white pieces first, and the set of legal moves for both players depends on the state of the board, and the other player moves.

These games are usually solved by *backward induction*, which is a technique that first considers the last possible outcomes in the game, and determines the best ones for the player in each case. Then, assuming those moves as future ones, it proceeds backwards in time, determining the best move for the player, until the beginning of the game is reached.

7.4 Two-Person Zero-Sum Games

As stated at the beginning of this chapter, John von Neumann, together with Oskar Morgenstern, published in 1944 his book *Theory of Games and Economic Behavior*; where he laid the foundations of game theory. The theory of von Neumann and Morgenstern is most complete for the class of games called 2-person zero-sum games. In a two-player zero-sum game, one player's gain is the other player's loss, so their interests are diametrically opposed. In general, a game is considered zero-sum if for any game outcome, the sum of the payoffs for all players is zero.

These games are usually represented in the strategic form, with a game matrix and two sets of strategies, one for every player. Matrix cells represent the payoff for each player (see Fig. 7.1). We can formally define a 2-person zero-sum game as a simultaneous game with a triplet (S_1, S_2, P), where:

1. S_1 is a nonempty set of strategies for Player one.
2. S_2 is a nonempty set of strategies for Player two.
3. The payoff matrix P is symmetric, and the winnings of player one are the losses of player two and viceversa.

When we choose among the set of *pure strategies* for player one in S_1 (or in S_2 for player two) with certain random probabilities, then we have a *mixed strategy* for such set. Besides, a 2-person zero-sum game (S_1, S_2, P) is a finite game if both strategy sets S_1 and S_2 are finite.

7.4.1 The Minimax Criterium

The minimax theorem, introduced by von Neumann, is one of the key results in game theory. It is a very defensive approach used to minimize the opponent's maximum payoff, which is equivalent in zero-sum games to maximize own's minimum gain.

For every finite two-person zero-sum game with finite strategies, there is a number V, called the value of the game, such that:

1. At least, there is a mixed strategy for Player one such that 1's average gain is at least V no matter what Player two does, and
2. At least, there is a mixed strategy for Player two such that 2's average loss is at most $-V$ no matter what Player one does.

In game theory literature, two-person zero-sum games are usually represented by a payoff matrix with a unique value in each cell, representing the earnings for player one, and this explains the asymmetric perspective of the minimax criterium for each player. Nevertheless, and for the sake of clarity, we will use the same matrix representation introduced before with positive and negative values to describe what each player earn in each outcome.

Figure 7.3 presents a simple example, considering only pure strategies. We have a symmetric payoff matrix representing the earnings (and losses) for player one (or player two). How does player one think in this case? Well, he knows that if he chooses the strategy A, then he earns a minimum of $+1$, while if he chooses strategy B, then he earns a minimum of $+2$ units. Then he maximize his minima, so his most secure option is to choose strategy B, i.e., to select the maximum of his minima. From player two point of view, if he chooses strategy C then his highest lost is -2, while if he chooses strategy D his highest lost is -4. Then, he will choose C in order to minimize the maxima of his opponent. The resulting outcome using both strategies (B, C) will be 2 units for player one.

Fig. 7.3 An example of two-person zero-sum game

Player II

Player I	C	D
A	+1,-1	+4,-4
B	+2,-2	+3,-3

Maximin is a term commonly used for non-zero-sum games to describe the strategy which maximizes one's own minimum payoff; which, in non-zero-sum games, it is not generally the same as minimizing the opponent's maximum gain.

7.5 Relevant Concepts

In this section we are going to mention some relevant concepts, from game theory, that will help us to analyze games in strategic form. Afterwards, we present one of the most relevant results from game theory, the Nash equilibrium.

7.5.1 Best Response

The best response a player may perform, assuming the other one is going to play a certain action, is the action that produces the most favorable outcome for him.

Consider Fig. 7.4 and suppose that player two plays C, then the best response that player one can do is to play A, winning 5 units, instead of playing B that only provides 2 units in the end.

Note that both players may play pure or mixed strategies as best responses, i.e., given a pure or mixed strategy for player two, there is a best response strategy for player one that could be either pure or mixed.

Fig. 7.4 Best response and dominance

Player II

Player I	C	D
A	5,1	4,2
B	2,3	3,5

7.5.2 Dominant Strategies

A strategy dominates another one from the same player if it always gives a better payoff, regardless of what the other player does. It weakly dominates the other strategy if it is always at least as good. A rational player will never play a dominated strategy.

A row (column) may also be removed if it is dominated by a probability combination of other rows (columns), i.e., by a mixed strategy.

In the example of Fig. 7.4 we see that for such game, strategy A dominates strategy B for player I, and strategy D dominates strategy C for player II. Therefore the only rational outcome for such a game is to play (A, D) resulting in a $(4, 2)$ payoff for the players.

7.5.3 Pareto Optimality

Pareto optimality, or Pareto efficiency, is a criterion that desirable solutions should satisfy. We say that an outcome is Pareto optimal if there is no other outcome that gives a better utility for a player, without providing a worse utility for the other player. When players stay at a Pareto inefficient outcome, rational players should accept to move to a Pareto one, as none of them lose anything; and at least one of them would gain more (but in real world scenarios, rationality not always rules).

In the matrix in Fig. 7.4, we can see that the outcomes resulting from the strategies (A, C), (B, D) and (A, D) are Pareto optimal.

7.5.4 Nash Equilibrium

Intuitively, a Nash equilibrium is a situation, where once you assume that the other player is going to play something, you can not do better than playing a particular strategy, and viceversa. As an example, suppose that you want to meet a friend with whom you have lost contact with, and you know that she will be at a certain pub today; then you have no more options than to go there if you want to find her.

Here, the equilibrium concept is very relevant, since rational agents will not have incentives to deviate from a Nash equilibrium.

Formally, a pair of strategies s_1 and s_2, for players 1 and 2 respectively, are in *Nash equilibrium* when

- assuming that player two is going to play s_2, then player one can not do better than playing s_1, and
- assuming that player one is going to play s_1, then player two can not do better than playing s_2.

When the previous two conditions happen, then those strategies are also the *best response* to each other. The opposite is also true and is a simple way to find the pure Nash equilibria in a matrix. Unfortunately, not every matrix have a pure Nash equilibrium, and some matrices have more than one Nash equilibrium. Fortunately, John F. Nash have found a nice result to clarify the scenario, which is that *any finite strategic-form game has an equilibrium in mixed strategies*. Besides, in 2-person zero-sum games, the minimax solution is the same as the Nash equilibrium. If you review now the matrix in Fig. 7.3 you will see how the minimax solution is also the Nash equilibrium of the matrix.

As another example, consider the matrix in Fig. 7.4, where we see that the pair of strategies (A,D) are in Nash equilibrium. Assuming that player one is going to play A, then player two best response is D, and assuming that player two is going to play D, then player one best response is A. In many game examples there are more than one pure Nash equilibrium.

7.6 Games in Coalitional Form

In previous non-cooperative games we assume that players may not establish agreements, and that utility has only sense for each individual as a result of its own actions. In *cooperative games*, we accept that players may reach agreements in order to decide how to play, and how to share the resulting common payoff. Hence, we also assume that there is a transferrable utility (TU) that allows these side payments. Therefore, given the use of side payments, there will be a tendency for certain players, with similar objectives, to create alliances and to establish coalitions. In reality, this type of games usually happen when there are coalitions, whose players have similar objectives and are usually linked by a contract (like companies or football teams). They play cooperatively games against other teams, and the team gets the payoff for the whole contest; so some players or professionals may specialize and/or scarify its own payoff for the benefit of the whole group. In this section, we describe the coalitional form in order to adequately study these games.

7.6.1 N-Person TU Games

Let $n \geq 2$ denote the number of players in the game, numbered from 1 to n, and $N = \{1, 2, \ldots, n\}$ the set of players. A coalition C is defined as a subset of N, $C \subset N$, and the power set of all coalitions is denoted by S_C. By convention, we also consider the empty coalition $\emptyset$, and the set N as the grand coalition.

For example, just considering two players $n = 2$ we have 4 possible coalitions, $S_C = \{\emptyset, \{1\}, \{2\}, N\}$. With $n = 3$ players, there are 8 possible coalitions, $S_C = \{\emptyset, \{1\}, \{2\}, \{3\}, \{1, 2\}, \{1, 3\}, \{2, 3\}, N\}$. In general, with n players, the set S_C has 2^n elements.

Definition 7.1 (*The Coalitional Form*) The coalitional form of an n-person game is given by the pair (N, v), where N is the set of players and v is a real-valued function, called the *characteristic function* of the game, defined over the set S_C of all the potential coalitions $v : 2^n \rightarrow \mathbb{R}$, and satisfying:

1. $v(\emptyset) = 0$, and
2. if P and Q are disjoint coalitions ($P \cap Q = \emptyset$), then $v(P) + v(Q) \leq v(P \cup Q)$

The first condition states that the empty set has value zero, and the second states a superadditivity property, i.e., a synergy effect: two coalitions working together have a value at least equal or greater than working apart. A game in coalitional form where $v(S) + v(\bar{S}) = v(N)$ for all coalitions $S \in S_C$ is denoted as a *constant-sum game*. In addition, if $v(N) = 0$ then the game is denoted as *zero-sum*.

Observe that we are not deciding how this value $v(C)$ is divided among the members of coalition C, and that also the value itself depends on every game.

7.6.2 Stages for Cooperating

What are the actions needed to start cooperating and create a coalition? Sandholm in [32] describe three key stages for cooperation to appear:

1. *Coalition structure generation*: The first thing that a player has to decide is if he joins a coalition or not, and in the first case, to which coalition it joins. Then, given a player has joined a coalition also appears the problem if it is better to leave it to be alone again, or if it is better to join a more useful coalition to maximize its own utility. Hence, this concept is related with stability, so the first thing we need to know is what coalitions are stable, and for this we will describe soon the *Core* concept.
2. *Maximizing the utility of each coalition*: Then, once created, it is needed to optimize the behavior of each coalition deciding the collective plan to play in order to get the highest reward. Here we talk about a team to describe a coalition having a tactic or a plan to play the game.
3. *Dividing the payoff in each coalition*: Finally, each coalition needs to be fair with its members, in order to divide the resulting payoff in a way that takes into account the individual contributions. We will introduce the Shapley value to analyze this last stage.

7.6.3 Imputations

Considering the previous definition of the coalitional form it seems reasonable for the agents to form a grand coalition, since its synergy makes that $v(N)$ is as large as any amount obtained by disjoint coalitions. The problem is then to agree on how this amount should be split among the players.

Given a payoff vector $x = (x_1, x_2, \ldots, x_n)$, where x_i is the individual payoff received by player i, we define:

Definition 7.2 A payoff vector x is **group rational** if $\sum_{i=1}^{n} x_i = v(N)$

Definition 7.3 A payoff vector x is **individually rational** if $x_i \geq v(\{i\}), \forall i = 1, \ldots, n$

Definition 7.4 An **imputation** is a payoff vector that is both: group and individually rational.

The set of imputations is never empty due to superadditivity $\sum_{i=1}^{n} v(\{i\}) \leq v(N)$. A game in coalitional form is said to be inessential if $\sum_{i=1}^{n} v(\{i\}) \leq v(N)$, and essential if $\sum_{i=1}^{n} v(\{i\}) < v(N)$. If a game is inessential, then the unique imputation is $x = (v(\{1\}), \ldots, v(\{n\}))$, every player expects its safety level, and there is no tendency to form coalitions. Two-person zero-sum games are all inessential.

7.6.4 The Core

Suppose that an imputation x is proposed to split $v(N)$ among the players. If there is a coalition C, whose return from x is less than what its members can achieve acting isolatelly, then such imputation has an inherent instability.

Definition 7.5 An imputation x is unstable if there is a coalition C such that $\sum_{i \in C} x_i < v(C)$.

An unstable imputation will never be valid, because some of its coalitions will never form.

Definition 7.6 The set of stable imputations is called the Core,
$Core = \{x = (x_1, \ldots, x_n) : \sum_{i \in N} x_i = v(N) \text{ and } \sum_{i \in C} x_i \geq v(C), \forall C \in S_C\}$

The core concept is useful as a measure of stability, but it provides a set of imputations without establishing preferences. Besides, the core can be also empty, but if it is not, then there must be a way for the imputations to cooperate, and to distribute the payoff in an acceptable allocation from its members point of view.

The concept of the core is close to the concept of Nash Equilibrium in non-cooperative games. Recall that a Nash Equilibrium is a strategy profile such that no player has incentive to deviate from it unilaterally. Now, a *strong Nash Equilibrium* in cooperative games is a strategy profile such that any subset of players has no incentive to deviate collectively. Any strong Nash Equilibrium is a Nash Equilibrium but the inverse is not necessarily true.

The core concept has not any notion of fairness in its definition, as it basically deals with stability; so even if it is not empty, it provides a hint about what are the possible imputations, what are the fair ones or which ones are more probable to be used. Next we introduce the Shapley value that deals with these considerations.

7.6.5 *The Shapley Value*

The major concept in cooperative games is the Shapley value, introduced by Nobel prize winner Lloyd Shapley in 1953. We will assign to each game in coalitional form a unique vector of payoffs, called the *value*. The entry i of the value vector may be considered as a measure of the power of player i in the game. The idea is to allocate to each participant a payoff portion according to its relative contribution to the coalition: the higher the contribution of an agent in the coalition, the higher the payoff received.

Definition 7.7 A value function ϕ assigns to each possible characteristic function of an n-player game v an n-tuple, $\phi(v) = (\phi_1(v), \phi_2(v), \ldots, \phi_n(v))$ of real numbers, where $\phi_i(v)$ represents the worth of player i in a game with characteristic function v.

The Shapley Axioms for $\phi(v)$ are:

1. *Efficiency*: $\sum_{i \in N} \phi_i(v) = v(N)$
2. *Symmetry*: If i and j are such that $v(C \cup \{i\}) = v(C \cup \{j\})$ for every coalition C not containing i and j, then $\phi_i(v) = \phi_j(v)$.
3. *Dummy Axiom*: If i is such that $v(S) = v(S \cup \{i\})$ for every coalition C not containing i, then $\phi_i(v) = 0$.
4. *Additivity*: If u and v are characteristic functions, then $\phi(u + v) = \phi(u) + \phi(v)$.

The first axiom states group rationality, i.e., the total value of the players is the value of the grand coalition. The second axiom says that if players i and j contribute the same to a coalition, then the payoffs assigned to i and j should be equal. The third axiom says that if player i is a dummy in the sense that he neither helps nor harms any coalition he may join, then his value should be zero. The fourth axiom is the stronger one, and it states that the arbitrated value of two games played at the same time should be the sum of the arbitrated values of the games if they are played at different times.

But, how do we calculate the Shapley value? Suppose we form a coalition C by entering its players sequentially, i.e., one at a time. As each player enters the coalition, it demands a fair compensation related with how its entry increases the value of the coalition $[v(C \bigcup i) - v(C)]$. But, the payoff a player receives by this scheme depends on the order in which such player enters the coalition. Therefore, each player should take the average of its marginal contribution over all the possible permutations in which the coalition can be formed, and this brings us to the concept of the Shapley value.

Theorem 7.8 (The Shapley value) *Given a coalitional game (N, v), the Shapley value given by $\phi = (\phi_1, \ldots, \phi_n)$ where for $i = 1, \ldots, n$, describes what player i gets and can be obtained by:*

$$\phi_i(v) = \sum_{C \subseteq N \setminus \{i\}} \frac{|C|!\,(n - |C| - 1)!}{n!} (v(C \cup \{i\}) - v(C)) \tag{7.1}$$

Therefore, the Shapley value is a mathematical concept to state that an agent should get the average marginal contribution it gives to a coalition. Of course, it satisfies the fairness axioms described above, and, besides Shapley proved that it is the unique value satisfying those axioms.

But, the problem is how to compute it, as in practice it requires exponential time depending on the number of players, because we must consider every possible permutation of the players in each coalition; thus as soon as the number of players in a game grows, the procedure becomes computationally complex.

7.7 Popular Games

In this section we analyze, using the game theoretical concepts discussed in the previous section, some popular games in strategic form, and we finish the section with the iterated version of the famous Prisoner's Dilemma.

7.7.1 Stag Hunt

The stag hunt is a game that describes a conflict between safety and social cooperation. The scenario comes from the philosopher Jean-Jacques Rousseau in his *Discourse on Inequality*.

Two hunters go out for a hunt, and each one must choose to hunt a stag or to hunt a hare. Each hunter must decide without knowing the decision taken by the other. If one decides to hunt a stag, he needs the cooperation of the other in order to succeed. Any hunter can hunt a hare by himself, but the worth of the hare is less than the stag. This game can be described by the matrix in Fig. 7.5.

In this game there are two pure strategy Nash equilibria: when both players choose the same strategy, either to hunt a stag or to hunt a hare. In addition to the pure strategy Nash equilibria, there is one mixed strategy Nash equilibrium, which depends on the payoffs.

Fig. 7.5 The stag hunt game

		Player II	
		Stag	Hare
Player I	Stag	3,3	0,2
	Hare	2,0	1,1

7.7.2 The Battle of Sexes

The battle of sexes belongs to a more general category of coordination games. Imagine a couple who has to decide what to do tonight. The girlfriend has mentioned the opera, while the boyfriend has suggested a football match, and assume that then both left to work without deciding the final place. The rest of the day they are away from each other, and they cannot communicate. But, they have to meet that night, and obviously each one prefers the place he has proposed; but both would prefer to go to the same place rather than different ones. A possible payoff matrix for this game appears in Fig. 7.6.

This game has two pure strategy Nash equilibria, one where both go to the opera and another where both go to the football game. Observe that if both players decide by any random method what to do (p.e., flipping a coin), then assuming the action done by the other, there is no incentive to change. There is also a mixed Nash equilibrium, where the players go to their preferred event more often than the other, with probability 3/5.

7.7.3 Hawks and Doves

This game models the behavior of two players that may fight (figuratively) like a hawk, or like a dove, when they try to obtain a certain resource. When both players act likes doves, they gently share the resource. When one player act like a hawk and the other like a dove, the hawk gets the resource as the dove flees. But if both players act like hawks, then they fight, and both get hurt at a dangerous level.

The symbolic payoff matrix describing this game appears in Fig. 7.7, where V is the value of the resource under contest, and C is the cost of fighting. In the hawk-dove

Fig. 7.6 The battle of sexes game

		Player II	
		Opera	Football
Player I	Opera	3,2	0,0
	Football	0,0	2,3

Fig. 7.7 The hawk-dove game

		Player II	
		Hawk	Dove
Player I	Hawk	$\frac{V-C}{2}, \frac{V-C}{2}$	$V, 0$
	Dove	$0, V$	$\frac{V}{2}, \frac{V}{2}$

Fig. 7.8 An example of the hawk-dove game

		Player II: Hawk	Player II: Dove
Player I	Hawk	-1,-1	2,0
	Dove	0,2	1,1

game it is usually assumed that the value of the resource is less than the cost of a fight ($C > V > 0$), otherwise it becomes the one-shot prisoner's dilemma game described next. Figure 7.8 provides an example of the Hawk-Dove game for $V = 2$ and $C = 4$.

The hawk-dove game has two pure Nash equilibria corresponding to the strategies (Hawk, Dove) and (Dove, Hawk), and has been extensively used to model male contests in Biology and in nuclear warfare scenarios. From game theory point of view, this game is also known as the *snowdrift* or the *chicken* game, which has its origins in a competition in which two players drive their cars towards each other on a collision course. At least one of the players must swerve, or both players may die in the crash, but if one driver swerves and the other does not; the one who swerved will be called a "chicken" meaning a coward. This game was supposedly popular among juvenile delinquents in America during the 50s. A version of this game was made immortal by James Dean in the film *Rebel without a cause*, where the contenders drive towards a cliff, and the first to swerve was the loser.

7.7.4 The Prisoner's Dilemma (PD)

Two members of a criminal band are arrested, imprisoned, isolated and accused for a crime. Due to this solitary confinement, each one has no means communicate with the other. The Police knows it does not have enough evidence to accuse them of murder. Hence, the Police offers each prisoner a deal, where each one has the opportunity to betray his mate, by testifying that the other committed the crime. Otherwise, he can cooperate with his mate by remaining silent. These are the outcomes:

1. If both prisoners betray (defect) the other, each ones stays 4 years in prison.
2. If the prisoner A defects his mate B, while B keeps silent, then A will be set free and B will serve 5 years in prison (and vice versa).
3. If both prisoners remain silent (cooperate), both of them will only remain 2 year in prison for a minor charge.

This game is represented in Fig. 7.9, which has negative values representing the years lost in prison. In order to use positive values (utilities), such matrix is usually substituted by the one appearing in Fig. 7.10, which models the same game by just

Fig. 7.9 The prisoner's dilemma game with years in jail (negative values)

Player I \ Player II	Cooperate	Defect
Cooperate	-2,-2	-5,0
Defect	0,-5	-4,-4

Fig. 7.10 The prisoner's dilemma game with positive utility units

Player I \ Player II	C	D
C	3,3	0,5
D	5,0	1,1

Fig. 7.11 The symbolic payoffs for the prisoner's dilemma game

Player I \ Player II	C	D
C	R,R	S,T
D	T,S	P,P

adding 5 units to every payoff in the previous matrix. In this new matrix cooperation is substituted by C and defection by D as usually happens in the game theory literature.

How to play this game, and why is it a dilemma? Looking at the payoff matrix we realize that the best outcome for both players occurs when both cooperate, as they earn 3 units each, but if one of them defects then he gets 5 units and the other zero. Hence, the temptation to defect is very strong, and the same happens with the other player as we have again a symmetric matrix. But, if both players defect they just get 1 unit each, which is the worst global outcome. The values inside the payoff matrix are usually represented by the variables that appear in Fig. 7.11, where the meanings are: (T) is the temptation payoff to defect, (S) is the sucker's payoff for cooperating when the other defects, (P) is the punishment payoff when both defect, and finally (R) is the reward payoff when both agents cooperate. In order to have a valid PD matrix the next rule is required: $T > R > P > S$.

The way to understand the game is to analyze it using the tools we have seen in the previous section. Player one should think, if the other player cooperates my best response is to defect, and if the other player defects my best response is to defect too; therefore player one defects. The same accounts for player two and the rational result is that both players defect obtaining the worst global payoff. In fact, the defection (D) strategy dominates the cooperation (C) one, and the outcome (D, D) is the unique

Nash equilibrium. At the same time (D, D) is the only point which is not Pareto optimal.

Thus, both players realize that it would be much better to cooperate, but as the temptation is too high, then here comes the dilemma.

The prisoner's dilemma models multiple interactions in nowadays global world. From nuclear negotiations to economic or social scenarios. Take for instance the *Tragedy of Commons*, that describes a scenario where a set of self-interested players may abuse of a common resource. In 1833 the English economist William Forster Lloyd published a scenario describing the use of a common piece of land by the village farmers, where they can let their cows graze. In English villages, shepherds also had sometimes grazed their sheep in common areas, and sheep ate grass more severely than cows do. Under this scenario, a problem of overgrazing could result, because shepherds will get the benefits of his sheep grazing in the common land, while the whole group share the damage. If all the shapers behave in the same way, the result is that the common resource is altered and finally removed. "Commons" has here multiple significances in common resources like atmosphere, oceans, rivers, fish stocks, tax payment, public funding, peer-to-peer networks, or any other shared resource which is not formally regulated.

Usually, prisoner's dilemma is understood as a nightmare for human cooperation as the rational result is pessimistic.

7.7.5 The Iterated Prisoner's Dilemma (IPD)

Looking at the prisoner's dilemma game, and the inherent philosophical conclusions, perhaps we could end up with a pessimistic perception about the future of the humankind concerning common shared resources in the planet. But, the future is not necessarily so dark, as we will see now considering a more realistic, and time dependent application of the prisoner's dilemma game. What happens when you play the PD game more than once? Then you get the iterated PD version of the game. In this case, besides the basic PD rule $T > R > P > S$; another rule, $2R > T + S$, is required to play the game, preventing to alternate cooperation and defection, and obtaining a greater reward than mutual cooperation.

Imagine now that both players keep playing the PD game many times and both know and remember what the opponent has played in the previous rounds. If you can meet again your PD opponent in the future, then the incentive to defect is reduced, mainly for two reasons:

1. If a player defects, then the opponent can punish such behavior defecting during the next rounds.
2. If you try to cooperate, and get a defection, then such loss of utility is not so relevant if the game extends during multiple rounds. On the other hand, the potential for achieving mutual cooperation, and the best common outcome, increases.

In fact, if you play the PD an infinite number of times then the *shadow of the future* forces cooperation as a rational outcome ([5], p. 358). We do not interact with others an infinite number of times, but one-shot games are also rare with people in your social network, so social and rational cooperation can emerge. But we must be careful yet: imagine that you play a certain number of rounds known by both players, p.e., ten rounds. Then on round 10 you know that it will be the last one, so the rational option is to defect as in the one-shot game. Then round 9 becomes the last *real* round, and again the rational option is to defect. Continuing and applying backward induction leads to the conclusion that in the IPD, with a known fixed number of rounds, defection is the rational dominant strategy ([5], p. 354).

So now, coming back to the real world interactions, we know that if we always play a finite number of interactions with other players, so does it mean that we are doomed to defection? Well, things are not as bad as they could look like, even the number of interactions is finite we can rely in the hope of cooperation as long as we expect to meet the other player again, i.e., if we are not sure that a certain interaction is the last one. Thus, rational cooperation is still possible if both players hope to meet again in the future. Besides, even we can meet natural born defectors along our lives, we can get benefitted by finding people more oriented to cooperation and interacting with them. Finally, real games have mechanisms to enforce defection or at least to reduce it to a certain acceptable level.

Axelrod's Tournaments

In 1980, Robert Axelrod organized a public tournament to play the iterated prisoner's dilemma among several players, and he invited game theorist in economics, sociology, political science, and mathematics. Axelrod, as a political scientist, was interested in see how cooperation can emerge in societies of self-interested agents. The contenders have to submit a computer program to play a multiplayer finite IPD game. He received 14 entries, and he added a totally random strategy. All of them were paired with each other in a round robin tournament [1]. The rules for the game were the next:

- The game length was 200 rounds.
- Each program had available the previous choices made by its opponent, in order to play the next round.
- The winning program was the one that got the best accumulated payoff score, considering all the games played among all its opponents.

The computer programs, sent to the contests, ranged from a few lines of coding to hundreds of lines. As an example, in one of the programs, the player models the behavior of the other player as a Markov process, and then uses Bayesian inference to select what seems the best choice for the long run. Alternatively, some examples where very simple strategies to play the game:

- RANDOM: This strategy selects C or D with equal probability.
- ALL-D: This is the rational strategy in the one-shot PD, i.e., to always defect.
- ALL-C: This is the most naive strategy, and it cooperates all the time.

- SPITEFUL: Cooperates until the opponent defects, then defects all the time.
- TIT-FOR-TAT (TFT): This strategy cooperates in the first round, and then next rounds it does what the opponent did in the previous round.

TFT was the simplest strategy submitted to the tournament, with only a few lines in Fortran, and developed by Anatol Rapoport, a mathematical psychologist. Surprisingly, TFT was the winner, and even more significant than TFT's initial victory, perhaps, is the fact that TFT again won Axelrod's second tournament done the next year, where 62 entries were submitted from six countries. Robert Axelrod in his book *The Evolution of Cooperation* (1984) [2] describes these two tournaments, and he compiled a set of rules to succeed in the multiplayer IPD:

1. Be nice: try to cooperate, and never be the first to defect. This was a success predictor in the game, as 'nicer' programs succeed as a general rule.
2. Be reactive to opponent actions: TFT has a perfect balance between retaliation and forgivingness. It is initially cooperative, but it responds to opponents defection by punishing, but without revenge; and forgiving opponent's defections as soon as possible, rewarding cooperation.
3. Don't be envious: be fair with your opponent, your objective is to earn as much payoff as possible, not to beat the others.
4. Don't be too clever: the winner was the simplest program, while other programs showed a behavior much more complex, it was difficult to understand by the opponents.

Noise and Communication

From the previous rules, we can deduce that there is no *best* strategy for the iterated prisoner's dilemma game, as it depends on the type of players participating in a tournament, and its rules to play. In general, each player must figure out his opponent's strategy, and then pick a strategy that is best suited for the situation. Observe also that the Axelrod's competition rewarded the highest overall payoff obtained adding all the encounters. If, for instance, the winner was the strategy that wins more encounters, then no one can beat the ALL-D strategy. Therefore, depending on the players of a game, and its context, there can be more chances to get good results with a particular strategy. For instance, when an IPD tournament introduces noise (errors or misunderstandings) TFT strategies can get trapped into a long string of retaliatory defections. In 1992 Martin Nowak and Karl Sigmund shown [27] that a strategy called *Pavlov* performs better in these circumstances. Pavlov cooperates at the first iteration, and whenever both players do the same at the previous round, but it defects when both players behave different at the previous round. Pavlov, also known as win-stay, lose-switch, resembles a common human behavior that keeps the present strategy while winning, or change to another one when losing.

A team from the University of Southampton in England (led by Professor Nicholas Jennings) introduced a new strategy at the 20th-anniversary iterated prisoners' dilemma competition, which became more successful than TFT. They submitted 60 programs to the competition, among 223 total entries, which were designed to recognize each other through a series of five to ten moves at the beginning. On the one hand, if the recognition was made, one program would always cooperate and the other would always defect, assuring the maximum number of points for the defector. On the other hand, if the program realized that it was playing a non-Southampton player, it would continuously defect in an attempt to minimize the score of the competing program. As a result, Southampton players ended up taking the top three positions in the competition, as well as multiple positions towards the bottom. Southampton strategy took advantage of the fact that multiple entries were allowed from a team in this particular IPD competition. Nevertheless, this idea of communicating among players in a team was pointed out time before by Richard Dawkins in a latter edition of his book The Selfish Gene [9]. In any case, such competition just reinforces the value of communication and cooperation among players acting as a team.

7.7.6 Similar Games and Mechanisms for Enforcing Cooperation

Previously, we followed a discussion about the 2-player version of the prisoner dilemma, then we extended the basic game repeating the interactions for a finite number of times (n-step PD), afterwards we paired the players in a 2-player n-step tournament (Axelrod's tournament). The natural extension is to consider what could happen if we model a n-player prisoner's dilemma version with all the agents playing at the same time (denoted as the NIPD) [7]. Strategies that play well in the IPD do not get so good results in the NIPD, and, in general, it seems to be more difficult to evolve cooperation as the group size increases.

What happens if the players want to keep a nice perception from other players point of view to have more advantages? If there is some way to recognize a player with bad reputation (i.e., a defector) then players can adapt to such circumstance and take a more defensive strategy. Under this perspective, cooperation in games can consider prior interactions with other players, i.e., reputation allows evolution of cooperation by indirect reciprocity. In a highly simplified example, the donation game is used to show how the mechanism of indirect reciprocity operates using players' reputation to promote cooperation [28]. Unlike the case of direct reciprocity, whereby any altruistic act of helping to another player is returned, in indirect reciprocity the altruistic act of helping others is perceived by the community as helpful, providing good reputation, and receiving help in return by other players. Indirect reciprocity is also associated with interactions having short encounters (e.g., one-shot interactions) whereby the effects of direct reciprocity on the interaction outcome are minimized. However, strategies can evolve to use reputation as a mechanism to estimate behaviors of future partners and to elicit cooperation right from the start of interactions. Cooperation occurs when strategies evolve to maintain high reputation scores.

Similar to the NIPD, but from a different perspective, there are a set of games known as *Public Good Games (PGG)* have got attention from the scientific community along the last years. The PGG is a standard of experimental economics, where players choose to put a certain number of their private tokens (cooperate) into a public pot, or to do not put them (defect). The tokens in the pot are multiplied by a factor (greater than one and less than the number of players) and this *public good* payoff is evenly divided among players. In the PGG, the global payoff is maximum when all the player contribute in the stated value, however, the Nash equilibrium in this game is just zero; because any rational agent gets the benefit of secrecy to do not contribute. In PGG games, the average contribution typically depends on the multiplication factor [16], and similar to the IPD, there is an iterative version of the PGG where players do it a certain number of rounds.

In games like the NIPD or PGG, when there is low level of transparency and players can benefit from the secrecy of their actions, hidden in the global group behavior, defective strategies tend to become more frequent. Curiously, these games model much better many social, economic and human behaviors that the PD or the IPD. For instance, tax payment, the free-rider users in Peer-to-Peer networks, and many other scenarios where users may enjoy some level of anonymity may be well modeled by these games. Therefore, there is a need for external mechanisms to increase the level of cooperation.

A classical mechanism, for instance used by governments to achieve tax compliance, is *inspection*; where a central entity inspect with a certain probability the behavior of the players in the last round and, if the contribution is inadequate, impose a penalty, i.e., a certain cost for such actions. Varying the inspection probability we can tune the level of game transparency. Of course, if we set a inspection probability of 100% then we can identify all defectors, but this probably needs a lot of resources from the inspecting entity. So the other parameter that becomes relevant is the penalty for defecting: the bigger it is, the lower interest in non contributing behaviors. Here, rational players determine the resulting game form, taking into account the penalty and the inspecting probability, and then take their decisions accordingly.

Other interesting mechanism is to allow cooperators to punish defectors. *Punishment* has some differences, compared with inspection: firstly, there is no central entity, and secondly, punishment is usually performed a certain cost to the punishers. This cost models, for instance, that the relation with such player becomes weakened. Axelrod used this model in his interesting paper from 1986 to model norm emergence [3].

7.7.7 Social Altruism

Social interactions are strongly related with concepts like altruism that has puzzled biologist and anthropologist for decades. Martin Nowak, in his article from 2006 titled *Five Rules for the Evolution of Cooperation*, put some light to explain the emergence of altruism in human societies from a game theory point of view. Nowak describe five mechanisms by which natural selection in Evolution can lead to cooperation [26]:

- Kin selection, when natural selection can favor cooperation if the donor and the recipient of an altruistic act are genetic relatives.
- Direct reciprocity, happening in games with repeated interactions, like the IPD, where it pays off to cooperate to receive future cooperations.
- Indirect reciprocity, as theoretical and empirical studies of indirect reciprocity show that people who are more helpful are more likely to receive help.
- Network or social reciprocity, which relies on geographical or social factors to increase the interactions with nearer neighbors.
- Group selection, which determines that groups with a higher percentage of cooperators are more successful as a whole, and grown faster that groups with a high percentage of defectors.

7.8 Evolutionary Game Theory (EGT)

Game theory was initially conceived for human interaction, but it turned out as a very active field of research in biology and related sciences. The reason was the seminal work done by John Maynard Smith and George Price in 1973 in his article *The logic of animal conflict* [22]. The quest was to find a realistic model to predict how animals behave when competing for a resource, and the answer led to evolutionary game theory, as a way to describe animal contests as games with strategies, and the underlying mathematical criteria needed to predict the evolution of such competing strategies. EGT has become a major vehicle to help to understand some fundamental questions in biology like group selection, sexual selection, altruism, parental care, co-evolution, and ecological dynamics.

EGT does not need that players act rationally, instead, the notion of rationality is replaced with the much weaker concept of reproductive success. Players only must have a strategy, and the evolutionary game will show how good it is. In fact, Evolution by natural selection tests alternative strategies for their ability to survive and reproduce. In Biology, strategies are genetically inherited and control individual's actions, just like a computer program, by means of their selfish genes [9]. The success of a strategy does not depend only on how good it is isolated, it depends how good such strategy plays against other alternative strategies, depending on their relative frequencies within a competing population. Besides, it is also relevant how good a strategy plays against itself, if eventually it dominates a population. Thus, EGT differs from classical game theory by focusing more on the dynamics of strategy change, depending on the quality of the competing strategies, and also on the effect of the frequency of those competing strategies over the whole population. Under this model, the payoff utility is measured in fitness units, that describe the reproductive success: strategies that are successful on average will be used more frequently, and prevail in the end.

7.8.1 Replicator Dynamics

According to Charles Darwin's paradigm of Evolution by Natural Selection, the species that live today are the ones that better fit to the dynamic environments and competitors faced along millions of years. In EGT, the *Replicator Equation* models mathematically such reasoning describing the ability to reproduce in terms of the average offspring. This ability is related with the advantage of a certain population in terms of fitness.

The dynamics of the game assume that each strategy is played by a certain fraction of individuals. Then, given a strategy distribution, individuals with better payoff in average will be more successful than others from a natural selection point of view, i.e., they will have a greater offspring, so their proportion in the population increases over time. This, in turn, affects the global popularity and distribution of the strategies, which determine what strategies are more successful than others in the next round. These dynamics can be described mathematically, using the most general continuous form, which is given by the differential equation:

$$\dot{x_i} = x_i[f_i(x) - \phi(x)], \quad \phi(x) = \sum_{j=1}^{n} x_j f_j(x) \tag{7.2}$$

where x_i is the proportion of type i in the population, $x = (x_1, \ldots, x_n)$ is the vector of the distribution of types in the population, $f_i(x)$ is the fitness of type i (which is dependent on the population), and $\phi(x)$ is the average population fitness (given by the weighted average of the fitness of the n types in the population). In many cases, particularly in symmetric games with only two possible strategies, the dynamic process evolves to an equilibrium.

7.8.2 Evolutionary Stable Strategies (ESS)

EGT assumes that there is a large population of individuals,[1] that sometimes meet and compete for a certain resource: a piece of land, a couple, food, etc. Under such set of contests, each individual can play one of the strategies presented. For instance, take a look to the hawk-dove game presented again in Fig. 7.12. Remember that a Hawk is a fighter strategy, while a Dove is a fight-avoider one. Remember also that the resource is valued as V, and that the damage for losing a fight is given by cost C.

Thinking qualitatively, if there are too many hawks they will kill each other, so the doves' ratio in the population will grow. On the opposite, if there are too many doves, hawks will also grow in ratio as they usually conquer doves' resources. In an imminent contest, the actual payoff that an individual may expect, depends on the probability of meeting a Hawk or a Dove opponent, which also depends the

[1]The model can be applied to different species or even to males at the same specie.

Fig. 7.12 The hawk-dove game

		Player II: Hawk	Player II: Dove
Player I	Hawk	$\frac{V-C}{2}, \frac{V-C}{2}$	$V, 0$
	Dove	$0, V$	$\frac{V}{2}, \frac{V}{2}$

percentage of Hawks and Doves in the population. Hence, we can match the ratios of hawks and doves with a mixed strategy, where the probability of selecting an action is equal to the ratio of each strategy. In the end, the ratios of hawks and doves will converge to an equilibrium that depends on the payoff matrix, and this point of convergence is denoted as an Evolutionary Stable Strategy (ESS).

The ESS is similar to a Nash equilibrium[2] in classical game theory. An ESS is instead a state of game dynamics where, in a very large (or infinite) population of competitors, another mutant strategy cannot successfully enter the population to really disturb such equilibrium.

To be part of an ESS, a strategy must be effective against competitors, when it invades a population, but also be successful later, when having a dominant position has to compete against itself, or against new invaders. For population dynamics, we can understand an ESS by saying that if the ratios of population match an ESS, then any mutation of populations, which leads to a deviation from the ESS, will not succeed, i.e., the deviation will disappear in time. Hence, mutations cannot invade the populations of an ESS.

For the sake of clarity, in the previous descriptions we have skipped some mathematical details about how to obtain the replicator dynamics equation, and also how are the population dynamics when an ESS is invaded by a mutant population. We refer to Ken Binmore's excellent book [5] for a more detailed introduction and discussion.

Given a strategy distribution x, for any $n \times n$ symmetric game, we have that [11]:

- Every Nash equilibrium is a steady state for the replicator dynamics.
- A stable steady state of the replicator dynamics is a Nash equilibrium.
- An ESS is an asymptotically stable steady state of the replicator dynamics.

where a stable steady state is one that, after suffering a small perturbation, is pushed back to the same steady state by the system's dynamics. The last result comes naturally from the definition of an ESS, but observe that the converse is not true, i.e., a stable state in the replicator dynamics does not need to be an ESS.

Therefore, an ESS state can be obtained solving the replication dynamics differential equations, or alternatively by solving the equations to determine the stable stationary points and analyzing if they are robust against perturbations. This is illustrated over the next example.

---

[2]Remember that a Nash equilibrium is a game equilibrium where it is not rational for any player to deviate from the present strategy they are applying.

Considering the Hawk-Dove game with $C > V$, we try to find the conditions for a static population, where the fitness of Hawks will be exactly the same as the fitness of Doves, so having the same growth rates.

Assume a particular strategy distribution such that the chance of meeting a Hawk player is p, so the chance of meeting a Dove player becomes $(1 - p)$. Now we must search for an equilibrium point where the payoff of a hank P_h and the one of a dove P_d are equal. We have:

$$P_h = p.\frac{V - C}{2} + (1 - p).V \tag{7.3}$$

$$P_d = p.0 + (1 - p).\frac{V}{2} = (1 - p).\frac{V}{2} \tag{7.4}$$

Now equating both fitnesses, we have:

$$p.\frac{V - C}{2} + (1 - p).V = (1 - p).\frac{V}{2} \tag{7.5}$$

and solving for p we obtain a steady state with a Hawk distribution $p = \frac{V}{C}$, which is, after analyzing its stability under mutant invasions, an ESS for such game.

This simple example shows that when the risk of losing a fight (the cost C for injury or death) is greater than the value of winning a reward V, which is the normal situation in the natural world; then the population ends in an ESS (a mixed strategy) where the population of Hawks is V/C. The population will progress back to this equilibrium point if any new Hawks or Doves make a temporary perturbation in the population. The solution of the Hawk-Dove game explains behaviors, actually observed in Nature; for instance why most animal contests involve only *ritual fighting* rather than fatal battles.

7.8.3 Cyclic Behavior

In order to explore cyclic behaviors we will introduce the popular game of rock-paper-scissors. The rules are simple: rock beats scissors (blunts them), scissors beats paper (cut it), and paper beats rock (wraps it up). The payoff matrix is represented in Fig. 7.13, and the Nash equilibrium for this game is to play a mixed strategy with equal probability for each pure one. If this game is played only with the pure strategy (either Rock, Scissor or Paper) then the evolutionary game is dynamically unstable, as mutant strategies can invade pure populations, triggering a cyclic invasion behavior over the three pure strategies.

This game resembles some situations in Nature, where more than two species are present. For instance, competition between two species might indirectly help a third one to enter into the ecosystem. In such cases, the distribution of species can be cyclic over time.

Fig. 7.13 The rock-paper-scissors game

Player I \ Player II	Rock	Paper	Scissors
Rock	0,0	−1,+1	+1,−1
Paper	+1,−1	0,0	−1,+1
Scissors	−1,+1	+1,−1	0,0

7.8.4 Coevolution

We have seen two types of dynamics up to now, one where the evolutionary game reach a stable situation denoted as an evolutionarily stable strategy, and another one, where the evolutionary game exhibit a cyclic behavior, and the proportions of strategies continuously cycle over time. A third dynamic can be found in Nature, which contain not only intra-species competition but also inter-species competition as well, and it is denoted as *coevolution*. We can define two different types of coevolution, one is competitive (as predator-prey or host-parasite competitions) and another one is mutualistic (as the relations of some insects or birds with some flower plants).

In competitive coevolutionary systems, adaptions that are better for competing against another counter specie are promoted. But, the *counterstrategy* of the competitor specie is similarly affected, creating an overall competitive arms race. The global effect is denoted as a *Red Queen* dynamics where, as in Alice in Wonderland, the protagonists must *run as fast as they can, just to keep at the same place*.

Several EGT models have been produced to encompass coevolutionary situations, and use complex mathematical models to describe them [29].

7.8.5 Extensions of the Evolutionary Game Theory Model

Following Maynard Smith's seminal work in EGT, there has been a variation in the models to extend the results, and to simulate different conditions changing the parameters and the topology where the different elements interact. Some of these key extensions to EGC are:

Finite Populations

Evolutionary games have been modeled and simulated considering finite populations rather than infinite ones. In most cases this does not significantly change game dynamics, but in others there are significant differences, for example in relation with the ratios in mixed strategies.

Spatial Games

Spatial game models describe topologically the interactions among the players by using a regular set of connections, usually a lattice of cells over a two o three dimensional grid to represent this spatial component. The local dimension limits interactions to immediate neighbors, and successful strategies propagate over these immediate neighborhoods, and then go further with adjacent ones over the next generations. These models has been especially interesting to analyze the spatial interactions among defectors and clusters of cooperators in the IPD [25], where Tit for Tat (TFT) is a Nash Equilibrium but not an ESS.

Memetics

The memetics concept comes from the book The Selfish Gene [9], where Dawkins proposes that social ideas, what he calls *memes* are a non-organic replicator form. In Dawkins' view, the fundamental characteristics of life are replication and evolution. In biological life, genes serve as the fundamental replicators; while in human culture, memes are the fundamental ones. Memetics belong to evolutionary games because the evolutionary process is essentially a scenario of replication dynamics based on survival of the fittest [23].

In the memetics model, less successful individuals and groups within a population imitate the behavior of the more successful ones in order to improve their competence for resources. Accordingly, the better an individual is perceived, the more others copy his behavior. As a result, the population establishes, and self-enforces over time, standards of *normal* behavior. Memes got popular as internet media pieces, jokes, ideas, etc. but also include tunes, catch-phrases, taboos, beliefs, scientific ideas, and fashions among others.

Complex Network Games

The basic principle of the regular structures in spatial games can be abstracted into a more general and complex network of interactions. This is the foundation of evolutionary graph theory [25].

7.9 Behavioral Game Theory

There are differences between humans and game theoretical computer programs, when they play games, because the concept of rationality is not necessarily used most of the time by human players. When trying to model human behavior, there is an experimental branch of game theory, namely behavioral game theory, that studies human player decisions. While traditional game theory focuses on mathematical equilibriums, perfect rationality and utility maximizing; behavioral game theory uses experimental psychology and experimental economics. In general, the results obtained show that human choices are not always rational, neither always try to maximize utility.

This experimental branch of game theory started in the fifties with the Allais paradox, and afterwards in the sixties with the Ellsberg paradox [12]. Both paradoxes show that choices made by participants in a game do not reflect the benefit they expect to receive from making those decisions. Later on, in the seventies, Vernon Smith shown the advantages of considering an experimental perspective in the analysis of economic markets, rather than only using a theoretical one. Within this economic framework, other economists conducted experiments that discovered variations of traditional decision-making models such as regret theory, prospect theory, and hyperbolic discounting [12]. In the eighties researchers started considering the factors that influence decisions. The ultimatum game, the dictator game and bargaining games examined how good are humans to predict opponent's behavior. In the 2000s researchers considered new models based on the rational choice theory, but adapted to reflect decision maker preferences, and to attempt to rationalize choices that did not maximize utility [12].

Traditional game theory uses rational choice theory and players' perfect knowledge in order to predict utility-maximizing decisions and also the opponents' strategies. Therefore, it is a normative theory whose aim is not to explain the reasons underlying human decisions. Behavioral game theory is a positive theory rather than a normative one [8]. Therefore, it seeks to describe phenomena rather than prescribe a correct actions to take. Positive theories must be testable and can be proven true or false.

To sum up, behavioral game theory attempts to explain strategies making using experimental data, in order to find the factors that influence real world decisions. The discoveries found in behavioral game theory show that human decision makers consider many factors when making choices including regret, emotions, bounded rationality, cultural influences, and reputation among others. Economy is a natural area where it has been successfully applied, leading to the appearance of Behavioral Economics that questions classical assumptions, p.e., rational approaches and equilibria, at least at a micro-scale level.

7.10 Mechanism Design

Mechanism design is a relatively recent sub-field of game theory that study solution for a kind of games with a *reverse approach* than classical game theory. In traditional approaches, the game, its rules and *mechanisms* are already given, and the aim is to find the best strategies to play and the equilibrium solution. In mechanism design, the known information is the objective function, while the mechanism is the unknown. In 2007 Leonid Hurwicz, Eric Maskin, y Roger Myerson won the Nobel Prize in economy for stating the basis for the theory of mechanism design.

These games aim to provide incentives for players when they behave as the designer pretends, i.e., the game has been designed accordingly to the expected result. Two key characteristics of these games are:

- that designers choose the game design rather than inherit it, and;
- that the designer is mainly interested in a certain outcome of the game.

A classical example is the auction designing problem. In auctions, the main player is the seller, who wants to sell its good at the highest possible price. There is also a set of buyers, who have their own valuation for the good. Then, the seller's mechanism design problem is to define an auction system that provides him the higher payoff. For this problem there are several types of auctions adapted to several conditions and scenarios. The most popular type of auction is the English one, where buyers bid prices, and the buyer that provides the highest bid effectively buys the good. One problem with this mechanism is that buyers have incentives to bid undervaluations of the good to obtain it at the lowest possible price. One alternative is the Vickrey auction, where the good goes to the highest bidder, but the price is defined by the second-highest bid. This auction gives incentives to players to be truthful in their bid considering their good valuation. In fact, no matter what others do, each player has the incentive to be truthful, because truthfulness is a dominant strategy. There are many other examples of mechanism design in many other scenarios like market regulation, voting schemes or sport competitions.

7.11 Heuristic Game Coalitions

The use of coalitions is strongly related with cooperative game theory, but in this book we are interested in how coalitions can help to introduce cooperation among selfish agents that play in competitive environments. The idea behind is to evaluate how a number of players can self-organize into groups in a way that helps to improve their own benefit; but, at the same time, also maximizing the reward for the whole game community.

As we have seen in Sect. 7.6, the coalition structure generation (CSG) problem is computationally challenging, even for small scenarios. This happens because finding a certain bounded solution close to the optimum in a CSG becomes a computationally too hard problem when the number of agents increases, or when the relations among those agents and their environment are: partially observable, real-time, dynamic, uncertain, and/or even noisy. This have made the research community to develop a set of heuristics for solving the problem under different assumptions and approaches. Thus, in practice, the objective becomes to improve the coalition formation process, rather than to find an optimal solution, and to design adaptable agents that self-improve over time. In these cases the only practical approach is to use metaheuristics, which do not guarantee optimal solutions, but usually can be applied to very large problems, i.e., consisting of *many* agents, from a computational point of view. A detailed survey of many of those approaches, mainly centered in dynamic-programming and anytime algorithms, appears in [31] from a multi-agent point of view. In this last section, we review some metaheuristics approaches to deal with coalitions in games.

From the point of view of the CSG, among the first researches to analyze coalition formation, we find Shehory and Kraus [34], that considered a decentralized, greedy algorithm for coalition structure generation. The algorithm ignores coalitions when they contain more than a certain number of agents, and it constructs the coalition in a greedy manner at every iteration; joining the best remaining coalition member to the coalition structure using a distributed search by means of agent negotiation. Mauro et al. [21] proposed another greedy algorithm based on GRASP, a general purpose greedy technique, which after each iteration performs a quick local search to try to improve its solution and construct iteratively the coalition structure. Another greedy algorithm, named C-Link, was proposed by Farinelli et al. [10]. It starts at the top node in the coalition structure graph, and then moves downwards in a greedy fashion, i.e., searching for the highest immediate reward, without taking into consideration the future consequences of this choice. All these heuristic algorithms for CSG can not guarantee solution quality.

Another heuristic approach to CSG is to use Genetic Algorithms (GA), and one example was proposed later by Sen and Dutta [33]. As usually in this approaches, it starts with a random population composed by a set of coalition structures; and at every iteration it uses the typical three steps of genetic algorithms: (i) evaluating every coalition member, (ii) selection some coalitions, and (iii) recombining their agents. In [37], Yang and Luo also present a GA-based algorithm for coalition structure formation. Other work using evolutionary algorithms is [15], where Gruszczyk amd Kwasnicka introduce an evolutionary algorithm for creating agent coalitions and solving assigned tasks.

Keinanen [17] proposed the use of Simulated Annealing, which is a stochastic local search technique. Starting from a random initial structure, at every iteration, the algorithm moves from the current coalition structure to another one in its neighborhood, where neighborhoods can be defined using a variety of criteria. In [20] Li et al. propose a Quantum Evolutionary Algorithm for solving coalition formation of multi-robots in dynamic environments, where a skillful quantum probability representation of chromosome coding strategy is designed to adapt to the complexity of the coalition formation problem. In [35], Shen et al. present a coalition structure optimization algorithm in MAS based on Particle Swarm Optimization (PSO). The aim is to maximize the summation of the coalition values, and to search for an optimized coalition structure in a minimal searching range.

Optimizing n-skill games is also another way to find the best coalition structure to improve gains, when each agent has to perform a certain skilled task. The Coalitional Skill Games (CSGs) are a restricted form of coalitional games, where each agent has a set of skills, and each task requires a certain set of skills in order to be completed. Then, a coalition can accomplish a task if the coalition's agents cover the set of required skills for such task. In [4] Bachrach and Rosenschein consider the computational complexity of several problems in CSGs.

Traditional models assume that each agent participates in exactly one coalition. However, it is common that in real-life one agent can participate in various groups, and perform one task in each of them. Overlapping Coalition Formation (OCF) games are cooperative games where the players can simultaneously participate in several coalitions. Chalkiadakis et al. propose in [6] a game theoretic model for overlapping coalition formation.

Many of the previous works follow one of the two classical approaches to CSG depending on the assumption of how agents behave: self-interested agents or altruistic agents. The former is the classical approach from game theory, where agents are assumed to be self-interested, i.e., each agent joining a coalition tries to maximize its own benefit, and the payoff obtained by the coalition will be somehow divided among the member agents. The latter are based on economic team theory, also called teamwork [30], where agents are assumed to be altruistic, i.e., all the agents participating in a coalition have a common goal, and they do not have any other private interests. Sometimes this can be assumption imposed to the model in order to achieve the designer goal [14]. However, in many real-world scenarios, agents have to maximize their individual interests, and use coalitions as a way to achieve such private goals, while respecting coalition rules. This is a legitimate behavior that results in environments where self-interest agents sometimes behave altruistically inside coalitions. One of the first approaches within this paradigm was done by Xin Li in [19], where agents form coalitions by hybrid negotiation.

The next chapters contain a mix of different approaches to several problems and scenarios, and the common nexus is the use of self-organized coalitions, usually under a game theoretical framework; to guide the population of self-interested agents to reach a better outcome for the whole population, and for the goals of the game designer. Therefore, this book does not fall within the classical stereotypes of self-interested or altruistic agents; otherwise it fits within the hybrid paradigm, where coalitions are the game tool to introduce cooperation among competitive agents.

7.12 Conclusion

This chapter provides an introduction to game theory, as a research field for studying the mathematical models of conflict and cooperation among self-interested agents. First, we introduced three basic game representations: strategic, extensive and coalitional forms. Then we have described several types of games, and also presented some relevant concepts to analyze the conditions to find solutions in a particular game. We also introduced the basic concepts to deal with games in coalitional form. Then we have described several classical and popular games, and analyze them using the concepts described previously. Later on, we have considered three relevant branches derived from the classical game theoretical core. First, evolutionary game theory, where the notion of rationality is replaced by the concept of reproductive success: strategies that are successful on average will be used more frequently, and prevail in the end. Second, behavioral game theory that tries to explain human decision making, and uses experimental data in order to find the factors that influence such decisions. Third, mechanism design, as a way to create games or interaction models according to a set of rules that allow to emerge a predictable set of strategies and game results.

Finally, we have explored several heuristic approaches for dealing with the coalition structure generation problem in practice, as this is the framework where this book is built.

7.13 Further Reading

As an overall introduction to the topic, I would recommend the excellent book *Fun and Games: A Text on Game Theory* [5] from Ken Binmore. A classical work about evolutionary game theory is the book *Evolution and the Theory of Games* [23] by John Maynard-Smith. Finally, I would also recommend the book *The evolution of Cooperation* [2] by Robert Axelrod about the famous Prisoner's Dilemma experiment performed by the author in the 80s.

References

1. Axelrod, R., Hamilton, W.: The evolution of Cooperation. Science **211**, 1390–1396 (1981)
2. Axelrod, R.: The Evolution of Cooperation. Basic Books, New York (1984)
3. Axelrod, R.: An evolutionary approach to norms. Am. Polit. Sci. Rev. **80**(4), 1095–1111 (1986)
4. Bachrach, Y., Rosenschein, J.S.: Coalitional skill games. In: Proceedings of the 7th International Joint Conference on Autonomous Agents and Multiagent Systems (AAMAS'08), pp. 1023–1030. Estoril, Portugal (2008)
5. Binmore, K.: Fun and Games: A Text on Game Theory. D.C. Heath and Company, Lexington (1992)
6. Chalkiadakis, G., Elkind, E., Markakis, E., Polukarov, M., Jennings, N.: Cooperative games with overlapping coalitions. J. Artif. Intell. Res. (JAIR) **39**, 179–216 (2010)
7. Colman, A.M.: Game Theory and Experimental Games. Pergamon Press, Oxford (1982)
8. Colman, A.M.: Cooperation, psychological game theory, and limitations of rationality in social interaction. Behav. Brain Sci. **26**(02), 139–153 (2003)
9. Dawkins, R.: The Selfish Gene. Oxford University Press, Oxford (1976)
10. Farinelli, A., Bicego, M., Ramchurn, S., Zucchelli, M.: C-link: a hierarchical clustering approach to large-scale near-optimal coalition formation. In: Proceedings of the Twenty-Third International Joint Conference on Artificial Intelligence, IJCAI, pp. 106–112 (2013)
11. Fudenberg, D., Levine, D.K.: The Theory of Learning in Games. MIT Press, Cambridge (1998)
12. Gintis, H.: Behavioral game theory and contemporary economic theory. Anal. Krit. **27**(1), 48–72 (2005)
13. Gardner, M.: Mathematical games the fantastic combinations of John Conway's new solitaire game "life". Sci. Am. **223**, 120–123 (1970)
14. Gleizes, M.P., Camps, V., Glize, P.: A theory of emergent computation based on cooperative self-organisation for adaptive artificial systems. In: Fourth European Congress of Systems Science, Valencia (1999)
15. Gruszczyk, W., Kwasnicka, H.: Coalition Formation in multi-agent systems; an evolutionary approach. In: International Multiconference on Computer Science and Information Technology (IMCSIT 2008), pp. 125–130 (2008)
16. Gunnthorsdottir, A., Houser, D., McCabe, K.: Dispositions, history and contributions in public goods experiments. J. Econ. Behav. Organ. **62**(2), 304–315 (2007)

17. Keinanen, H.: Simulated annealing for multi-agent coalition formation. In: Proceedings of the Third KES International Symposium on Agent and Multi-agent Systems: Technologies and Applications, KES-AMSTA 09, pp. 30-39. Springer, Berlin/Heidelberg (2009)
18. Langer, P., Nowak, M.A., Hauert, C.: Spatial invasion of cooperation. J. Theor. Biol. **250**, 634–641 (2008)
19. Li, X.: Improving multi-agent coalition formation in complex environments. Doctoral Dissertation, The University of Nebraska-Lincoln (2007)
20. Li, Z., Xu, B., Yang, L., Chen, J., Li, K.: Quantum evolutionary algorithm for multi-robot coalition formation. In: Proceedings of the first ACM/SIGEVO Summit on Genetic and Evolutionary Computation, pp. 295–302. Shanghai, China (2009)
21. Mauro, N.D., Basile, T.M.A., Ferilli, S., Esposito, F.: Coalition structure generation with grasp. In: Proceedings of the 14th International Conference on Artificial Intelligence: Methodology, Systems, and Applications, AIMSA 10, pp. 111–120. Springer, Berlin/Heidelberg (2010)
22. Maynard-Smith, J., Price, G.: The logic of animal conflicts. Nature **246**, 15–18 (1973)
23. Maynard-Smith, J.: Evolution and the Theory of Games. Cambridge University Press, Cambridge (1982)
24. Nash, J.: Equilibrium points in n-person games. Proc. Natl. Acad. Sci. United States Am. **36**(1), 48–49 (1950)
25. Nowak, M.: Evolutionary Dynamics: Exploring the Equations of Life, pp. 152–154. Harvard University Press, Cambridge (2006)
26. Nowak, M.A.: Five rules for the evolution of cooperation. Science **314**, 1560–1563 (2006)
27. Nowak, M.A., Sigmund, K.: Tit for tat in heterogenous populations. Nature **355**(6016), 250–253 (1992)
28. Nowak, M.A., Sigmund, K.: Evolution of indirect reciprocity by image scoring. Nature **393**, 573–577 (1998)
29. Perc, M., Szolnoki, A.: Coevolutionary games - a mini review. BioSystems **99**, 109–125 (2010)
30. Pynadath, D., Tambe, M.: The communicative multiagent team decision problem: analyzing teamwork theories and models. J. Artif. Intell. Res. **16**, 389–423 (2002)
31. Rahwan, T., Michalak, T.P., Wooldridge, M., Jennings, N.R.: Coalition structure generation: a survey. Artif. Intell. **229**, 139–174 (2015)
32. Sandholm, T.: Distributed rational decision making. Multiagent Systems, pp. 201–258. The MIT Press, Cambridge (1999)
33. Sen, S., Dutta, P.: Searching for optimal coalition structures. In: ICMAS00: Sixth International Conference on Multi-Agent Systems, pp. 286-292 (2000)
34. Shehory, O., Kraus, S.: Methods for task allocation via agent coalition formation. Artif. Intell. **101**(12), 165–200 (1998)
35. Shen, Y., Guo, B., Wang, D.: Optimal coalition structure based on particle swarm optimization algorithm in multi-agent system. In: The Sixth World Congress on Intelligent Control and Automation (WCICA 2006), pp. 2494–2497 (2006)
36. von Neumann, J., Morgenstern, O.: The Theory of Games and Economic Behavior. Princeton University Press, Princeton (1947)
37. Yang, J., Luo, Z.: Coalition formation mechanism in multi-agent systems based on genetic algorithms. Appl. Soft Comput. **7**(2), 561–568 (2007)

Part II
Self-Organizing Algorithms

Chapter 8
Optimization Models with Coalitional Cellular Automata

Juan C. Burguillo and Bernabé Dorronsoro

Abstract This chapter analyzes the use of adaptive neighborhoods based on coalitions in evolutionary optimization frameworks. First, we introduce the concepts of evolutionary algorithms, population topologies and coalitions. We integrate all these topics to study how to avoid some of the drawbacks of previous evolutionary algorithms and to remove their typically required parameters. The main contribution of the chapter is a redefinition of the Evolutionary Algorithm with Coalitions (EACO), which uses cellular approaches with neighborhoods, allowing the formation of coalitions among cells as a way to create islands of evolution in order to preserve diversity. This idea speeds up the evolution of individuals grouped in high-quality coalitions that are quickly converging to promising solutions. In the results section, we successfully compare EACO with a canonical cGA (Cellular Genetic Algorithm), and provide evidences about the statistical significance of our results. We also analyze the influence of parameters in order to tune them up accordingly; and finally, we evaluate the performance of EACO under different complex network topologies.

8.1 Introduction

Evolutionary algorithms (EAs) are well-known population based metaheuristics [9, 30, 54] to solve complex optimization problems. They work on a set of *individuals* (called *population*), evolving them simultaneously towards (hopefully) better ones by applying some stochastic operators (typically called *evolutionary operators*, e.g., selection, recombination, and mutation). These individuals represent tentative solutions to the problem, and to measure their aptitude a fitness value is assigned to the individuals. This value is a quantitative metric of the quality of the solution

J. C. Burguillo (✉)
Department of Telematics Engineering, School of Telecommunications Engineering, University of Vigo, 36310 Vigo, Spain
e-mail: J.C.Burguillo@uvigo.es

B. Dorronsoro
Computer Science Engineering Department, University of Cadiz, 11519 Cadiz, Spain
e-mail: bernabe.dorronsoro@uca.es

J. C. Burguillo, *Self-organizing Coalitions for Managing Complexity*, Emergence, Complexity and Computation 29, https://doi.org/10.1007/978-3-319-69898-4_8

represented by the individual to the problem being solved, and it is used by the algorithm to guide the search. This way, those individuals with better fitness will survive for future generations more likely than those with a bad one.

Depending on the characteristics of the algorithm, as the problem representation, the population size, or the evolutionary operators applied, there are a number of different families of evolutionary algorithms. The most classical ones being Genetic Algorithms (GA) (originally using binary representations), Evolutionary Strategies (ES), Evolutionary programming (EP) and Genetic programming (GP), but new approaches have emerged recently.

The tradeoff between exploration of new areas of the search space and exploitation of good solutions accomplished by this kind of algorithms is one of the key factors for their high performance with respect to other metaheuristics. This exploration/exploitation balance can be sharpened with some different parameters of the algorithm such as the use of adaptive techniques [5], the design of hybrid or memetic approaches [53], or decentralizing the population [3, 4, 6, 21], among others.

In this chapter, we focus on the last approach, that we find very interesting because decentralizing the population helps in keeping the diversity of solutions for longer, because it slows down the convergence speed, and therefore increases the chance to find better solutions. Therefore, it allows balancing the exploration-to-exploitation ratio performed by the algorithm in the search space at very low computational cost, compared to the other existing techniques, and additionally; it can be implemented in any population based evolutionary algorithm.

Therefore, this chapter proposes several contributions considering recent approaches in the areas of Evolutionary Algorithms, Spatial Cellular Automata, Complex Networks, Evolutionary Game Theory and Coalitions. First, we consider the integration of all these topics in a way to obtain a synergy in the development of Evolutionary Algorithms, with the advantages of Distributed Evolutionary Algorithms (dEAs) and Cellular Evolutionary Algorithms (cEAs), but avoiding their particular drawbacks, and removing their typically required parameters; like the neighborhood to use (cEAs), the islands topology, the migration frequency, and the policies to exchange and discard individuals (in dEAs). This can be done by using cellular approaches with neighborhoods, but allowing the formation of coalitions among cells as a way to create dynamic islands of evolution in order to preserve diversity. Besides, we rely on Evolutionary Game Theory to consider every cell as a player of a game arranged in a two dimensional torus. Altogether, the main contribution of the chapter is a redefinition of the Evolutionary Algorithm with Coalitions (EACO), which uses cellular approaches with neighborhoods, allowing the formation of coalitions among cells as a way to create islands of evolution in order to preserve diversity. In this way, cells will be able to evolve depending on their payoff with respect to their neighbors, and having also the support provided by their coalition members. This approach allows the payoff of a given solution to be defined in terms of how much such solution has improved in the last generations. The idea is to speed up the evolution of individuals grouped in high-quality coalitions that are quickly converging to promising solutions.

The chapter is structured as follows. Section 8.2 introduced the concept of Evolutionary Algorithms, and Sect. 8.3 describes the particular case of Decentralized Evolutionary Algorithms. Then, Sect. 8.4 introduces the concept of population topology in Evolutionary Algorithms. Then, Sect. 8.5 introduces Evolutionary Algorithms, and describes the EACO algorithm. Afterwards, Sect. 8.6 presents a set of complex combinatorial problems we used in our simulations to compare EACO versus the canonical cGA. Section 8.7 presents the promising results obtained by EACO, and finally in Sect. 8.8 we outline our conclusions.

8.2 Evolutionary Algorithms

The evolutionary computation framework [9] stands for a wide set of families of techniques for solving the problem of searching optimal values by using computational models, most of them inspired by evolutionary processes (evolutionary algorithms). Evolutionary algorithms (EAs) are population based optimization techniques designed for searching optimal values in complex spaces. They are loosely based on some biological processes that can be appreciated in Nature, like natural selection [17] or genetic inheritance [50] of good traits. Part of the evolution is determined by the natural selection of different individuals competing for resources in the environment. Inevitably, some individuals are better than others, and those that are better are more likely to survive, learn, and propagate their genetic material.

Sexual reproduction allows some shuffling of chromosomes, producing offspring that contain a combination of information from each parent. This is known as the *recombination* operation, which is often referred to as crossover because of the way that biologists have observed strands of chromosomes crossing over during the exchange. Recombination happens in an environment where the selection of the mating pool is largely a function of the fitness of individuals, i.e., how good each individual is at competing in its environment.

As in the biological case, individuals can occasionally mutate. *Mutation* is an important source of diversity for EAs. In an EA, a large amount of diversity is usually introduced at the start of the algorithm by randomizing the genes in the population. The importance of mutation, which introduces further diversity while the algorithm is running, is a matter of debate. Some refer to it as a background operator, simply replacing some of the original diversity which may have been lost, while others view it as playing a dominant role in the evolutionary process (e.g., avoiding getting stuck in local optima).

In Fig. 8.1 we show the functioning of a typical EA. As it can be seen, an EA proceeds in an iterative way by successively evolving the current population of individuals. This evolution is usually a consequence of applying stochastic variation operators such as selection, recombination, and mutation on the population in order to compute a whole generation of new individuals. The initial population is usually generated randomly, although it is also usual to use some seeding technique in order to speed up the search by starting from good quality solutions. A fitness evaluation

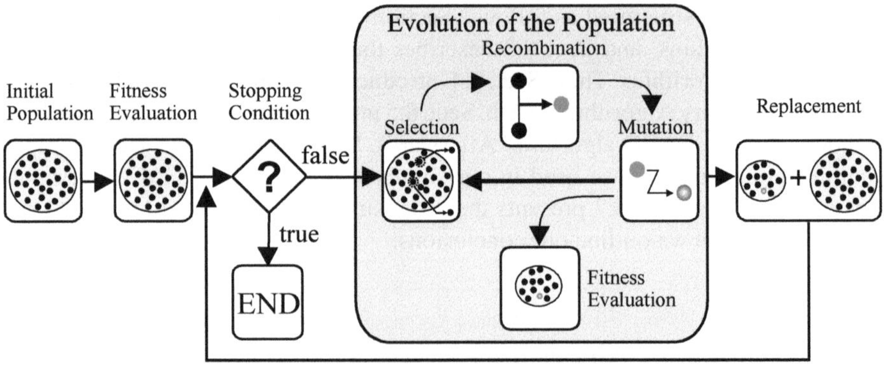

Fig. 8.1 Basic working principles of a typical EA

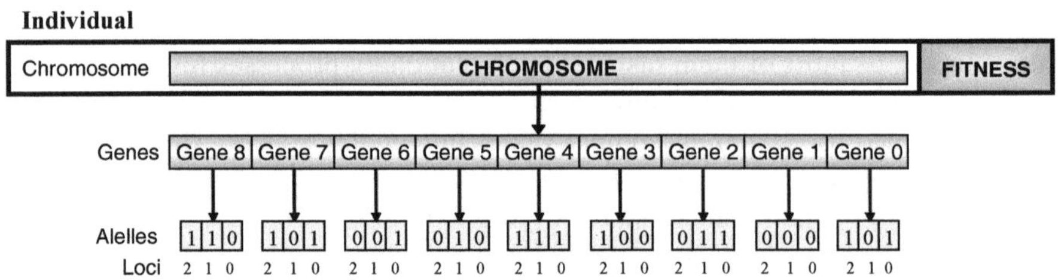

Fig. 8.2 Typical structure of individuals in an EAs

assigns a value to every individual, which is representative of its suitability to the problem in hand. This evaluation can be performed by an objective function (e.g., a mathematical expression or a computer simulation) or by a subjective opinion, in which the best solutions are selected by an external agent (e.g., expert design of furniture or draws using interactive EAs). The stopping criterion is usually set as reaching a preprogrammed number of iterations of the algorithm, or as finding a solution to the problem (or an approximation to it, if it is known beforehand).

Individuals encode tentative solutions to the problem at hands usually in the form of strings (of binary, decimal, or real numbers) or trees. Every individual has assigned a fitness value as a measure of its adequacy, so better fitness values represent better individuals. This fitness value is used for deciding which individuals are better and which ones are worse. We present in Fig. 8.2 an example of a typical individual structure in EAs.

The chromosome encodes the problem variables in some manner. More than one chromosome could possible exists inside the same individual (e.g., diploid representation like in humans [32]). A single fitness value is allocated to the solution encoded in the chromosome(s) of one single individual after decoding them appropriately (e.g., binary to decimal). Every problem variable is a gene, and usually (but not forcedly) every gene encodes a value. The individual positions inside every gene are named loci, and in general alleles are said to be the values stored in every loci.

Problem: maximize f(x) = x^2

String#	String	Fitness	% of the Total
1	01101	169	14.4
2	11000	576	49.2
3	01000	64	5.5
4	10011	361	30.9
Total		1170	100.0

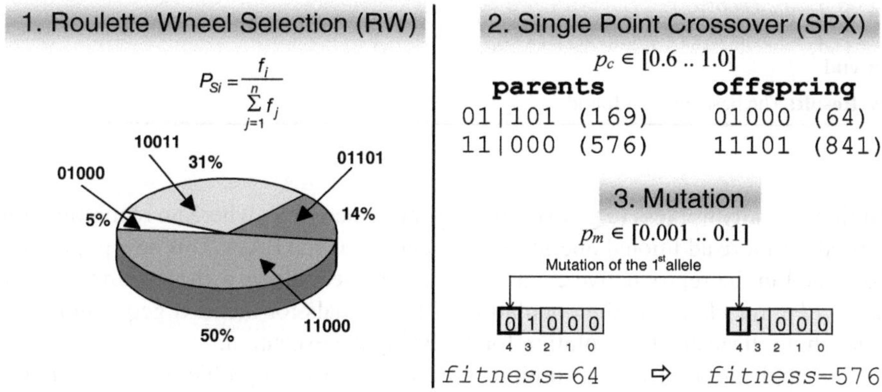

Fig. 8.3 Example of the application of the variation operators in an EA

In Fig. 8.3 we show an example of the application of some specific variation operators in a population composed of 4 individuals for the problem of maximizing $f(x) = x^2$. As it can be seen, the 'String' column in the upper table is the binary codification of the problem variable x using one single gene in one single chromosome in binary. The selection operator chooses the parents with probability equal to the percentage of their fitness values with respect to the sum of the fitness values of all the individuals in the population. The recombination operator (single point crossover) splits the chromosomes of the two individuals into two different parts in a randomly chosen location, and joins the parts of the different individuals in order to generate two new offsprings. Finally, the mutation in this example flips the value of a random loci (the first one in this case), in order to introduce some more diversity, and hopefully getting a better individual, as it is the case in the figure.

Now we analyze the functioning of an evolutionary algorithm in detail. Its pseudocode is shown in Procedure 2. As it was said before, evolutionary algorithms work on populations of individuals which are tentative solutions to the problem. The initial population is usually composed by randomly created individuals, although problem knowledge can help creating faster EAs (e.g., by using a greedy initial feeding of solutions). After the generation of the initial population, the fitness value of each individual is computed, and the algorithm starts the reproductive cycle. This step lies in generating a new population through the selection of the parents, their recombination, the mutation of the offsprings obtained and then, their evaluation. These three variation operators are typical of most EAs, specially GAs, although many EA

Procedure 2: Pseudo-code of an Evolutionary Algorithm

$P \leftarrow$ *GenerateInitialPopulation*();
Evaluate (P);
while *Not StopCondition* **do**
  $P' \leftarrow$ *SelectParents* (P);
  $P' \leftarrow$ *ApplyVariationOperators* (P');
  Evaluate (P');
  $P \leftarrow$ *SelectNewPopulation* (P, P');
end
Result: The best solution found

families usually use less (e.g., evolutionary strategies –ES– where no recombination is used) or more additional operators (e.g., decentralized EAs). This new population generated in the reproductive cycle (P') will be used, along with the current population (P), for obtaining the population of individuals for the next generation. The algorithm returns the best solution found during the execution.

The application of EAs to optimization (and learning) problems has been very intense during the last decades [9]. In fact, it is possible to find this kind of algorithm applied to solving complex problems like constrained optimization tasks, problems with a noisy objective function, or problems which have high epistasis (high correlation between the values to optimize) and multimodality [2, 5]. The high complexity and applicability of these algorithms has promoted the emergence of innovative new optimization models.

Initially, four kinds of evolutionary algorithms [36] could be clearly differentiated. These four families of algorithms were simultaneously developed by different research groups around the world:

- *Genetic algorithms (GAs)* were initially studied by J.H. Holland [34, 35] in Ann Arbor (Michigan), H.J. Bremermann [11] in Berkeley (California), and A.S. Fraser [27] in Sidney (Australia).
- *Evolutionary strategies (ES)* were proposed by I. Rechenberg [58, 59] and H.-P. Schwefel [60] in Berlin (Germany).
- *Evolutionary programming (EP)* was firstly proposed by L.J. Fogel [26] in San Diego (California).
- *Genetic programming (GP)* appeared two decades later, in 1985, as an adaptation of N. Cramer [16] of a genetic algorithm which worked with tree shaped genes, instead of the strings of binary characters traditionally used in GAs, and it is now widely used thanks to the leading works of Koza [44].

Nowadays, the evolutionary algorithms field is growing and evolving itself. An evidence is the number of new families that recently emerged, such as particle swarm optimization (PSO) [10], ant colony optimization (ACO) [19], estimation of distribution algorithms (EDAs) [45], analytic programming [73], and differential evolution [57], among others.

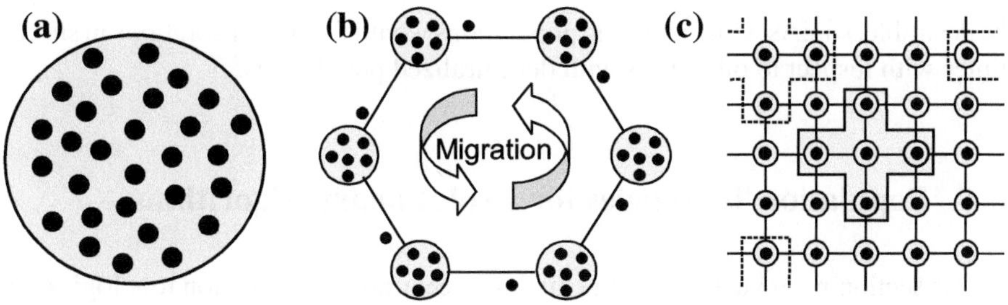

Fig. 8.4 In figure **a** appears a panmictic EA with all its individuals (black points) in the same population. Structuring the population usually leads to distinguishing between distributed EAs in figure (**b**) and cellular EAs in (**c**)

8.3 Decentralized Evolutionary Algorithms

Most EAs use a single population (panmixia) of individuals and apply operators on them as a whole (see Fig. 8.4a). In contrast, there is also some tradition in using EAs with structured, or decentralized, populations. The use of parallel distributed populations is based on the idea that the isolation of populations enables a higher genetic differentiation [71]. In many cases [4], these algorithms using decentralized populations provide a better sampling of the search space, and thus improve both the numerical behavior and the execution time of an equivalent panmictic algorithm.

There are two main ways for decentralizing the population in EAs, namely island or coarse-grained EAs (dEAs) [3], and cellular or fine-grained EAs (cEAs) [6]. On the one hand, in dEAs, the population is split into several smaller sub-populations that are independently evolved by EAs, and exchanging some information (typically the best solution found so far) with other sub-populations. Thus, the EAs in the different islands will perform a fast convergence to hopefully distinct regions of the search space, preserving this way the overall population diversity. The information exchange among islands allows them to benefit from the exploration performed by the others and to introduce diversity in the local sub-populations. On the other hand, in the case of cEAs, the individuals composing the population are arranged in a usually 2-dimensional toroidal mesh, and only next individuals can interact in the breeding loop. Diversity is preserved thanks to the isolation by distance introduced with the use of neighborhoods. The effect is the formation of niches in the population exploring different regions of the search space. At the same time, these niches are not isolated, since due to the neighborhood overlapping, individuals located at the borders of the niches can interact and exchange information.

Therefore, the goal of structuring the population in EAs is to somehow preserve the population diversity for longer, typically at the cost of slowing down the convergence speed of the algorithm. This could be a good strategy for complex multimodal and/or epistatic problems for which a too exploitative behavior results in quick diversity loss in the population, so the algorithm is stuck in some local optimum from which it cannot escape. However, for some other problems, this fast convergence speed could be desirable, so EAs with panmictic populations would find better solutions in shorter times with respect to other EAs with decentralized populations.

8.4 Population Topologies for Evolutionary Algorithms

In this section we review the main existing decentralized population topologies that have been proposed for EAs. There is a large number of papers in this topic, and it is not the scope of this section to revise all of them, but only some of the most outstanding ones.

The influence of using fine and coarse-grained populations in EAs has been deeply investigated in the literature [1, 3, 6, 13, 52, 63, 66, 67]. As we already mentioned, they are only two boundary cases of decentralized populations. Recently, several works appeared studying new population topologies that share properties of both models. They are discussed below.

8.4.1 Cellular Evolutionary Algorithms

Cellular EAs [6, 49, 69] are structured population algorithms with a high explorative capacity. The individuals composing their population are arranged into a (usually) two dimensional toroidal mesh, and only neighbor individuals (i.e., the closest ones measured in Manhattan distance) are allowed to interact during the breeding loop (see Fig. 8.5). This way, we are introducing some kind of isolation in the population that depends on the distance between individuals. Hence, the genetic information of a given individual can be spread slowly through the grid (since neighborhoods are overlapped), and it will need a high number of generations to reach distant individuals (therefore preventing the population from premature convergence). Structuring the population in this way we can achieve a good exploration/exploitation trade off on the search space, thus improving the capacity of the algorithm for solving complex problems [3].

A canonical cEA follows the pseudo-code included in Procedure 3. In this basic cEA, the population is usually structured in a regular grid of d dimensions ($d = 1, 2, 3$), and a neighborhood is defined on it. The algorithm iteratively considers as current each individual in the grid (line 3), and individuals may only interact with individuals belonging to their neighborhood (line 4), so parents are chosen among the neighbors (line 5) with a given criterion. Crossover and mutation operators are applied to the individuals in lines 6 and 7, with probabilities P_c and P_m, respectively. Afterwards, the algorithm computes the fitness value of the new offspring individual (or individuals) (line 8), and inserts it (or one of them) instead of the current individual in the population (line 9) following a given replacement policy. This loop is repeated until a termination condition is met (line 2).

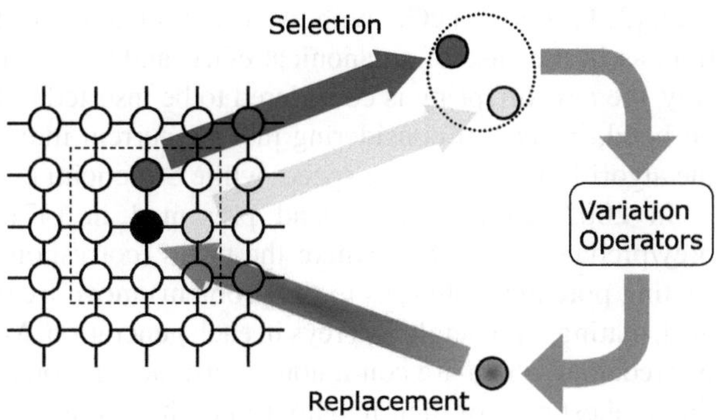

Fig. 8.5 In cellular EAs, individuals are only allowed to interact with their neighbors

Procedure 3: Pseudocode for a canonical cEA

```
//Algorithm parameters in 'cea'
while Not StopCondition do
    for individual in 1 to cea.popSize do
        n_list ← GetNeighborhood (cea, position(individual));
        parents ← Selection (n_list);
        offspring ← Recombination (cea.Pc, parents);
        offspring ← Mutation (cea.Pm, offspring);
        Evaluation (offspring);
        SelectNewIndividual (position(individual), individual, offspring);
    end
end
```

The cEA described here is asynchronous, since the population is updated with the next generation individuals just after creating them. This way, these new individuals can interact with those belonging to their parents' generation. On the contrary, there is also the possibility of storing all the offspring individuals in an auxiliary population, and then replace all the individuals in the population at the same time. This last version matches with the synchronous cEA model. As it was studied in [6, 7], the use of asynchronous policies allows faster convergence of the population than in the case of the synchronous one.

8.4.2 Enhanced Cellular Topologies

In this section we address several papers that propose enhancements on cGAs that modify the algorithm dynamics, but keeping the original cellular population topology. Simoncini et al. [61] propose the use of an anisotropic selection that is giving higher priority to some of the individuals in the neighborhood to be chosen against others. Specifically, those individuals at the north and south positions will be more likely to be chosen than the individuals in the east and west locations. Ishibuchi et al. [37] propose a cGA with two different neighborhood structures: one for selection, as in the case of a canonical cGA, and the second one for replacement. This way, the new offspring is considered to be inserted in the whole replacement neighborhood, instead of considering just the current individual. This model accelerates the algorithm convergence speed, while Simoncini et al. one is reduces it.

In 2002, Li and Sutherland presented in [47] a variation of cGA called prey/predator algorithm, where the preys (corresponding to the individuals representing potential solutions to the problem) move freely around the positions of the grid, mating with neighbor preys in each generation. Moreover, there exists a number of predators which are continuously displacing around the population, and they kill the weakest prey of their neighbor in each generation.

Finally, Alba et al. propose in [8] a new Estimation of Distribution Algorithm (EDA) with a cellular population in which small subpopulations are located in every location of the lattice, instead of only one single individual. This is done because the EDA needs large populations to get enough information to estimate the distribution of the solutions of the variables, therefore adding small subpopulations in every lattice will multiply the number of solutions in the neighborhood.

8.4.3 Hierarchical Populations

Janson et al. proposed in [38] a hierarchical Particle Swarm Optimization method (H-PSO) in which individuals are arranged in a tree hierarchy. Therefore, better solutions move towards the highest levels of the hierarchy, exchanging their position with worse ones. In this hierarchy, particles are influenced by their personal best solution and by its parent in the hierarchy. Later, in [39], the authors propose a new cellular Genetic Algorithm (cGA) with a pyramidal hierarchy into the population, such that the best individuals will be placed in the highest levels of the hierarchy. Therefore, individuals interact in this case with more than one individual from the hierarchy, as determined by the neighborhood defined in the cellular population. The effect is that the exploitation of the most promising solutions is enhanced, since they are located next to each other in the population thanks to the hierarchy, and the cellular population promotes the interaction of neighboring individuals. At the same time, the diversity of solutions in the population is maintained due to the evolution of worse individuals at the lower levels of the hierarchy, promoting the exploration of other regions of the search space different than the ones where the most promising current solutions are.

8.4.4 Population Structures Based on Social Networks

There exist some works analyzing new fine-grained topologies that are not as connected as the panmictic population (which is fully connected), but have shorter characteristic path length (i.e., the maximum distance between any two solutions) than the cellular model. In particular, it is worth mentioning the studies made by Giacobini et al. to both theoretically [29] and empirically [28] analyze the behavior of different GAs using random, scale-free, and small-world topologies. Additionally, Payne et al. addressed in [55] another theoretical study on the behavior of GAs with scale-free and small-world topologies, and they later extended it in [56] to analyze the effects of some characteristics of these kinds of networks, like the scale and assortativity (the degree of connections of nodes). They arrived to the conclusion that increasing the assortativity leads to shorter convergence times, while high scale networks provide longer convergence rates. However, the main conclusion of these studies by Giacobini et al. and Payne et al. is that small-world populations are competitive, but the potential utility of scale-free population structures is still unclear. In [22], the influence on the behavior of the algorithm under different small-world topologies generated in several different ways is analyzed.

Despite random graphs were not the best performing ones in the studies made by Giacobini et al. for GAs [28, 29], they have become popular for PSO algorithms. Indeed, this is the population model used in the Standard PSO 2007 (SPSO 07) [62]. Kennedy and Mendes [41, 42] have deeply investigated on the use of fine-grained population topologies (e.g., ring, von Neumann, random, pyramid, or star graphs, to name a few) for PSO algorithms. As a conclusion of their studies, they recommend the use of von Neumann topologies. In [51], the same authors present a new fully informed PSO algorithm in which the information of all the neighbors is taken into account in the generation of new solutions.

In [21], Dorronsoro and Bouvry compare the behavior of several Differential Evolution (DE) algorithms using different population topologies, like panmictic, cellular, island, hierarchical, random, or small-world. Among the compared topologies, the island and small-world populations were the best performing ones for the considered continuous problems.

8.4.5 Dynamic Topologies

There also exist several approaches using dynamic topologies. In this sense, Suganthan presented in 1999 [65] probably the first EA with a dynamic population topology: a PSO algorithm with variable neighborhood sizes. The idea is that every particle starts with a neighborhood of size one (the particle itself), and then the size of the neighborhood is increasing during the run. One year later, Kennedy [40] proposed splitting the population into clusters (groups of swarms), where the centroid of every cluster is the one influencing the other particles in its cluster. A similar concept is the clubs-based PSO [25], but in this case particles can belong to different clubs (or clusters), and they can also join to or quit from a club. A third algorithm we would like to mention here is the species-based PSO (SPSO), proposed by Li in 2004 [46]. In this case, the population is split into swarms of the same species, i.e., containing particles with some similarities. Only those particles of the same species interact in the creation of the next generation of solutions.

In 2008, Whitacre et al. proposed in [68] a dynamic population structure for GAs that automatically evolves by following a set of simple rules to mimic the interaction networks of complex systems. Individuals are dynamically created or removed from the population in terms of their quality by following some rules. The population is modeled as a graph, and therefore its structure is self-evolving when adding or removing solutions. As a result, this kind of population performs longer convergence times compared to cGAs. However, as a consequence, they can provide better results after a high number of evaluations. Similar topologies were proposed for PSO by Godoy et al. [31] and Clerc [15].

Dorronsoro et al. proposed different self-adaptive cGAs that automatically modify the population topology to a more suitable one according to the convergence speed of the population. In [5], the shape of the population is changed based on the principle that narrower populations provide with more exploration capabilities than square ones, that are more exploitative. More recently, a new cGA that is using different neighborhood sizes according to the quality of individuals was proposed in [20]. In [23], the authors propose an adaptive technique that automatically chooses the neighborhood size of each individual and the population shape, designing the first decentralized EA that does not require any additional parameters.

Finally, there are two recent papers [12, 72] that incorporate a novel idea in which the evolution process of the cGA is guided by a cellular automaton (CA). In particular, the CA is used to activate or deactivate the evolution of the individuals in the population topology at each generation according to the cellular automaton state. This is done to slow down the convergence speed of the algorithm, thus keeping diversity for longer.

8.5 Evolutionary Algorithms with Coalitions

In this section, we introduce an informal description of an Evolutionary Algorithm with Coalitions (EACO), previously introduced in [24], which is a new class of EA with dynamic population topology that aims at taking profit of the benefits of the two main existing population structures, namely cellular and island populations. This is achieved by introducing the concept of coalitions. Individuals are arranged in a toroidal lattice, as in cellular EAs, but they form coalitions according to some policies that will be defined next. All the individuals belonging to the same coalition can interact among them, like the subpopulation of an island in EA.

In EACO, individuals can only belong to one single coalition, but they can leave it and/or join another coalition from which they expect to get a *better benefit*. Therefore, when an individual leaves a coalition, it can either remain independent, join another existing coalition or even create a completely new one. Thus, we apply a game theoretic perspective, where individuals are considered selfish entities; able to collaborate but looking for its own benefit. Here, the main goal of every cell is to take the actions needed to produce successful offspring and reach the global optimum.

In order to further evolve to better solutions, individuals will be interested in mating with other *good quality* individuals. Therefore, it would be beneficial for them to belong to coalitions composed by a large number of high quality members with genetic diversity. In order to evaluate this, the quality of a coalition will be valued in terms of its size, the average quality of the solutions forming it, and their diversity. In this way, belonging to a high quality coalition will generally be beneficial for individuals. Here, the term quality has a broader meaning than classical fitness, as, besides fitness, it also includes other components to measure a group of individuals.

Some desirable characteristics for an EACO algorithm can be:

- Larger coalitions are desirable, as cells have more candidates for mating.
- Coalitions with higher genetic diversity are desirable.
- Coalitions with better average fitness are desirable.
- Every coalition or independent cell will have a quality value, that integrates the size, diversity and fitness values.
- Cells from a given coalition can interact with any other cell from the same coalition, so the coalition behaves like a panmictic island.
- Cells can leave a coalition, remain independent, join other neighboring coalitions or join another neighboring independent cell (de facto creating a new coalition), according to their selfish behavior to maximize its success.
- Each independent cell mates with all its neighbors.
- Inside a coalition, offspring generation is performed as in regular EAs; typically, parents are selected from the population using some scheme, and then variation operators are applied in order to generate the offspring that replace the parents, following some policy to keep the population size constant.

8.5.1 Algorithmic Description of EACO

Procedure 4 describes the pseudocode of the EACO algorithm, which is relatively similar to the one described before for the cEAs case. In EACO, the population is also structured in a regular grid of d dimensions ($d = 1, 2, 3$), and a neighborhood is defined on it. The algorithm first evaluates the quality of every coalition (line 3). At the beginning, all the cells are independent, so their quality differences will be based only in the fitness of those cell solutions. Then, the algorithm iteratively considers each individual in the grid (line 4), and individuals may only interact with individuals belonging to their neighborhood (line 5), in the independent case, or within the coalition neighborhood. Then, parents are chosen among those neighbors (line 6) with a given criterion. Crossover and mutation operators are also applied to the individuals in lines 7 and 8, with probabilities P_c and P_m, respectively. Afterwards, the algorithm computes the fitness value of the new offspring individual (or individuals) (line 9), and inserts it (or one of them) instead of the current individual in the population (line 9) following a given replacement policy. Finally, the cell can join, leave or change from one coalition to another (line 10) depending on its present context and its particular solution. All this procedure is addressed through the quality concept, and the loop is repeated until a termination condition is met (line 2).

Procedure 4: Pseudocode for EACO

```
1  //Algorithm parameters in 'eaco'
2  while Not StopCondition do
3      coa_Q_list ← CalculateCoalitionsQuality (coa_list);
4      for individual in 1 to eaco.popSize do
5          n_list ← GetCoalition_or_Neighborhood (eaco, coa_list, position(individual));
6          parents ← Selection (n_list);
7          offspring ← Recombination (eaco.Pc, parents);
8          offspring ← Mutation (eaco.Pm, offspring);
9          eaco.pop ← SelectNewIndividual (position(individual), individual, offspring);
10         individual.coa ← ChangeCoa (position(individual), coa_Q_list);
11     end
12 end
```

In line 3, the algorithm calculates the quality of the different coalitions. As said before, every coalition (C_i) has a quality value, which is determined with the formulae,

$$Quality(C_i) = \alpha \cdot RelSize(C_i) + \beta \cdot AvgDiv(C_i) + \gamma \cdot AvgFitness(C_i) \ , \quad (8.1)$$

where α is the weight given to the relative size of the coalition, β is the weight given to the diversity among the coalition members, and γ is the weight given to the average quality of the solutions. The relative size (*RelSize*) of the coalition is determined dividing its number of members by the total number of population cells. The diversity (*AvgDiv*) is measured comparing the average hamming distance between the best element in the coalition with the rest of its members. Finally, the average fitness (*AvgFitness*) is determined using all coalition members.

Procedure 5 describes how a cell decides if it will be independent or a coalition member along the next generation. In (line 3) it considers if its quality is worse than the worst cell in its neighborhood, then it joins its best neighbor, either in a coalition or de facto creating a new one. If it already belongs to a coalition (line 6) three things may happen. First it checks if its neighbors belong to the same coalition (line 7), otherwise the cell considers itself as isolated and gets independence. Otherwise, it considers if it is the worst in its neighborhood, and in such case it tries to join its best neighbor (implicitly changing coalition if they are not in the same one). Finally, if

Procedure 5: Pseudocode for the procedure ChangeCoa

```
ChangeCoa (position(individual), coa_Q_list)
if isIndependent() then
    if (c(i).Quality() <= WorstNeighbor) then
        JoinBestNeighborCell();
    end
else
    if (not isCoaNear()) then
        GetIndependence();
    else if (c(i).Quality() <= WorstNeighbor)) then
        JoinBestNeighborCell();
    else if (Ramdom() < P_reb)) then
        GetIndependence();
    end
end
```

a certain probability of rebellion P_{reb} happens, we consider that the cell decides to rebel and leave the present coalition.[1]

8.6 Set of Problems

In this section, we present the set of problems chosen for this study. The benchmark is representative because it contains many different interesting features in optimization, such as epistasis, multimodality, deceptiveness, use of constraints, parameter identification, and problem generators. These are important ingredients in any work trying to evaluate algorithmic approaches with the objective of getting reliable results, as stated by Whitley et al. in [70].

We will experiment with the reduced set of problems studied in [2], which includes the massively multimodal deceptive problem (MMDP), the multimodal problem generator P-PEAKS, the error correcting code design (ECC), maximum cut of a graph (MAXCUT), and the minimum tardy task problem (MTTP).

The set of problems selected for this study is justified by both their difficulty and their application domains (combinatorial optimization, continuous optimization, telecommunications, scheduling, etc.). Therefore, this selection guarantees a high level of confidence in the results.

[1] Like mutations in genetic evolution, P_{reb} alters the state of a cell within a coalition; and implicitly the whole coalition structure.

Unitation	Subfunction value
0	1.000000
1	0.000000
2	0.360384
3	0.640576
4	0.360384
5	0.000000
6	1.000000

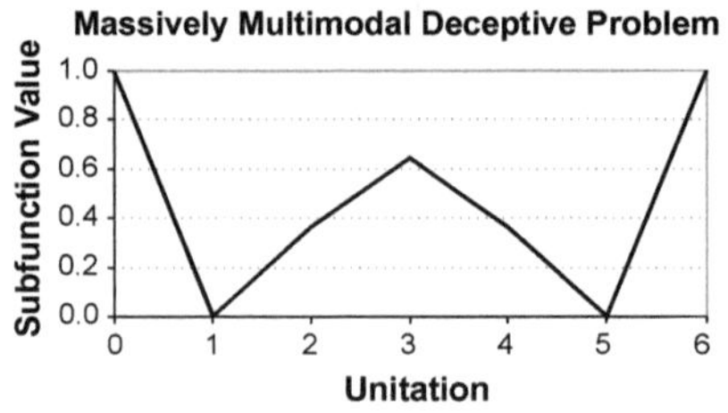

Fig. 8.6 Basic deceptive bipolar function (s_i) for MMDP

8.6.1 Massively Multimodal Deceptive Problem (MMDP)

The MMDP is a problem that has been specifically designed to be difficult for an EA [33]. It is made up of k deceptive subproblems (s_i) of 6 bits each one, whose value depends on the number of ones (*unitation*) a binary string has (see Fig. 8.6). It is easy to see (graphic of Fig. 8.6) that these subfunctions have two global maxima and a deceptive attractor in the middle point.

In MMDP, each subproblem s_i contributes to the fitness value according to its *unitation* (Fig. 8.6). The global optimum has a value of k and it is attained when every subproblem is composed of zero or six ones. The number of local optima is quite large (22^k), while there are only 2^k global solutions. Therefore, the degree of multimodality is regulated by the k parameter. We use here a considerably large instance of $k = 40$ subproblems. The instance we try to maximize for solving the problem is shown in Eq. 8.2, and its maximum value is 40.

$$f_{MMDP}(\mathbf{s}) = \sum_{i=1}^{k} fitness_{s_i} \tag{8.2}$$

8.6.2 Multimodal Problem Generator (P-PEAKS)

The P-PEAKS problem [18] is a multimodal problem generator. A problem generator is an easily parameterizable task which has a tunable degree of epistasis, thus admitting to derive instances with growing difficulty. Also, using a problem generator removes the opportunity to hand-tune algorithms to a particular problem, therefore allowing a larger fairness when comparing algorithms. With a problem generator we evaluate our algorithms on a high number of random problem instances, since a different instance is solved each time the algorithm runs, then the predictive power of the results for the problem class as a whole is increased.

The idea of P-PEAKS is to generate P random N-bit strings that represent the location of P peaks in the search space. The fitness value of a string is the number of bits the string has in common with the nearest peak in that space, divided by N (as shown in Eq. 8.3). By using a small/large number of peaks we can get weakly/strongly epistatic problems. In this work we have used an instance of $P = 100$ peaks of length $N = 100$ bits each, which represents a medium/high epistasis level [2]. The maximum fitness value for this problem is 1.0.

$$f_{P-PEAKS}(\mathbf{x}) = \frac{1}{N} \max_{1 \le i \le p} \{N - HammingD(\mathbf{x}, Peak_i)\} \tag{8.3}$$

8.6.3 Error Correcting Code Design Problem (ECC)

The ECC problem was presented in [48]. We will consider a three-tuple (n, M, d), where n is the length of each codeword (number of bits), M is the number of codewords, and d is the minimum Hamming distance between any pair of codewords. Our objective will be to find a code which has a value for d as large as possible (reflecting greater tolerance to noise and errors), given previously fixed values for n and M. The problem we have studied is a simplified version of that in [48]. In our case we search for half of the codewords $(M/2)$ that will compose the code, and the other half is made up by the complement of the codewords computed by the algorithm.

The fitness function to be maximized is:

$$f_{ECC}(C) = \frac{1}{\sum_{i=1}^{M} \sum_{\substack{j=1 \\ i \neq j}}^{M} \frac{1}{d_{ij}^2}}, \tag{8.4}$$

where d_{ij} represents the Hamming distance between codewords i and j in the code C (made up of M codewords, each of length n). We consider in the present work an instance where $M = 24$ and $n = 12$. The search space is of size $\binom{4096}{24}$, which is approximately 10^{87}. The optimal solution for $M = 24$ and $n = 12$ has a fitness value of 0.0674 [14].

8.6.4 Maximum Cut of a Graph (MAXCUT)

The MAXCUT problem is to look for a partition of the set of vertices (V) of a weighted graph $G = (V, E)$ into two disjoint subsets V_0 and V_1 so that the sum of the weights of the edges with one endpoint in V_0 and the other one in V_1 is maximized. For encoding the problem we use a binary string $(x_1, x_2, \ldots, x_n)$ of length n where each digit corresponds to a vertex. If a digit is 1 then the corresponding vertex is in set V_1; if it is 0 then the corresponding vertex is in set V_0. The function to be maximized [43] is:

$$f_{MAXCUT}(\mathbf{x})=\sum_{i=1}^{n-1}\sum_{j=i+1}^{n} w_{ij}\cdot\left[x_i\cdot(1-x_j)+x_j\cdot(1-x_i)\right] \tag{8.5}$$

Note that w_{ij} contributes to the sum only if nodes i and j are in different partitions. We have considered in this study three different instance graphs. Two of them are randomly generated graphs of moderate sizes: a sparse one "cut20.01", and a dense one "cut20.09"; both of them are made up of 20 vertices. The other instance is a scalable weighted graph of 100 vertices. The maximum fitness values for these instances are 10.119812 for "cut20.01", 56.740064 in the case of "cut20.09", and 1077 for "cut100".

8.6.5 *Minimum Tardy Task Problem (MTTP)*

MTTP [64] is a task-scheduling problem wherein each task i from the set of tasks $T=\{1,2,\ldots,n\}$ has a length l_i —the time it takes for its execution—, a deadline d_i —before which a task must be scheduled, and its execution completed—, and a weight w_i. The weight is a penalty that has to be added to the objective function in the event that the task remains unscheduled. The lengths, weights, and deadlines of tasks are all positive integers. Scheduling the tasks of a subset S of T is to find the starting time of each task in S, such as at most one task at time is performed and such that each task finishes before its deadline.

We characterize a one-to-one scheduling function g defined on a subset of tasks $S\subseteq T: S\mapsto \mathbb{Z}^+\bigcup\{0\}$, so that for all tasks $i, j\in S$ has the following properties:

1. A task can not be scheduled before any previous one has finished: $g(i)<g(j)\Rightarrow g(i)+l_i\leq g(j)$.
2. Every task finishes before its deadline: $g(i)+l_i\leq d_i$.

The objective function for this problem is to minimize the sum of the weights of the unscheduled tasks. Therefore, the optimal scheduling minimizes Eq. 8.6:

$$f_{MTTP}(\mathbf{x})=\sum_{i\in T-S} w_i \tag{8.6}$$

The schedule of tasks S can be represented by a vector $\mathbf{x}=(x_1,x_2,\ldots,x_n)$ containing all the tasks ordered by its deadline. Each $x_i\in\{0,1\}$, where if $x_i=1$ then task i is scheduled in S, while if $x_i=0$ means that task i is not included in S. The fitness function is the inverse of Eq. 8.6, as described in [43]. We have used in this study two different instances [43] for analyzing the behavior of our algorithms with this function: "mttp100", and "mttp200", with sizes 100 and 200, and maximum fitness values of 0.005 and 0.0025, respectively.

8.7 Results

In this section, we compare the performance of EACO versus the canonical cellular GA, which was demonstrated to outperform other GAs with panmictic and decentralized populations in [6]. Then, we explore the behavior of EACO evaluating the influence of the different parameters under several scenarios.

The results, unless otherwise stated, were obtained after averaging over 1000 independent runs, using a square grid with 400 cells (20×20), and the basic parameters used by the EACO algorithm are: $P_{reb} = 0.5$, $\alpha = 1$, $\beta = 1$ and $\gamma = 1$.

8.7.1 Selecting the Population Size

The first question we address is to decide the population size. On the one hand, smaller populations will produce a lower percentage of hits when searching for the optimum. On the other hand, higher population sizes increase the time needed to do the calculations for each generation. In order to analyze this, five significant sizes are considered for a spatial square lattice as described in Fig. 8.7 for the cGA algorithm and Fig. 8.8 for EACO. The horizontal axis shows the problem number,[2] while the vertical axis shows the percentage of runs in which the optimum has been found. The different bars refer to the five different population sizes considered: 25, 100, 400, 900 and 1.600; corresponding to square latices with sides of 5, 10, 20, 30 and 40 cells, respectively. We can clearly appreciate in the figures that increasing the number of cells leads to a higher efficacy of the algorithms. We remark that EACO needed, in average, a lower number of generations to find the optimum, and outperforms cGA in all cases.

In the simulations performed to generate Figs. 8.7 and 8.8, the number of generations gets reduced when increasing the population size. These is due to the effect of having a larger population, that allows to better explore the optimization scenario in order to find the global optimum. However, the reduction in the number of generations is accompanied by a raise in the execution time, as more cells need to be evaluated in each generation.[3] We are interested in those configurations offering a good balance between execution time and accuracy for the different problems. Therefore, from now on we present the results considering a population size of 400 cells, which provides a reasonable compromise.

---

[2]The problem order is: 0. ECC, 1. MAXCUT100, 2. MAXCUT20_01, 3. MAXCUT20_09, 4. MMDP, 5. MTTP100, 6. MTTP200 and 7. P-PEAKS.

[3]The most computation demanding problem is P-PEAKS, which takes around 100 milliseconds per run to be solved by both algorithms with a population of 25 cells, and using a standard Intel dual-core computer with 8 GB of RAM. However, the same problem takes about ten times longer with a population of 1600 cells.

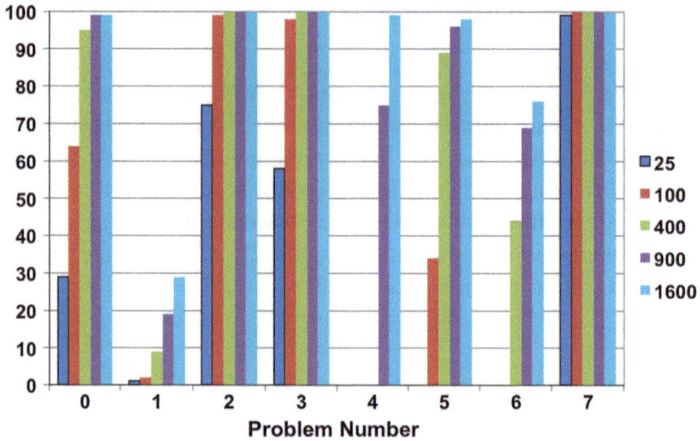

Fig. 8.7 cGA: Average percentage of optimum values found when considering different population sizes with a spatial topology (1000 runs)

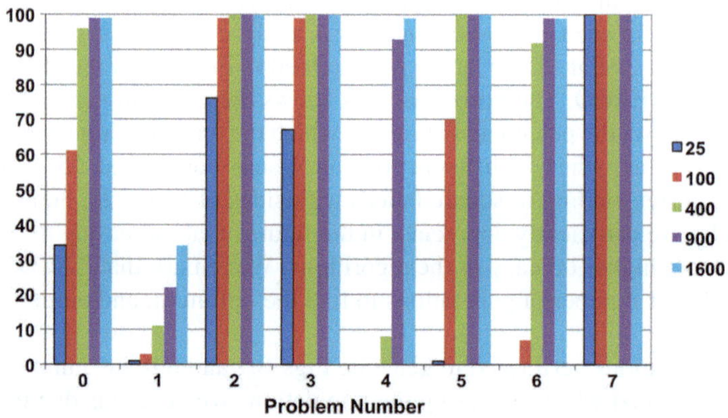

Fig. 8.8 EACO: Average percentage of optimum values found when considering different population sizes with a spatial topology (1000 runs)

8.7.2 Comparing cGA Versus EACO

We present in Table 8.1 the results reported by both algorithms for every considered problem. The table shows the percentage of executions (over 1000 runs) in which the optimum was found, or the average value found when such optimum was not achieved. In smaller font size we also present the standard deviation of the results found by the algorithm in every problem. We applied the Wilcoxon unpaired signed rank sum statistical test to assess statistical confidence in the comparison of the two algorithms. The statistically best results, with 95% confidence level, are emphasized with a gray background in the table.

Table 8.1 Comparison between cGA and EACO (1000 runs)

	cGA	EACO
0. ECC	$6,7332E-002(96\%)$ $\pm 4,4183E{-}004$	$6,7318E-002(95\%)$ $\pm 4,7349E{-}004$
1. MAXCUT100	$1,0433E003(9\%)$ $\pm 1,8350E001$	$1,0445E003(11\%)$ $\pm 1,8513E001$
2. MAXCUT20_01	$1,0120E001(100\%)$ $\pm 0,0000E000$	$1,0120E001(100\%)$ $\pm 0,0000E000$
3. MAXCUT20_09	$5,6740E001(100\%)$ $\pm 0,0000E000$	$5,6740E001(100\%)$ $\pm 0,0000E000$
4. MMDP	$3,8516E001(0\%)$ $\pm 6,6758E{-}001$	$3,9190E001(8\%)$ $\pm 5,1082E{-}001$
5. MTTP100	$4,9071E-003(90\%)$ $\pm 2,9155E{-}004$	$4,9991E-003(99\%)$ $\pm 2,7980E{-}005$
6. MTTP200	$2,2914E-003(44\%)$ $\pm 2,2591E{-}004$	$2,4713E-003(92\%)$ $\pm 1,1239E{-}004$
7. P-PEAKS	$1,0000E000(100\%)$ $\pm 0,0000E000$	$1,0000E000(100\%)$ $\pm 0,0000E000$

We can see in the table that the two algorithms find always the optimal solution for MAXCUT20_01, MAXCUT20_09 and P-PEAKS. In the ECC case both obtain similar results, while in the other four problems EACO finds the optimum with a higher probability.

We are also interested on the computational effort required by the algorithms to find the optimum. To analyze that, Table 8.2 presents the average number of generations needed by every algorithm to find the optimum, and below that the standard deviation. It can be seen that EACO requires significantly less generations than cGA to find the optimum, for all problems.

Summarizing, the redefined EACO algorithm clearly outperforms the cGA in terms of efficiency (i.e., the number of generations required to find the optimum), accuracy (i.e., the quality of solutions when the optimum is not found), and mainly in efficacy (i.e., the percentage of runs in which the optimum is found).

8.7.3 Influence of Parameters

Now, we analyze the effect of the four main parameters used in EACO algorithm, namely the probability P_{reb} and the weights α, β and γ. These weights can take any value, and only the relative influence compared with the two others is relevant. Taking this into account, we have considered values in the interval [0,1] for these three weights. We have evaluated the four parameters in all the problems described, and here we only show some representative cases.

Table 8.2 Generations needed by cGA and EACO (1000 runs)

	cGA	EACO
0. ECC	$1,2580E002$ $\pm 2,2967E001$	$1,1673E002$ $\pm 2,1449E001$
1. MAXCUT100	$1,7440E002$ $\pm 3,6190E001$	$1,4129E002$ $\pm 2,8096E001$
2. MAXCUT20_01	$8,1420E000$ $\pm 2,9097E000$	$5,5890E000$ $\pm 1,9302E000$
3. MAXCUT20_09	$1,2856E001$ $\pm 4,5499E000$	$8,9810E000$ $\pm 3,0819E000$
4. MMDP	$2,1189E002$ $\pm 1,7656E001$	$1,6777E002$ $\pm 1,7196E001$
5. MTTP100	$1,3400E002$ $\pm 1,7521E001$	$9,3726E001$ $\pm 1,0551E001$
6. MTTP200	$2,6691E002$ $\pm 2,5388E001$	$1,9056E002$ $\pm 1,6884E001$
7. P-PEAKS	$5,6787E001$ $\pm 4,0381E000$	$5,0436E001$ $\pm 3,6344E000$

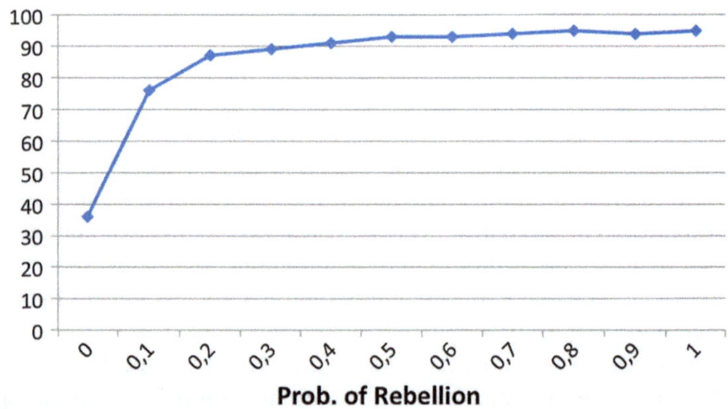

Fig. 8.9 Average percentage of optimum values found when changing the probability of rebellion for the MTTP200 problem (1000 runs)

Figure 8.9 shows the average results of the influence of P_{reb} in a set of simulations performed over the MTTP200 problem. This problem is a representative example of our results, i.e., any value of $P_{reb} \geq 0.5$ is appropriate, as they all provide very similar results. Therefore, $P_{reb} = 0.5$ has been the value selected all along the rest of the simulations.

Concerning α, Fig. 8.10 shows the results when changing this parameter while β and γ are set to zero or one (four possible scenarios). As we can see we need $\alpha > 0$. However, this is not enough to reach the best results in the MTTP200 problem, as

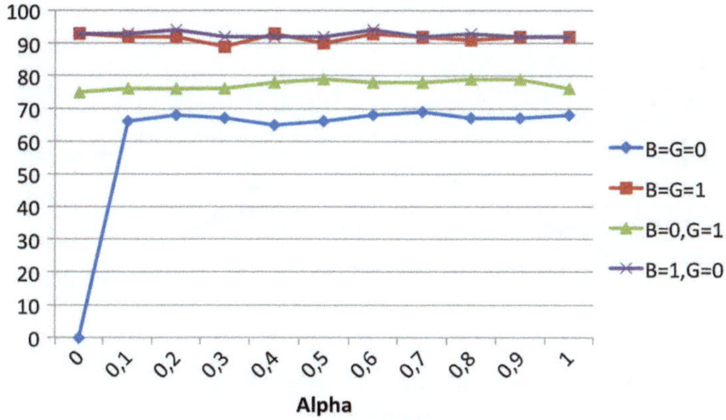

Fig. 8.10 Average percentage of optimum values found when changing the influence of α for the MTTP200 problem (1000 runs) and considering four possible scenarios for the other two parameters

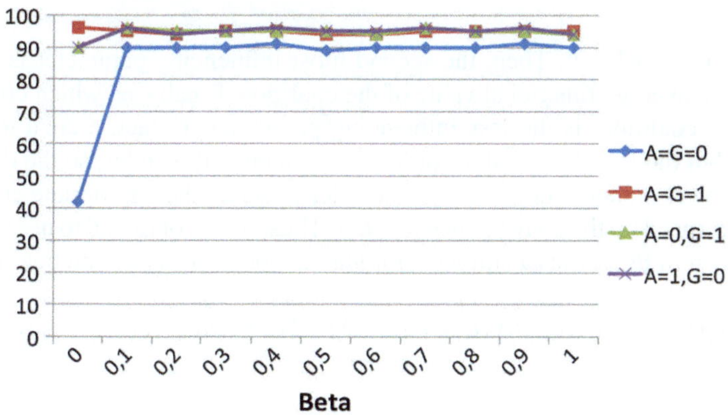

Fig. 8.11 Average percentage of optimum values found when changing the influence of β for the ECC problem (1000 runs) and considering four possible scenarios for the other two parameters

the figure also shows that at least setting $\beta = 1$ produces much better results, and that the influence of γ is low, but the one from α even lower.

Figure 8.11 shows the results when changing β while setting α and γ to zero or one, respectively. As we can see using β alone in the ECC problem is almost enough to get very good results. Nevertheless, results can be improved including the other two parameters.

Finally, Fig. 8.12 shows the variation of the γ parameter, when the other two ones are set to zero and one respectively. We observe again how γ is necessary but not enough to solve the MMDP problem, and again that the influence of α is minimal.

To sum up, we have found that β, which controls the weight of the divergence inside a coalition, is the parameter having the highest influence on the performance of the algorithm. We have observed that having a high divergence is really good for

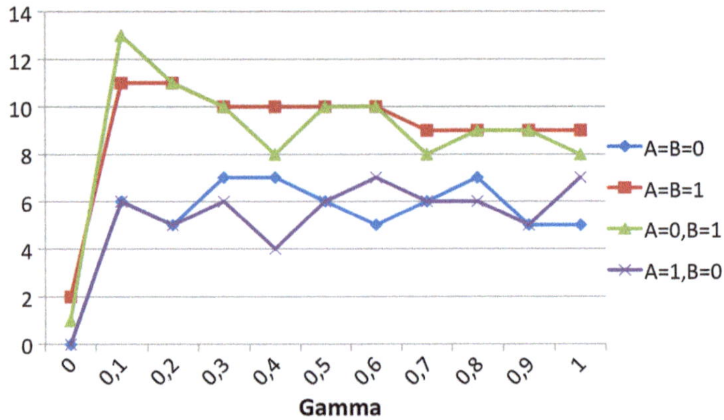

Fig. 8.12 Average percentage of optimum values found when changing the influence of γ for the MMDP problem (1000 runs) and considering four possible scenarios for the other two parameters

joining a new coalition. Then, the second most influencing parameter is γ, which controls the average functional value of the coalition. Finally, α, which controls the size of the coalition, is the less influencing parameter. In fact, a coalition can be very big, but the functional values of its cells can be very similar; so the divergence managed by β is much more relevant in order to select the appropriate coalition to join and to speed up the convergence process. Thus, we recommend to use the quality value shown in Eq. 8.7, as a simpler formulae to obtain the coalition quality.[4]

$$Quality_2(C_i) = \delta \cdot AvgDiv(C_i) + (1 - \delta) \cdot AvgFitness(C_i), \quad \delta \in [0, 1] \quad (8.7)$$

8.7.4 Complex Networks

In many real contexts like geography, biology, or telecommunication networks; each node interacts mainly with its neighbors. Therefore, we place agents in a complex network since they provide a realistic model of the topological features found in many nature, social and technological networks (see Chap. 3). In the simulations, we mainly focus on four types of well-known topologies such as spatial, random, small-world and scale-free networks; since they model the most common networks appearing in Nature and in our human societies. For more information about these four topologies, we redirect to the interested reader to Sect. 3.5 from Chap. 3.

The parameters used for building the networks are a von Neumann neighborhood for the spatial network (i.e., four neighbors), $W_{1600}^{4;0.1}$ for the small-world network, $S_{1600}^{4;-2}$ for the scale-free network and $R_{1600}^{0.0025}$ for the random network. These values

[4]Note that $Avgfitness$ in the formulae must be positive when we are maximizing, but negative if we minimize.

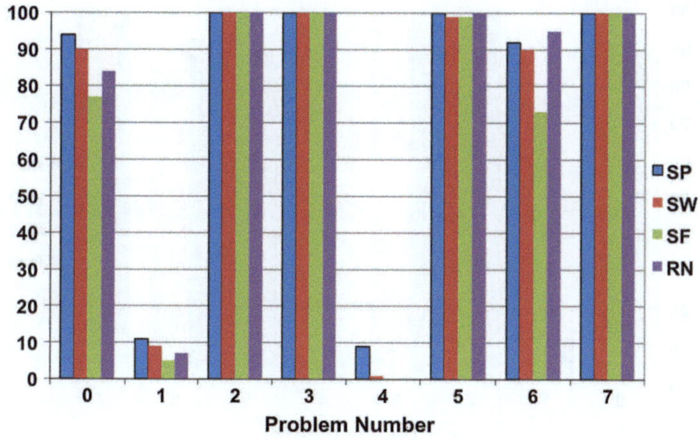

Fig. 8.13 Average percentage of optimum values found for different topologies and all the problems (1000 runs)

have been chosen to compare the different topologies in a fair condition, i.e., having four neighbors per cell in average.

Figure 8.13 shows the results obtained with EACO using different complex topologies with an average neighborhood of 4 cells. As we can see the best results are obtained with a spatial network, while the worst ones happen with the scale-free one. This means that homogenizing the size of the neighborhood is relevant for EACO to work well, and that is why the spatial topology is the most appropriate; since it is a pure regular network.

8.7.5 Changing the Neighborhood

Finally, we are going to analyze what happens when we modify the properties of the neighborhoods studied up to now. For that we consider two scenarios, firstly a higher neighborhood size using a spatial topology and increasing the radio, and secondly cells can dynamically modify their neighborhood during runtime by selecting successful new partner cells, and deleting the connections with the worst neighbors.

Increasing the Size of the Neighborhood

Figure 8.14 shows the average results, when changing the radio in a spatial topology. It becomes clear that bigger neighborhood radios produce worse results. The figure also shows four problems where differences are significative, while in the other four there is no difference at all. As expected, increasing the radio increases the exploitation capabilities of the network, but reduces the exploration ability to escape from local optima and in the end to find the optimum.

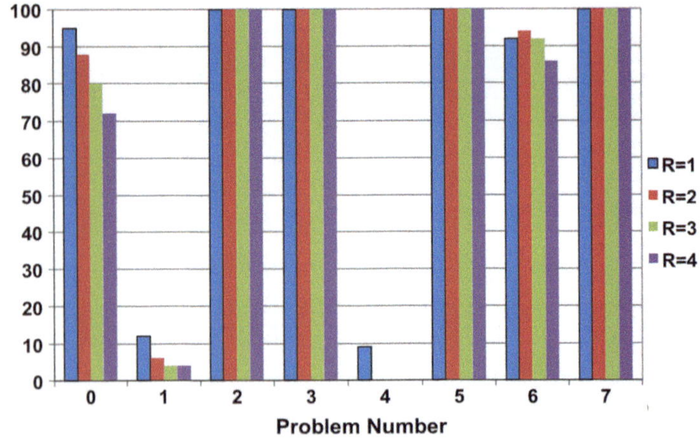

Fig. 8.14 Average percentage of optimum values found when increasing the neighborhood radio, from one to four, in a spatial topology (1000 runs)

Changing the Neighbors (Rewiring)

Finally, we show what is effect of changing the links in our network, i.e., switching the partners in a cell neighborhood (a.k.a. rewiring). Here, rewiring means deleting the link with the neighbor that has the worst functional value, and create a new link with a new cell of the network. For this, we considered two alternatives: rewiring to the best neighbor[5] of a neighbor, or using a random rewiring. However, we point out that, if the cell to delete has only one link (of course with the cell trying to rewire), then rewiring is aborted, i.e., we do not allow disconnected cells in our network. We also remark that the rewiring process does not affect the global number of links in the network, which remains stable; as for every new connection another one is deleted. Nevertheless, rewiring really affects the network structure, as some cells get more links; while others reduce their neighborhood depending on their success to obtain a better quality value.

Figure 8.15 shows the results when using a static spatial network (No RW), using rewiring-to-the-best (RW Best) or using random rewiring (RW RND). As we can see, there are no significative differences when using rewiring or not. The results obtained for the small-world, scale-free and random networks are much worse, in all problems, when compared with the ones obtained starting with a spatial topology; and we do not show them here.

8.8 Conclusions

In this chapter we have presented an introduction to the areas of evolutionary algorithms, population topologies and coalitions. We considered the integration of all

[5]The best neighbor means the cell with the best functional value.

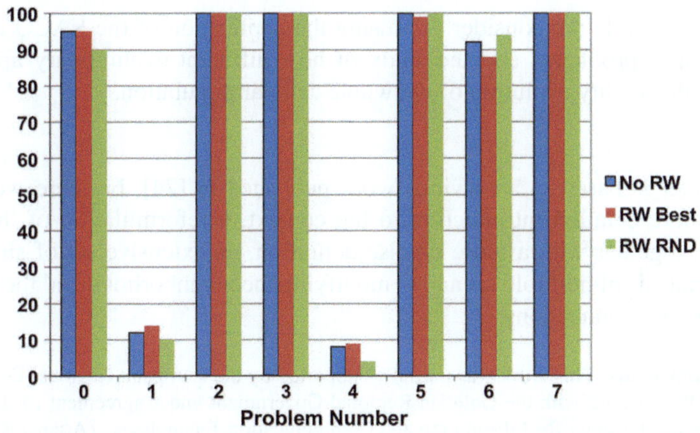

Fig. 8.15 Average percentage of optimum values found when considering rewiring in a spatial network (1000 runs). Three options are shown: **a** No rewiring, **b** rewiring to the best neighbor, and **c** rewiring randomly

these topics in a way to obtain a synergy in the development of new evolutionary algorithms, with the advantages of dEAs and cEAs, but avoiding their particular drawbacks, and removing their typically required parameters, like the predefined neighborhood (in cEAs), the islands topology, the migration frequency, and the policies to exchange and discard individuals (in dEAs).

The main contribution of the chapter is a redefinition of the EACO algorithm, which uses cellular approaches together with coalitions among cells, as a way to create islands of evolution in order to preserve diversity. Besides, we rely on evolutionary game theory to consider cells as players; who evolve depending on their payoff, with respect to their neighbors, and also the feedback provided by their coalition members. This approach allows the payoff of a given solution to be defined in terms of how much such solution has improved in the last generations. This idea speeds up the evolution of individuals grouped in high-quality coalitions that are quickly converging to promising solutions.

As the reader can see in the results section, EACO performs better than the canonical cGA concerning the efficacy (understood as how many times the optimum is found). Besides, EACO is more accurate than the cGA, and also a bit more efficient (understood as faster to find the optimum) for all the problems revised. We have also provided evidences about the statistical significance of all these results, and we have analyzed the little influence of the parameters used by EACO, that convert it in a parameter free algorithm. Finally, we have also evaluated the performance of EACO considering different complex topologies, and we have found that the best results come from static spatial networks with small neighborhoods, p.e., like the von Neumann neighborhood.

As future work, we consider promising the application of the EACO algorithm to continuous problems, and the study of how different evolutionary approaches influence the quality evolution of the whole cellular population.

Notes

This chapter is based in a previous work published in [24], but here we provide a much more detailed introduction to the context, a reformulation of the EACO algorithm together with a more precise definition, an extensive set of simulations and the analysis of multiple scenarios modifying the neighborhood and the topology of the network connections.

Acknowledgements This work was partially supported by the European Regional Development Fund (ERDF) together with the Galician Regional Government under agreement for funding the Atlantic Research Center for Information and Communication Technologies (AtlantTIC).

References

1. Alba, E.: Parallel Metaheuristics: A New Class of Algorithms. Wiley, New York (2005)
2. Alba, E., Troya, J.M.: Cellular evolutionary algorithms: evaluating the influence of ratio. In: Schoenauer, M. et al. (eds.) PPSN-6. Lecture Notes in Computer Science, vol. 1917, pp. 29–38. Springer, Berlin (2000)
3. Alba, E., Tomassini, M.: Parallelism and evolutionary algorithms. IEEE Trans. Evol. Comput. **6**(5), 443–462 (2002)
4. Alba, E., Troya, J.M.: Improving flexibility and efficiency by adding parallelism to genetic algorithms. Soft Comput. **12**(2), 91–114 (2002)
5. Alba, E., Dorronsoro, B.: The exploration/exploitation tradeoff in dynamic cellular evolutionary algorithms. IEEE Trans. Evol. Comput. **9**(2), 126–142 (2005)
6. Alba, E., Dorronsoro, B.: Cellular Genetic Algorithms. Operations Research/Compuer Science Interfaces. Springer, Heidelberg (2008)
7. Alba, E., Dorronsoro, B., Giacobini, M., Tomassini, M.: Decentralized cellular evolutionary algorithms. Handbook of Bioinspired Algorithms and Applications, pp. 103–120. CRC Press, Boca Raton (2006)
8. Alba, E., Madera, J., Dorronsoro, B., Ochoa, A., Soto, M.: Theory and practice of cellular UMDA for discrete optimization. In: Runarsson, T.P. et al. (eds.) Proceedings of the International Conference on Parallel Problem Solving from Nature IX (PPSN-IX), Reykjavik, Iceland. Lecture Notes in Computer Science, vol. 4193, pp. 242–251. Springer, Berlin (2006)
9. Bäck, T., Fogel, D.B., Michalewicz, Z. (eds.): Handbook of Evolutionary Computation. Oxford University Press, Oxford (1997)
10. Bonabeau, E., Dorigo, M., Theraulaz, G.: Swarm Intelligence. From Natural to Artificial Systems. Oxford University Press, Oxford (1999)
11. Bremermann, H.: Optimization trough evolution and resombination. Self-Organizing Systems, pp. 93–106. Spartan Books, Washington DC (1962)
12. Cantor, G., Gómez, J.: Maintaining genetic diversity in fine-grained parallel genetic algorithms by combining cellular automata, cambrian explosions and massive extinctions. In: Proceedings of the IEEE International Conference on Evolutionary Computation (CEC), pp. 1–8 (2010)
13. Cantú-Paz, E.: Efficient and Accurate Parallel Genetic Algorithms. Book Series on Genetic Algorithms and Evolutionary Computation, vol. 1, 2nd edn. Kluwer Academic Publishers, Dordrecht (2000)

14. Chen, H., Flann, N.S., Watson, D.W.: Parallel genetic simulated annealing: a massively parallel SIMD algorithm. IEEE Trans. Parallel Distrib. Syst. **9**(2), 126–136 (1998)
15. Clerc, M.: Particle Swarm Optimization. ISTE (International Scientific and Technical Encyclopedia) (2006)
16. Cramer, N.: A representation for the adaptive generation of simple sequential programs. In: Grefenstette, J. (ed.) Proceedings of the First International Conference on Genetic Algorithms and their Applications, Carnegie-Mellon University, Pittsburgh, PA, USA, 24–26 July 1985, pp. 183–187 (1985)
17. Darwin, C.: On the Origin of Species by Means of Natural Selection. John Murray, Londres (1859)
18. De Jong, K.A., Potter, M.A., Spears, W.M.: Using problem generators to explore the effects of epistasis. In: Bäck, T. (ed.) Proceedings of the 7th International Conference of Genetic Algorithms, Morgan Kaufman, pp. 338–345 (1997)
19. Dorigo, M., Stützle, T.: Ant Colony Optimization. The MIT Press, Cambridge (2004)
20. Dorronsoro, B., Bouvry, P.: Adaptive neighborhoods for cellular genetic algorithms. In: Nature Inspired Distributed Computing (NIDISC) Sessions of the International Parallel and Distributed Processing Symposium (IPDPS) 2011 Workshop, pp. 383–389 (2011)
21. Dorronsoro, B., Bouvry, P.: Improving classical and decentralized differential evolution with new mutation operator and population topologies. IEEE Trans. Evol. Comput. **15**(1), 67–98 (2011)
22. Dorronsoro, B., Bouvry, P.: On the use of small-world population topologies for genetic algorithms. In: EVOLVE 2011, A Bridge Between Probability, Set Oriented Numerics and Evolutionary Computation, e–proceedings (2011)
23. Dorronsoro, B., Bouvry, P.: Cellular genetic algorithms without additional parameters. J. Supercomput. **63**(3), 816–835 (2012)
24. Dorronsoro, B., Burguillo, J.C., Peleteiro, A., Bouvry, P.: Evolutionary algorithms based on game theory and cellular automata with coalitions. Handbook of Optimization, pp. 481–503. Springer, Berlin (2013)
25. Elshamy, W., Emara, H.M., Bahgat, A.: Clubs-based particle swarm optimization. In: Proceedings of the IEEE Swarm Intelligence Symposium (SIS), pp. 289–296 (2007)
26. Fogel, L.: Autonomous automata. Ind. Res. **4**, 14–19 (1962)
27. Fraser, A.: Simulation of genetic systems by automatic digital computers II: effects of linkage on rates under selection. Aust. J. Biol. Sci. **10**, 492–499 (1957)
28. Giacobini, M.T.M., Preuss, M.: Effects of scale-free and small-world topologies on binary coded self-adaptive CEA. In: Evolutionary Computation in Combinatorial Optimization (EvoCOP). Lecture Notes in Computer Science (LNCS), vol. 3906. Springer, Berlin (2006)
29. Giacobini, M., Tomassini, M., Tettamanzi, A.: Takeover time curves in random and small-world structured populations. In: Proceedings of the Genetic and Evolutionary Computation Conference (GECCO), Washington D.C. USA, 25–29 June 2005, pp. 1333–1340. ACM Press (2005)
30. Glover, F.W., Kochenberger, G.A. (eds.): Handbook of Metaheuristics. International Series in Operations Research Management Science. Kluwer, Dordrecht (2003)
31. Godoy, A., Von Zuben, F.J.: A complex neighborhood based particle swarm optimization. In: Proceedings of the IEEE International Conference on Evolutionary Computation (CEC), pp. 720–727 (2009)
32. Goldberg, D.: Genetic Algorithms in Search, Optimization and Machine Learning. Addison-Wesley Publishing Company, Reading (1989)
33. Goldberg, D., Deb, K., Horn, J.: Massively multimodality, deception, and genetic algorithms. In: Proceedings of the International Conference on Parallel Problem Solving from Nature II, pp. 37–46 (1992)
34. Holland, J.: Outline for a logical theory of adaptive systems. J. ACM **9**(3), 297–314 (1962)
35. Holland, J.: Adaptation in Natural and Artificial Systems. University of Michigan Press, Ann Arbor (1975)

36. Hussain, T.: An introduction to evolutionary computation. Tutorial presentation, CITO Researcher Retreat, 12–14 May 1998, Hamilton, Ontario (1998)
37. Ishibuchi, H., Sakane, Y., Tsukamoto, N., Nojima, Y.: Implementation of cellular genetic algorithms with two neighborhood structures for single-objective and multi-objective optimization. Soft Comput. **15**(9), 1749–1767 (2011)
38. Janson, S., Middendorf, M.: A hierarchical particle swarm optimizer and its adaptive variant. IEEE Syst. Man Cybern. Part B **35**(6), 1272–1282 (2005)
39. Janson, S., Alba, E., Dorronsoro, B., Middendorf, M.: Hierarchical cellular genetic algorithm. In: Gottlieb, J., Raidl, G. (eds.) Evolutionary Computation in Combinatorial Optimization (EvoCOP), Budapest, Hungary. Lecture Notes in Computer Science (LNCS), vol. 3906, pp. 111–122. Springer, Berlin (2006)
40. Kennedy, J.: Stereotyping: improving particle swarm performance with cluster analysis. In: Proceedings of the IEEE International Conference on Evolutionary Computation (CEC), vol. 2, pp. 1507–1512 (2000)
41. Kennedy, J., Mendes, R.: Population structure and particle swarm performance. In: Proceedings of the IEEE International Conference on Evolutionary Computation (CEC), pp. 1671–1676. IEEE Press (2002)
42. Kennedy, J., Mendes, R.: Neighborhood topologies in fully informed and best-of-neighborhood particle swarms. IEEE Trans. Syst. Man Cybern. Part C Appl. Rev. **36**(4), 515–519 (2006)
43. Khuri, S., Bäck, T., Heitkötter, J.: An evolutionary approach to combinatorial optimization problems. In: Proceedings of the ACM Press Computer Science Conference, Phoenix, Arizona, pp. 66–73. ACM Press (1994)
44. Koza, J.: Genetic programming. In: Williams, J., Kent, A. (eds.) Encyclopedia of Computer Science and Technology, vol. 39, pp. 29–43. Marcel-Dekker, New York (1998)
45. Larrañaga, P., Lozano, J. (eds.): Estimation of Distribution Algorithms. A New Tool for Evolutionary Computation. Kluwer Academic Publishers, Dordrecht (2002)
46. Li, X.: Adaptively choosing neighbourhood bests using species in a particle swarm optimizer for multimodal function optimization. In: Proceedings of the Genetic and Evolutionary Computation Conference (GECCO). Lecture Notes in Computer Science (LNCS), vol. 3102, pp. 105–116. Springer, Berlin (2004)
47. Li, X., Sutherland, S.: A cellular genetic algorithm simulating predator-prey interactions. In: Proceedings of the Third International Conference on Genetic Algorithms (ICGA), Morgan Kaufmann, pp. 416–421 (2002)
48. MacWilliams, F.J., Sloane, N.J.A.: The Theory of Error-Correcting Codes. North-Holland, Amsterdam (1977)
49. Manderick, B., Spiessens, P.: Fine-grained parallel genetic algorithm. In: Schaffer, J. (ed.) Third International Conference on Genetic Algorithms ICGA-3, pp. 428–433. Morgan-Kaufmann (1989)
50. Mendel, G.: Versuche über Pflanzen-Hybriden. Verhandlungen des Naturforschedes Vereines in Brünn 4 (1865)
51. Mendes, R., Kennedy, J., Neves, J.: The fully informed particle swarm: simpler, may be better. IEEE Trans. Evol. Comput. **8**(3), 204–210 (2004)
52. Nedjah, N., Alba, E., de Macedo Mourelle, L.: Parallel Evolutionary Computations. Studies in Computational Intelligence. Springer, Berlin (2006)
53. Neri, F., Cotta, C., Moscato, P.: Handbook of Memetic Algorithms. Studies in Computational Intelligence, vol. 379. Springer, Berlin (2012)
54. Olariu, S., Zomaya, A.Y. (eds.): Handbook of Bioinspired Algorithms and Applications. CRC Press, Boca Raton (2006)
55. Payne, J.L., Eppstein, M.J.: Emergent mating topologies in spatially structured genetic algorithms. In: Proceedings of the Genetic and Evolutionary Computation Conference (GECCO), Seattle, Washington, USA, pp. 207–214. ACM Press (2006)
56. Payne, J.L., Eppstein, M.J.: The influence of scaling and assortativity on takeover times in scale-free topologies. In: Proceedings of the Genetic and Evolutionary Computation Conference (GECCO), Atlanta, Georgia, USA, pp. 241–248. ACM Press (2008)

57. Price, K.V., Storn, R.M., Lampinen, J.A.: Differential Evolution - A Practical Approach to Global Optimization. Natural Computing Series. Springer, Berlin (2005)
58. Rechenberg, I.: Cybernetic solution path of an experimental problem. Technical report, Royal Aircraft Establishment, Library translation No. 1122, Farnborough, Hants., UK, 1965
59. Rechenberg, I.: Evolutionsstrategie: Optimierung technischer Systeme und Prinzipien der biologischen Evolution. Frommann-Holzboog, Stuttgart (1973)
60. Schwefel, H.-P.: Kybernetische Evolution als Strategie der Experimentellen Forschung in der Strömungstechnik. Ph.D. thesis, Technical University of Berlin, 1965
61. Simoncini, D., Verel, S., Collard, P., Clergue, M.: Anisotropic selection in cellular genetic algorithms. In: Proceedings of the Genetic and Evolutionary Computation Conference (GECCO), Seattle, Washington, USA, pp. 559–566. ACM Press (2006)
62. Standard Particle Swarm Optimization, Particle Swarm Central website
63. Stender, J.: Parallel Genetic Algorithms: Theory and Applications. IOS Press, Amsterdam (1993)
64. Stinson, D.R.: An Introduction to the Design and Analysis of Algorithms. The Charles Babbage Research Center, Winnipeg, Manitoba, Canada, 1985 (second edition, 1987)
65. Suganthan, P.N.: Particle swarm optimiser with neighborhood operator. In: Proceedings of the IEEE International Conference on Evolutionary Computation (CEC), vol. 3, pp. 1958–1962 (1999)
66. Talbi, E.: Parallel Combinatorial Optimization. Wiley, New York (2006)
67. Tomassini, M.: Spatially Structured Evolutionary Algorithms: Artificial Evolution in Space and Time. Natural Computing Series. Springer, Berlin (2005)
68. Whitacre, J.M., Sarker, R.A., Pham, T.T.: The self-organization of interaction networks for nature-inspired optimization. IEEE Trans. Evol. Comput. **12**(2), 220–230 (2008)
69. Whitley, D.: Cellular genetic algorithms. In: Forrest, S. (ed.) Fifth International Conference on Genetic Algorithms ICGA-5, California, CA, USA, p. 658. Morgan-Kaufmann (1993)
70. Whitley, D., Rana, S., Dzubera, J., Mathias, K.E.: Evaluating evolutionary algorithms. Artif. Intell. **85**, 245–276 (1997)
71. Wright, S.: Isolation by distance. Genetics **28**, 114–138 (1943)
72. Yumind, L., Ming, L., Ling, L.: Cellular genetic algorithms with evolutional rule. In: International Workshop on Intelligent Systems and Applications (ISA), pp. 1–4. IEEE (2009)
73. Zelinka, I., Oplatkova, Z., Nolle, L.: Analytic programming-symbolic regression by means of arbitrary evolutionary algorithms. Int. J. Simul. Syst. Sci. Technol. **6**(9), 44–56 (2005)

Chapter 9
Time Series Prediction Using Coalitions and Self-organizing Maps

Juan C. Burguillo and Juan García-Rois

Abstract In this chapter we consider the Time Series Prediction problem (TSP), and the use of Self-organizing Maps (SOM) as the basic neural network model to apply. We conduct a topology-based TSP performance analysis, based on extensive numerical simulations, taking into account different complex networks for connecting the neurons within the SOM. We introduce the use of coalitions, and a parameter free version of our Coalitional Algorithm for SOM (CASOM), which adapts the neuron neighborhood to the time series under analysis by means of dynamic coalitions. The results obtained by CASOM are better than the ones provided by SOM in all the topologies and time series considered in this chapter, even when the network structure changes along the training process. Besides, and more remarkable, the number of training epochs, needed by CASOM to stabilize the weights of its neurons, is much lower than SOM.

9.1 Introduction

Time Series Prediction (TSP) is a function approximation to estimate (predict, forecast) future values after a time sequence of ordered observations. The aim of TSP methods is to find the model that best fits with the empirical observations, and once a particular model has been selected for TSP, the next step concerns with estimating its parameters from the available data. Finally, a model evaluation is done concerning its predictive ability, usually over a new set of testing data.

In the last decades, a growing interest in TSP methods has been observed. Particularly, within the field of neural networks [21, 28], some well-known supervised

J. C. Burguillo (✉) · J. García-Rois
Department of Telematics Engineering, School of Telecommunications Engineering, University of Vigo, 36310 Vigo, Spain
e-mail: J.C.Burguillo@uvigo.es

J. García-Rois
e-mail: jgrois@gti.uvigo.es

J. C. Burguillo, *Self-organizing Coalitions for Managing Complexity*, Emergence, Complexity and Computation 29, https://doi.org/10.1007/978-3-319-69898-4_9

neural architectures have been considered, such as the Multilayer Perceptron (MLP), the Radial Basis Functions (RBF) networks and, more recently, the Self-organizing Maps (SOM) [3]. A Self-organizing Map (SOM) [13] is a well-known competitive learning neural network, that learns from examples a mapping (projection) from a high-dimensional continuous input space onto a low-dimensional discrete one (usually a lattice). After training, the neuron weights in the map can provide a model of the training patterns mainly through: vector quantization, regression and clustering. Using these techniques, self-organizing maps have been applied to multiple applications [15] in areas like automatic speech recognition, monitoring of plants and processes, cloud classification, micro-array data analysis, document organization, image retrieval, etc.

In this chapter we consider the use coalitions and complex network topologies for connecting the neurons within a SOM. Not many works have dealt with such framework in the past, and most of them only consider the use of complex topologies from a partial point of view. In [2] the authors explore growing SOMs with the possibility to expand areas of the network adequate for achieving hierarchical clustering. Yang et al. [31] implemented a SOM with small-world topology, in which neighborhood sizes are progressively reduced during the learning process. The resulting SOM seems to have a faster convergence and have a more reasonable weight distribution. Furao et al. study in [9] incremental SOM for online unsupervised learning with two layer network structures to identify relevant clusters. Jiang et al. [12] use SOM with dynamic complex networks topologies, obtained by means of evolutionary optimization algorithms, and apply them for classification purposes. Abinader et al. [1] analyze different heuristics for constructing hierarchical SOMs for time series prediction, evaluating their computational cost and forecast precision. In a series of works, related with SOM and forecasting, Yin et al. study alternative models in order to gain more precision in time series prediction, mainly applied to foreign exchange (FX) rate forecasting. For instance, in [19] the authors propose a network, denoted as self-organising mixture autoregressive (SOMAR) model, that can be used to describe and model nonstationary, nonlinear time series by means of a number of underlying local regressive models. Most recently, in [20], the authors study the use of neural gas and autoregressive models for forecasting (NGMAR).

This chapter introduces SOM as a basic neural network model to solve the TSP problem. We conduct an extensive topology-based TSP performance analysis, based on extensive numerical simulations, taking into account the classical rectangular 2-D lattice; and also different complex networks for connecting the neurons within the SOM. We also consider the use of coalitions, and we introduce a redesigned Coalitional Algorithm for SOM (CASOM), which adapts the neuron neighborhood to the time series under analysis by means of dynamic coalitions, and allows CASOM to become a parameter free algorithm. The performance of the CASOM algorithm is better than SOM for all the topologies and time series considered in the chapter. Besides, and more remarkable, the number of training epochs needed by CASOM to stabilize the weights of its neurons is much lower than SOM.

The rest of the chapter is organized as follows. First, we introduce the concepts of Time Series Prediction and Self-organizing Maps in the next two sections. Then, in Sect. 9.5 we analyze the conditions for using classical SOM for TSP, in order to obtain the best results. Then, we consider the use of complex networks for connecting the neurons in the map. Afterwards, we evaluate the results over Real Time Series. Section 9.8 defines precisely a new Coalition Algorithm for SOM (CASOM). Afterwards, Sect. 9.9 shows the results obtained after applying CASOM over the set of benchmarks previously evaluated with SOM. Section 9.10 presents a set of results using dynamic networks. Finally, we present our conclusions.

9.2 Time Series Prediction (TSP)

A time series is a collection of data recorded over a period of time. Analyzing a time series can be used to take current decisions and to do plans based on long-term forecasting. The background is that we usually assume that patterns experimented in the past will happen again in the future. Typical time series examples could be the assets value of a certain company along several days in the stock exchange market, the series of temperature values in a certain location along a time period, or the electricity consumption by hours in a certain company.

Time series prediction or forecasting is the use of a model to predict future values based on previously observed ones. Even the terms prediction and forecasting have slightly different meanings in the literature, we will consider them equivalent along this chapter. Applications of time-series forecasting include economic planning, sales forecasting, stock control, production planning, financial risk management, etc.

A forecasting procedure is an algorithm for forecasting future values based on present and past ones. Alternatively, perhaps we may identify a certain probability model for the given data, and we can find the optimal forecasts conditional on that model. Therefore, the two terms *forecast method* and *forecast model* should be considered different. Mathematical, forecasting methods may be broadly classified into:

- *Univariate methods*, where forecasts depend only on present and past values of a single time series.
- *Multivariate methods*, where forecasts of a given variable may depend on values of additional time series variables called predictors. Multivariate forecasts may depend on a multivariate model, where the variables are jointly dependent.

This chapter will deal exclusively with *univariate methods*, as our main aim is to analyze several conditions for using SOM to approximate time series and their capabilities to predict future values.

9.3 Self-organizing Maps (SOM)

Self-organizing Maps (SOM) [13, 14], also denoted as Self-organizing Feature Maps (SOFM) or Kohonen Neural Networks, were introduced by the Finnish professor Teuvo Kohonen in the 1980 s based on biologically neural models from the 1970s [26] and morphogenesis models dating back to Alan Turing in the 1950s [24]. SOM are a type of artificial neural networks (ANN) that are trained using unsupervised learning to self-organize a discrete representation of the input space, denoted as a map. In order to do that, SOM use a certain amount of classificatory resources, usually denoted as neurons, nodes, cells or units.

From a topological point of view in neural networks, a SOM is a single layer neural network, where the neurons are set along a d-dimensional grid. In most applications this grid is 2-dimensional and rectangular, but hexagonal grids or other dimensional spaces have been also used in the literature. Associated with each neuron is a weight vector of the same dimension as the input vectors (patterns) and a position in the map space (see Fig. 9.1). Therefore, the self-organizing map describes a mapping from a higher dimensional input space to a lower dimensional map space (also know as multidimensional scaling).

SOM are different from other ANNs as they apply competitive learning instead of error-correction, such as backpropagation with gradient descent methods, and in the sense that they use a neighborhood function to preserve the topological properties of the input space. The procedure for placing a vector from the input space onto the map is to find the neuron with the closest (smallest metric distance) weight vector. After several iterations, the result is an self-ordered network in which nearby neurons will share certain similarities. Therefore similar input patterns activate similar areas in the SOM producing a local specialization in the global self-organized network. When the map contains a large number of neurons, it is possible to perform cluster operations over that map itself, and it may show emergent properties [25].

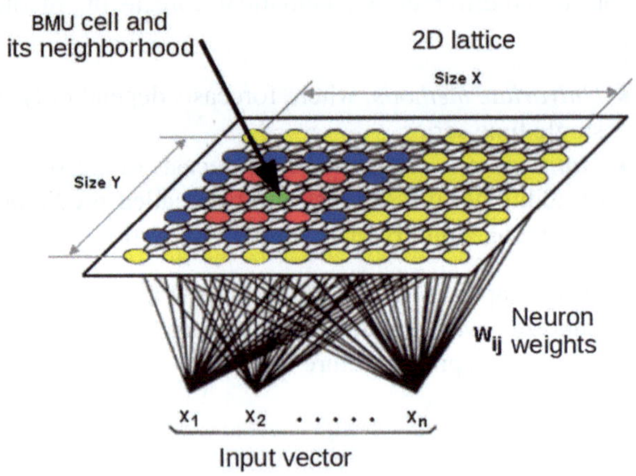

Fig. 9.1 A SOM map with an input vector, a winner neuron and its neighborhood

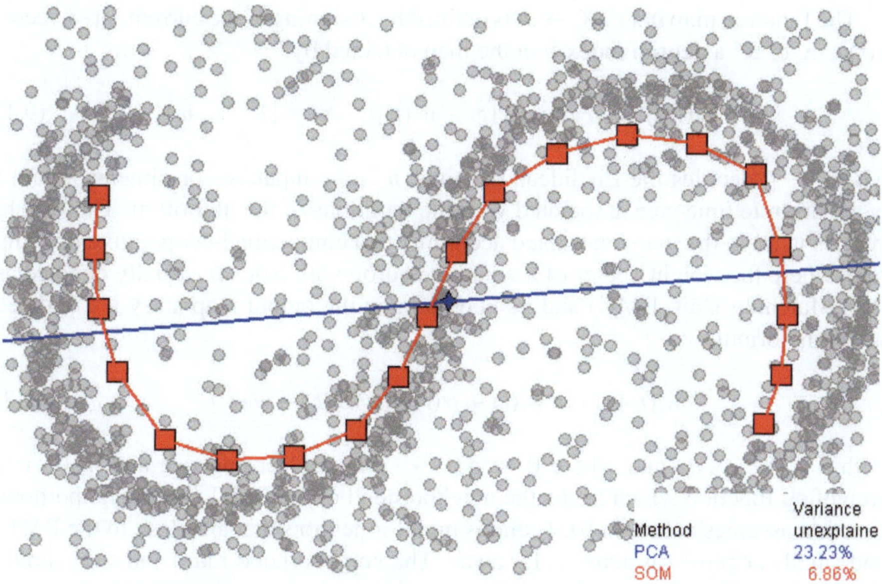

Fig. 9.2 The picture presents a comparison between SOM and PCA for data approximation (source Wikimedia Commons [29])

Before starting the training process, the neurons may be initialized randomly, and like most artificial neural networks, SOMs operate in two modes: training and mapping. *Training* builds the map using input examples (it is a competitive process, also called vector quantization), while *Mapping* automatically classifies a new input vector. The SOM network must be trained with a large number of sample vectors that represent, as close as possible, the kinds of vectors expected during mapping phase. Those training samples are usually iterated several times, also known as epochs.

Figure 9.2 presents a comparison between SOM and PCA methods for data approximation. The 2-dimensional input data points are represented by circles, while the SOM network contains 20 nodes and are represented by squares.

9.3.1 SOM Formal Definition

The SOM learns from the input patterns a mapping from a high-dimensional continuous input space X onto a low-dimensional discrete space L (the lattice) of m neurons which are arranged in fixed topological forms, e.g., as a rectangular 2-dimensional array. The number of neurons m in the map is usually defined after experimentation with the dataset.

The function map $b(x) : X \to L$ is defined by assigning to the current input vector $x(t) \in X \subset \mathbb{R}^n$ a neuron index b in the map obtained by:

$$b(x(t)) = argmin||x(t) - w_i(t)||, \quad \forall i \in \{1, \ldots, m\} \tag{9.1}$$

where $|| \cdot ||$ denotes the Euclidean distance, n is the input vector dimension and t is the discrete time step associated with the iterations of the algorithm. The weight vectors $w_i(t)$ in the map are trained according to a competitive-cooperative learning rule, where the weight vector of a winning neuron with index b (usually denoted as Best Matching Unit, BMU) and its neighbors in the output map array are updated with this formula:

$$w_i(t+1) = w_i(t) + \alpha(t)h_{bi}(t)[x(t) - w_i(t)], \tag{9.2}$$

with $i = 1, \ldots, m$, and where $0 < \alpha(t) < 1$ is the learning rate and $h_{bi}(t)$ is a weighting function which limits the neighborhood of the BMU. This neighborhood function assumes values in [0, 1] and is high for neurons that are close to the BMU, and small (or *zero*) for neurons far away. The neighborhood radio and $\alpha(t)$ should both decay with t to guarantee convergence of the weight vectors in the map to stable steady states.

The neighborhood function $h_{bi}(t)$ depends on the distance between the BMU (b) and a lattice neuron with a certain weight w_i. The simplest function considers $h_{bi}(t) = 1$ for all neurons close enough to the BMU and *zero* for the others, but we also could consider more complicated functions depending on iteration time or lattice space. Using a Gaussian function that shrinks with time is a common choice. Therefore, at the beginning the neighborhood produces a broader impact over the global lattice, but as soon as the training process advances the neighborhood shrinks to just a few neurons producing a local convergence.

Figure 9.3 presents an example with the training process of a SOM network. The blue blob is the distribution of the training data, the small white disc is the current input sample, and the yellow-two color disc is the BMU and its neighborhood. At first (left) the SOM nodes are arbitrarily positioned in the data space. After we get an input sample (white), we select the node (yellow) with the nearest weight to such input

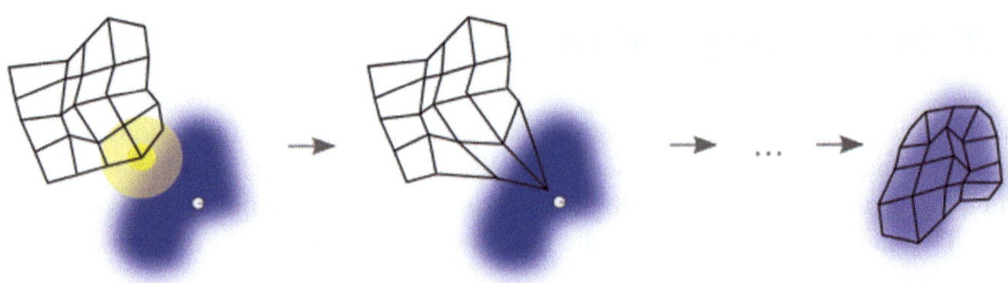

Fig. 9.3 Training a SOM (source Wikimedia Commons [30])

sample. Such node, together with its neighborhood, is moved towards the training sample. After many iterations, the SOM approximates the data distribution from the input set (right).

9.3.2 The VQTAM Model

In order to do TSP with SOM, we use the Vector-Quantized Temporal Associative Memory (VQTAM) model [4]. In a general formulation, the VQTAM models the input vector $x(t)$ at time step t by two parts: the first part, denoted $x^{in}(t) \in \mathbb{R}^p$, carries data about the input of the dynamic mapping to be learned. The second part, denoted $x^{out}(t) \in \mathbb{R}^q$, contains data concerning the desired output of this mapping. To do such estimation, the weight vector of neuron i, $w_i(t)$, has its dimension adapted as follows:

$$x(t) = \begin{pmatrix} x^{in}(t) \\ x^{out}(t) \end{pmatrix} \quad and \quad w_i(t) = \begin{pmatrix} w_i^{in}(t) \\ w_i^{out}(t) \end{pmatrix} , \tag{9.3}$$

where $w_i^{in}(t) \in \mathbb{R}^p$ and $w_i^{out}(t) \in \mathbb{R}^q$ are, respectively, the portions of the weight (prototype) vector which store information about the inputs and the outputs of the desired mapping. In general we have that $p > q$, and particularly for the univariate TSP that we consider in this chapter, we have that $p > 1$ and $q = 1$, so that the following definitions apply:

$$x^{in}(t) = [y(t) \quad y(t-1) \quad \dots \quad y(t-p+1)]^T \tag{9.4}$$

$$x^{out}(t) = s(t) , \tag{9.5}$$

where $s(t)$ is the value generated by the process at the time step t, $y(t)$ is a vector containing the p past samples, and T denotes the transpose vector. During learning, the winning neuron at time step t is determined based only on $x^{in}(t)$:

$$b(x^{in}(t)) = argmin||x^{in}(t) - w_i^{in}(t)||, \quad \forall i \in \{1, \dots, m\} . \tag{9.6}$$

For updating the weights, both $x^{in}(t)$ and $x^{out}(t)$ are used:

$$w_i^{in}(t+1) = w_i^{in}(t) + \alpha(t)h_{bi}(t)[x^{in}(t) - w_i^{in}(t)], \tag{9.7}$$

$$w_i^{out}(t+1) = w_i^{out}(t) + \alpha(t)h_{bi}(t)[x^{out}(t) - w_i^{out}(t)], \tag{9.8}$$

with $i = 1, \dots, m$, and these two learning rules perform topology-preserving vector quantization on the input and output spaces of the mapping. As training proceeds, the SOM learns to associate the input prototype vectors w_i^{in} with the corresponding output prototype vectors w_i^{out}. Once the SOM has been trained, its output $z(t)$ for a new input vector is estimated from the learned codebook vectors as follows:

$$z(t) = w_b^{out}(t) \ , \tag{9.9}$$

where $w_b^{out}(t)$ is the weight output vector of the corresponding winning neuron. From it we define the error function used in this chapter as:

$$e(t) = s(t) - z(t) = s(t) - w_b^{out}(t) \ . \tag{9.10}$$

9.4 Context of the Simulation Framework

All the simulations have been repeated 70 times (unless it is explicitly said) for average considerations. The particular VQTAM parameters used in all the next simulations are the initial standard ones in [4], i.e., $p = 2$ (input values) and $q = 1$ (output value). The case of $\alpha(t)$ function also follows standard variation, starting at $\alpha(0) = 1$ and decreasing linearly reaching zero at the end of the training phase.

Furthermore, in order to simplify the analysis as much as possible, and considering the particularities introduced by the complex topologies, our neighborhood function $h_{bi}(t)$ considers only the closer neuron's neighborhood, which means the four neurons conforming the von Neumman neighborhood (North, East, South, West) in the spatial case, and the direct neighbors in the case of the complex topologies used here (SW, SF and RN). The value defined for the BMU is $h_{bi}(t) = 1$ and for its neighbor neurons is $h_{bi}(t) = 0.5$. In addition, we also define a probability of neighbor updating $P_u \in \{0.25, 0.5, 0.75, 1\}$ as in our previous work [6], exploring in this chapter its behavior in a wider context.

For complex topology evaluation, in Sect. 9.6 we started by considering 4 initial neighbors in all the possible topologies, so the parameters used for building the RN, SW, and SF networks are $R_m^{0.0025}$, $W_m^{4;0.1}$ and $S_m^{4;-2}$ respectively.

We have tested several functions, benchmarks and real time series for TSP. We considered our previous SCSTS benchmark in [6], consisting in a trigonometric function $f(x) = 2sin(x) - cos(3x) + sin(5x)$, because it is continuous and its smooth variation sometimes induced higher error rates than classical discrete TSP benchmarks. We also have consider the BRTS benchmark presented in [3], which is a normalized chaotic time series from Lorenz dynamical system defined by the next system of equations:

$$\begin{aligned} dx(t)/dt &= a[y(t) - x(t)] \\ dy(t)/dt &= bx(t) - y(t) - x(t)z(t) \\ dz(t)/dt &= x(t)y(t) - cz(t) \end{aligned} \tag{9.11}$$

with default parameters $h = 0.01$, $a = 10$, $b = 28$, $c = {}^{8}/_{3}$ and random initial values in [0, 1]. We also used the well-known Mackey–Glass time series (MGTS) [16], which is a benchmark widely regarded for comparing the generalization ability of different methods, and corresponds to a chaotic time series generated from a time-delay ordinary differential equation. Concretely, for the Mackey–Glass equation, we followed the next non-linear time delay differential equation:

$$\frac{dx(t)}{dt} = \beta \frac{x(t-\tau)}{1+x(t-\tau)^n} - \gamma x(t), \quad \beta, \gamma, n > 0, \tag{9.12}$$

where $\beta = 0.2$, $\gamma = 0.1$, $\tau = 18$, $n = 10$, and $x(0)$ is a random real number in our simulations. Depending on the values of the parameters, this equation displays a range of periodic and chaotic dynamics.

Finally, the real time series considered in this chapter are a laser intensity time series (LITS) [23], and the classic Solar Spot Number time series (SSNTS), included in MATLAB R2012a. Further details are given in Sect. 9.7.

9.5 Analyzing SOM over Spatial Networks

In this section, we review the case of univariate time series prediction with 2-D grid SOM topologies (we named them as SP), for the benchmarks presented in Sect. 9.4. We subdivide the section accordingly with four main simulation scenarios. The evaluation method is the average Root Mean Squared Error (RMSE) between the predicted values and the real ones (i.e., the test set). Here, we review the prediction behavior depending on the probability of neighbor updating P_u, and how that behavior is influenced by changing the number of neurons m in the lattice. See [10] for a more detailed review considering different regular topologies, the impact of the training set and the use of using different $\alpha(t)$ functions during the learning process of the SOM. Even though the evolution of SOM, with all the mentioned parameters, is far from being deterministic in a wide sense, we analyze them fairly enough to provide useful understanding of how these parameters influence the results in TSP.

9.5.1 Evaluation of Regular Topologies

In this subsection, we study the influence of the $P \times Q$ lattice dimensions over the average RMSE. Depending on these dimension parameters, the resultant lattice can have slightly different shapes, such as a line ($1 \times m$), a rectangle or a square ($\sqrt{m} \times \sqrt{m}$). For completeness, we have also included in the analysis the case of a torus, where all neurons have degree equal to four, and no border neurons exists.

Simulation results are shown in Fig. 9.4 for both short (a) and long (b) Mackey Glass time series. The short case consists in 1000 training samples and a test set of

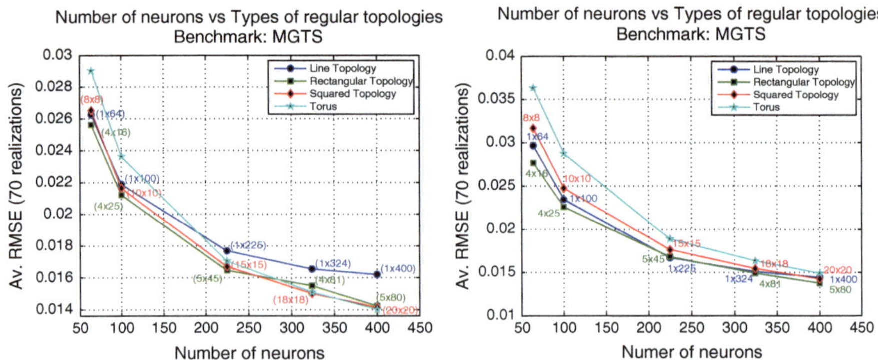

(a) Short MGTS (b) Long MGTS

Fig. 9.4 Performance of line, rectangular, square and torus topologies

500 samples. During the learning process, we repeat the training sequence 20 times (i.e., 20 epochs). The long MGTS case, consists of 19000 training samples and 1000 extra samples for the test phase, and no epoching is used. For further information about epoching, and the training process, we refer the reader to our previous work in [10].

Generally, results show good agreement among all topologies under analysis. We can observe that more regular topologies (p.e., going from line towards torus) does not necessarily mean a better prediction capability. In fact, rectangular lattices (something in between line and torus) provide the lower average RMSE in general. Henceforth, we will use rectangular 2-D grid topologies when we refer to spatial or regular networks in general.

9.5.2 Number of Neurons m and Updating Probability P_u

This subsection studies in a general context the Av. RMSE curve trends over the probability of neighbor update P_u and the number of neurons m. Hence, we used the three benchmarks, i.e., SCSTS, MGTS and BRTS, with 20000 samples each. The first 19000 samples are used to build the training set and the last 1000 ones for the test set. No epoching is used, i.e., the training set is not repeated during learning. This is mainly given that there are already a reasonable amount of samples for training. Figure 9.5 presents the results for a SCSTS benchmark, when the whole time series has only 10 periods (i.e., smooth variation). Figures 9.6 and 9.7 shows the results of the simulation using MGTS and BRTS benchmarks.

Simulation results show that the performance of an updating policy, which keeps a BMU's neighbor neuron unchanged based on a certain probability value $(1 - P_u)$ is, in general, highly dependent on the actual number of neurons m considered. The

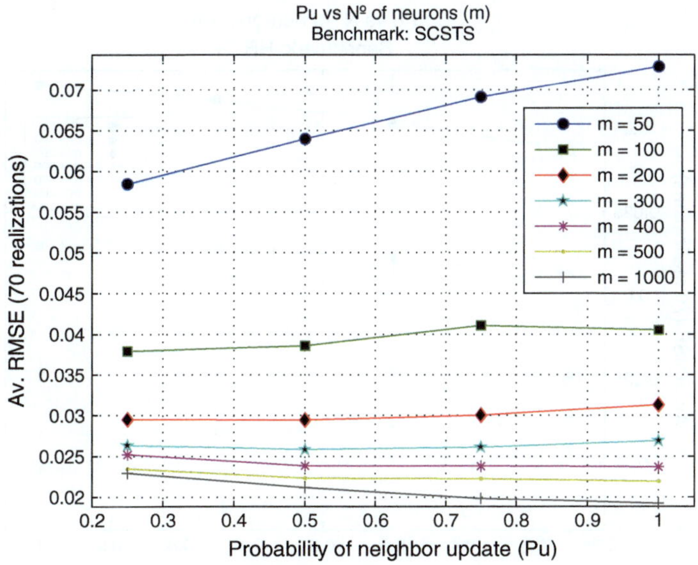

Fig. 9.5 SCSTS: Av. RMSE curves for different P_u values and number of neurons m

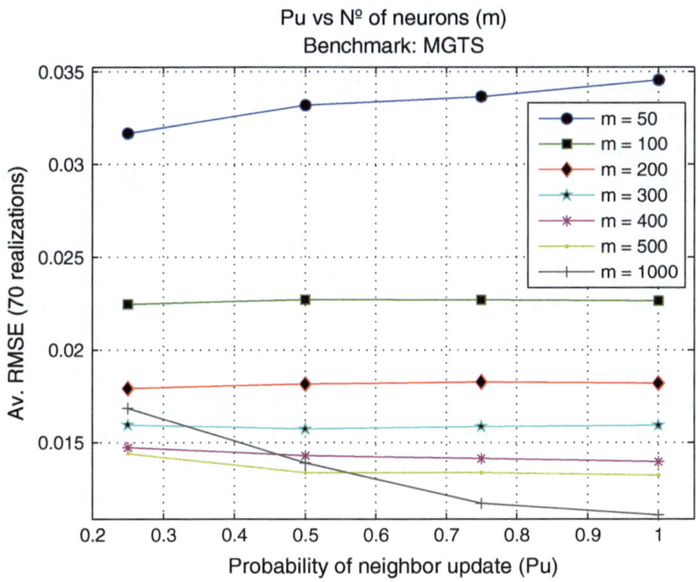

Fig. 9.6 MGTS: Av. RMSE curves for different P_u values and number of neurons m

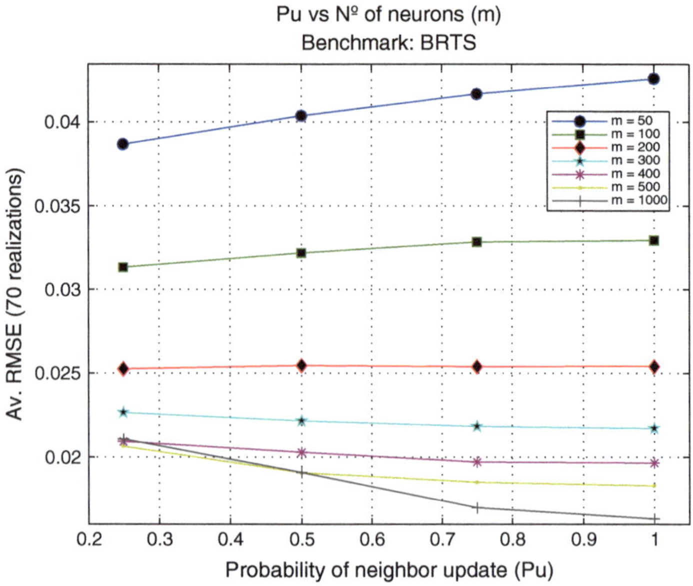

Fig. 9.7 BRTS: Av. RMSE curves for different P_u values and number of neurons m

results significantly agree no matter the benchmark being used (periodic or chaotic time series).

In order to quantify our results with previous works, we consider $m_h = 5.\sqrt{N_x}$ as the heuristic number of neurons usually applied in the SOM context [13], where N_x is the length of the training set used. In Figs. 9.5, 9.6 and 9.7, we observe two opposite trends over the P_u values as the number of neurons m increases.

Interestingly, low P_u values of the von-Neumman neighborhood provide a superior average RMSE performance, when $m \ll m_h$. Figures 9.5, 9.6 and 9.7 with $m = 50 \ll m_h \approx 680$, exemplify this result. When the situation is reversed, i.e., $m \gg m_h$, updating all BMU's neighbors is clearly advantageous. In Fig. 9.8a, b, we extended the previous analysis to epoching, considering both the case of $m = 50 \ll m_h \approx 680$ (a) and $m = 1000 \gg m_h \approx 680$ (b), which are the most representative scenarios of the opposite trends in our numerical example. It is worth noting the benefit of low P_u values when $m \ll m_h$, which improves with the number of epochs (Fig. 9.8a). On the contrary, the initial benefit of higher P_u, when $m \gg m_h$, happens to be significant only when no epoching is used (Fig. 9.8b).

Considering that the benefit of a high P_u is only clear when no epoching is used and $m \gg m_h$, together with the fact that time complexity increases with m, we can observe that if we do not update the whole BMU's neighborhood is more efficient, in general, from a practical point of view.

Intuitively, given a fixed training set with length L_{TS}, we can expect longer convergence time from the random initial position of the neurons to the output mapping as $m \gg m_h$. This convergence time also depends on the type of neighborhood, and

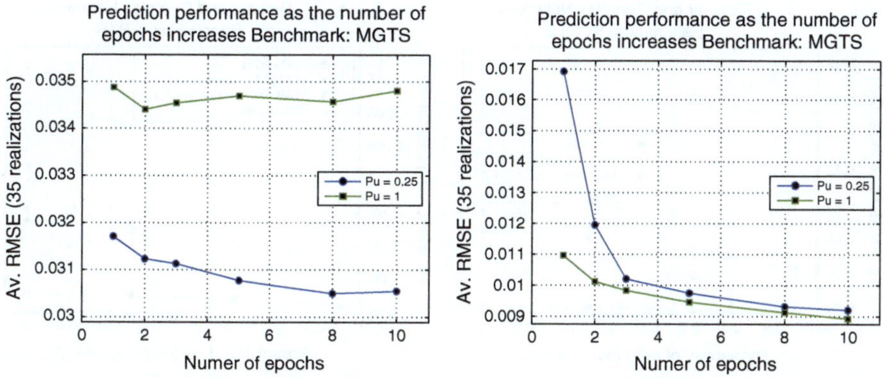

(a) $m = 50$ and $m_h \approx 680$ (b) $m = 1000$ and $m_h \approx 680$

Fig. 9.8 Performance gap between higher and lower P_u values for different scenarios

how the BMU influences its own neighborhood. In other words, it will take longer to the SOM to fit the output mapping if at each iteration a BMU is likely to affect only one neuron in average (i.e., $P_u = 0.25$). This is basically why in Fig. 9.8b we observe a significant difference in performance between $P_u = 0.25$ and $P_u = 1$, when no epoching is used, and why just by adding epoching the results get closer in both P_u cases.

On the other side, when $m \ll m_h$ the effect of updating all neighbors in each iteration becomes counter-productive. This is due to a constant moving of a big portion of the whole set of neurons available, and old BMUs are likely to loose an already good position. More formally, as the factor $\frac{m}{k}$ increases, being k the average neuron's degree, the higher P_u the better the performance, and viceversa.

Summing up, we have observed that the larger the neighborhood size (i.e., higher P_u, and thus the effective average neighbor degree k) with respect to the number of neurons m, the more likely a neuron will be constantly moving from one point within the space to another, which essentially prevents a precise tuning. This is why under similar training conditions (i.e., m, size of the training set, epoching, etc.), it is more practical to use lower P_u values (i.e., updating part rather than the whole neighborhood) for prediction. In addition, the finer tune allowed by lower P_u values, also suggest using P_u as a *second learning rate parameter* beyond the classic $\alpha(t)$ in Eq. 9.2.

9.6 Analyzing SOM Performance over Complex Networks

The analysis of complex networks in Self-Organizing Maps, for the task of time series prediction, has been first introduced in our previous work [6]. In that first approach, complex topologies were found to behave worse than the classic spatial

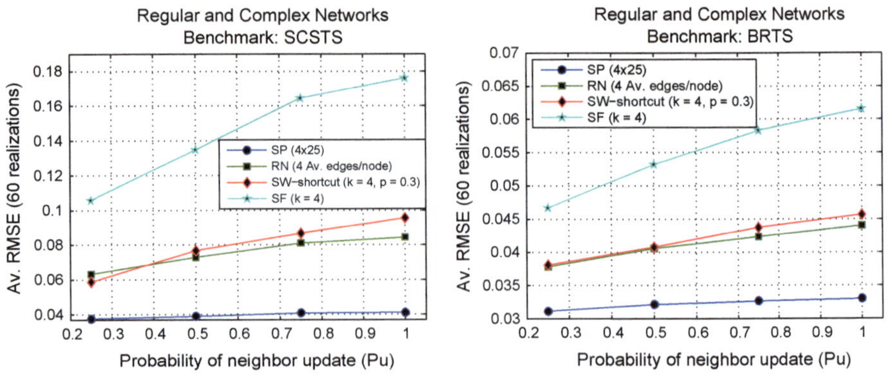

(a) SCSTS (b) BRTS

Fig. 9.9 Behavior using complex networks: $R_{100}^{4/100}$, $W_{100}^{4;0.3}$ and S_{100}^{4}

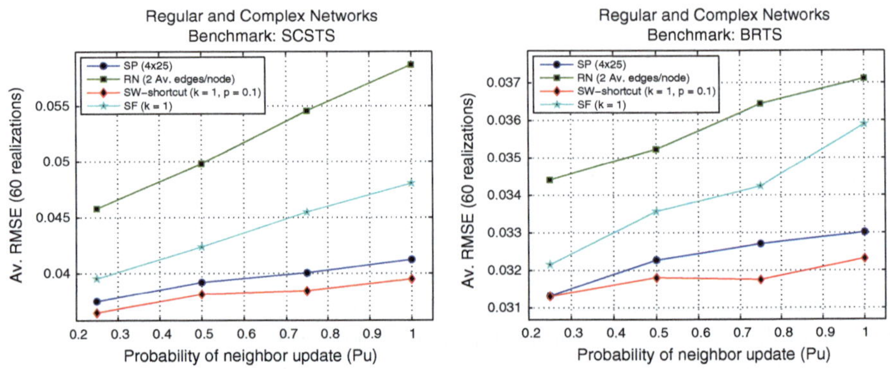

(a) SCSTS (b) BRTS

Fig. 9.10 Behavior using complex networks: $R_{100}^{2/100}$, $W_{100}^{1;0.1}$ and S_{100}^{1}

one. Indeed, we obtained a specially bad performance in prediction when scale-free (SF) models were used (see Fig. 9.9).

However, we believe that the complex topologies subject to that initial study (RN, SW and SF), deserve further analysis, given that each net can behave dramatically different by appropriately tuning its associated parameters. By doing so, we can find those parameters that make each net behave as its best for the benchmarks used.

In Fig. 9.10 we show the performance in prediction of complex networks for a different set of parameters, i.e., 2 average edges per node in RN, $k = 1$ and $p = 0.1$ in SW[1] with the link method in [11, 18], and $k = 1$ in the SF case. The values chosen for the previous parameters have been selected after an exhaustive testing in multiple experiments with the three benchmarks SCSTS, MGTS and BRTS.

[1]In the original Watts–Strogatz model, p stands for the rewiring probability of an existing edge, while p in [11, 18] refers to the probability of adding a new link.

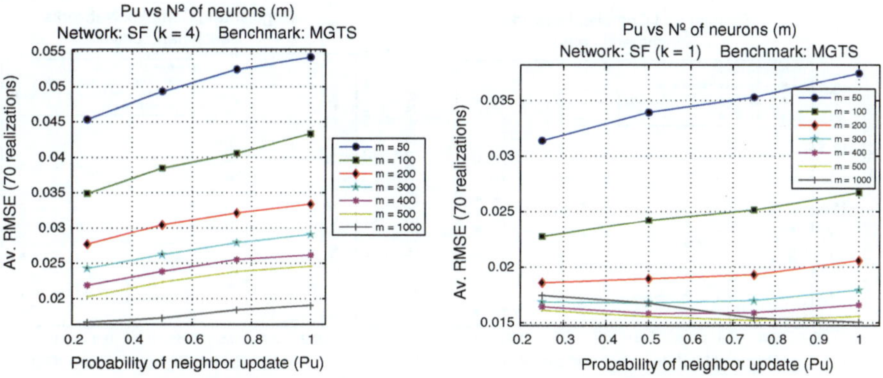

(a) SF with $k = 4$ (b) SF with $k = 1$

Fig. 9.11 Behavior of the neighborhood's updating rule with complex topologies, for different number of neurons

In random topologies R_m^p, the connection probability $p \in [0, 1]$, which determine the existence of a link between two neurons, should not be high. This is due to the fact that high p values in a RN topology cause to appear massive neighborhoods, which we have found to be specially disadvantageous. Concretely, we considered a relatively good practice following the rule $p = {}^{\Delta}/_{m}$, where $\Delta \in [1, 4]$ is the average number of edges per neuron.

Small-world nets, which are build upon the method of adding links in [11, 18], behave better than the classic Watts–Strogatz model in [27]. Concretely, setting $k = 1$, and a low probability of adding a new link $p \approx 0.1$, significantly improves the prediction (compare Fig. 9.9 with Fig. 9.10). Lastly, the specially disadvantageous performance of scale-free models can be significantly alleviated as k decreases, outperforming random networks.

Therefore, by changing the parameters of the complex networks, we allow the SW net to improve the spatial topology performance,[2] and a major improvement in SF. However, as the number of neurons m increases, the SP topology starts to take advantage over the other ones.

Finally, we have extended the analysis that combines the influence of the neighborhood updating rule P_u with the number of neurons m shown in Sect. 9.5.2, to the case of complex networks. Considering that all complex networks (RN, SW and SF) lead to similar results, we only show the case of SF. In the literature it has been shown that, in general, SOM's collaborative learning via neighborhoods brings topological orders and are fault tolerant [32, 33], but there can be scenarios where having a reduced effect over the neighborhood can be beneficial in terms of updating time and precision. In Fig. 9.11a, b we can observe, for different values of k, how the results are in line with the conclusions derived in Sect. 9.5.2. Thus, fixing a training set and hence a m_h value, the lower the ratio $\frac{m}{k}$ is, the better to update fewer neurons, and

[2]Except in the MGTS benchmark, although it reaches the best SP value when $P_u = 0.25$.

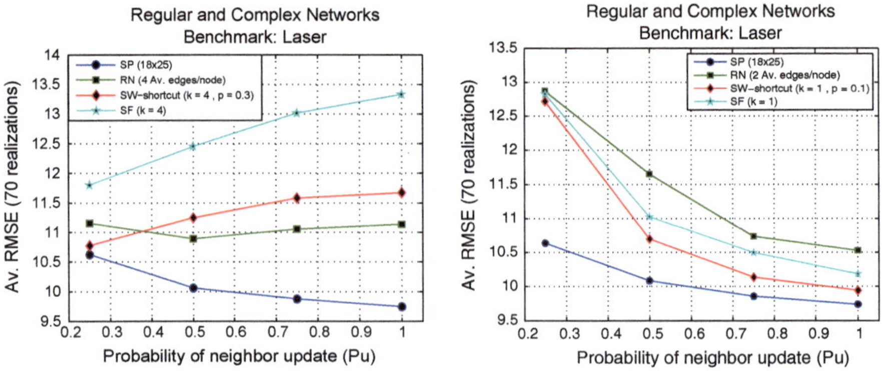

(a) $R_{450}^{4/450}$, $W_{450}^{4;0.3}$ and S_{450}^{4} (b) $R_{450}^{2/450}$, $W_{450}^{1;0.1}$ and S_{450}^{1}

Fig. 9.12 Laser intensity time series, 1 Epoch, 450 neurons

viceversa. Again, the main idea behind the probabilistic update of neighborhoods, is to help in the fine tune of neurons, avoiding a constantly moving of a relatively large amount of neurons. In addition, we also observe quite similar results between Figs. 9.11b and 9.6 in Sect. 9.5.2 (i.e., the 2D-lattice case), when the complex network becomes simpler and, therefore, its neighborhoods (likewise 2D-lattices have also simple von Neumman neighborhoods).

9.7 Analyzing SOM Performance over Real Time Series

In this section we evaluate the performance of regular and complex SOM topologies (SP, RN, SW and SF), in terms of the average RMSE, when real time series (RTS) are considered. As we also did in previous sections, we jointly consider the probability of neighbor updating P_u in the simulations. The RTSs, that we included in the analysis, are a *laser intensity* data over a period of chaotic activity, and the well-know *Solar Spot Number* (SSN).

The *laser intensity* univariate dataset, consisting of 10093 scalar values, represents the measurements of a far-infrared laser intensity over a period of chaotic activity. It was one of the data sets used in the Santa Fe Competition, directed by Neil Gershenfeld and Andreas Weigend, and can be found in [23]. We have used the first 8000 data samples for the training data, and the remaining data for the test. Given that there is a reasonable amount of data, we do not repeat the training set during learning (i.e., only one epoch). An example of this signal vs. our prediction output is given in Fig. 9.18 for $P_u = 1$.

As expected from previous Sect. 9.6, the simulation results show that scale-free models with $k = 4$, performs significantly worse than any other of the topologies under our study (Fig. 9.12a). However, this effect can be significantly reduced by

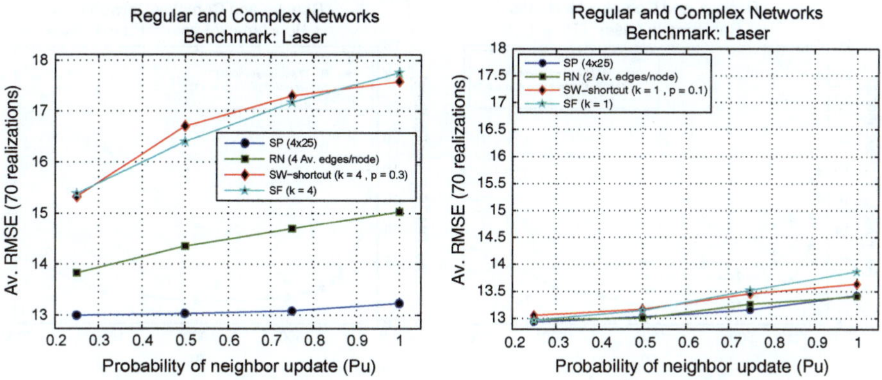

(a) $R_{100}^{4/100}$, $W_{100}^{4;0.3}$ and S_{100}^{4} (b) $R_{100}^{2/100}$, $W_{100}^{1;0.1}$ and S_{100}^{1}

Fig. 9.13 Laser intensity time series, 1 epoch, 1000 neurons

decreasing the parameter k, i.e., reducing the *hub-effect* of the network (Fig. 9.12b). Small-world topology also benefits from lower values of k and p. Similarly to results obtained in Sect. 9.5.2, the increasing/decreasing tendency of the average RMSE curves, with respect to P_u values, depends on the number of neurons m. However, in the case of complex topologies, it also jointly depends on their specific configuration parameters. For example, with $m = 100 \ll m_h \approx 450$ (Fig. 9.13), the tendency is the same, independently of the parameters of the complex nets, similarly to Sect. 9.6, and low P_u values are recommended. As m approaches the heuristic $m_h \approx 450$ the particular parameters of each type of network, can dramatically affect the tendency (Fig. 9.12).

Additionally, in Fig. 9.14 we show the performance of the different regular and complex topologies when the number of neurons m varies, for a fixed $P_u = 1$. From both Fig. 9.14a, b, we can see that the SP topology improves its average RMSE until $m \approx 1000$, a value after which the performance starts to deteriorate. In the case of complex networks, where the average node's degree is $k \approx 4$ (Fig. 9.14a), we observe that both SF and SW networks always improve with m, contrarily to SP and RN. However, SF and SW nets need a significant larger number of neurons to get started toward being competitive, which is not very practical indeed.

In case of decreasing k and p parameters (Fig. 9.14b), the initial left side of the graph (i.e., $m < m_h$) shows that all topologies have a similar performance, and thus complex topologies can easily approximate the result of the regular lattice SP. Nonetheless, when $m > m_h$ all networks start to deteriorate, including SF and SW.

Again, the main reason why SF and SW improve the RMSE with m in Fig. 9.14a but not in Fig. 9.14b is due to the ratio $\frac{m}{k}$, which is responsible of modifying the turning point to larger (>2000) or shorter (≈500) values of m. Note that SF and SW with $k = 1$ lead to smaller neighborhoods, and that increasing the number of neurons m starts to be counter-productive sooner for higher values of k.

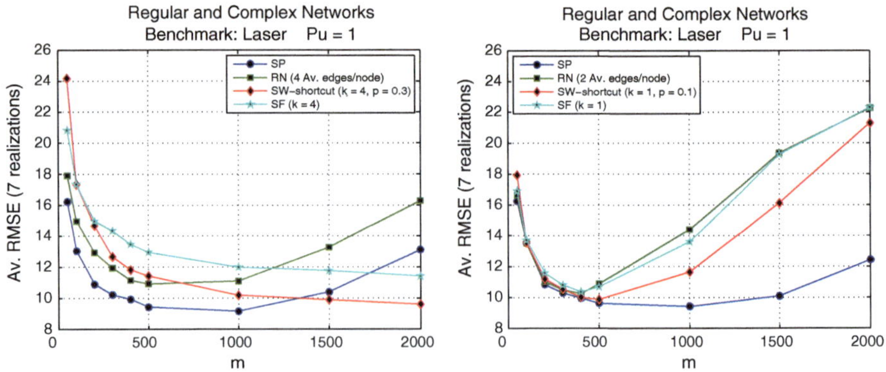

(a) $R_m^{4/m}$, $W_m^{4;0.3}$ and S_m^4 (b) $R_m^{2/m}$, $W_m^{1;0.1}$ and S_m^1

Fig. 9.14 Laser intensity time series, variation of m

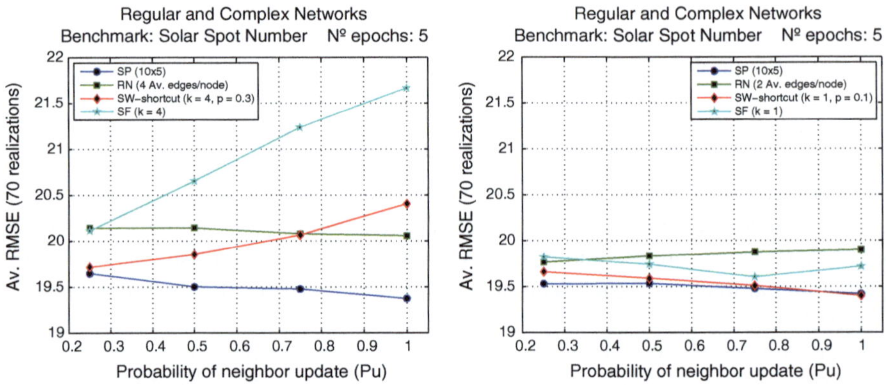

(a) $R_{50}^{4/50}$, $W_{50}^{4;0.3}$ and S_{50}^4 (b) $R_{50}^{2/50}$, $W_{50}^{1;0.1}$ and S_{50}^1

Fig. 9.15 Solar Spot Number time series, 5 epoch, $m = 50$

The *Solar Spot Number* series we have used is already included in MATLAB R2012a, and provides 2899 samples, which gives the monthly mean sunspot numbers for a 240 year period, that starts from January of 1749 and runs to July of 1990. We have selected the first 2600 samples for the training data and the remaining 299 for the test. A realization example is given in Fig. 9.19, where both the data for the test and the prediction behavior of the trained SOM is presented.

The SSN series is specially interesting, given its noisy variation and very short data sample. Therefore, the training set should be repeated several times, until we observe no further improvement in the error. In general, results show that few epoching ($\approx$5) is needed to reach a minimum value of the average RMSE, which is around 19.5. Neither doubling epoching nor increasing the number of neurons significantly improve the

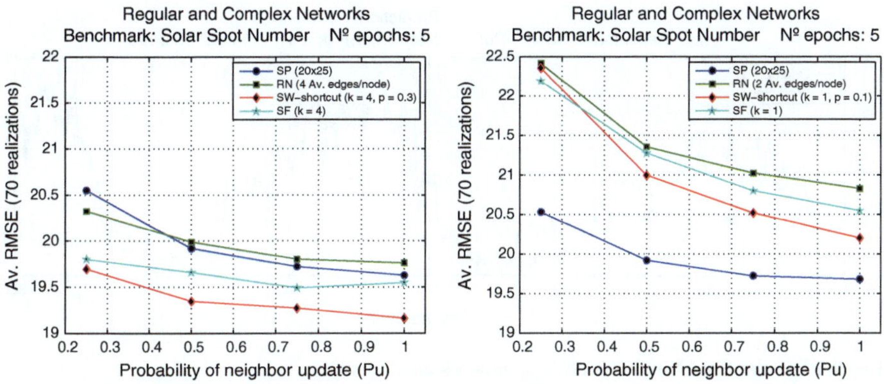

(a) $R_{500}^{4/500}$, $W_{500}^{4;0.3}$ and S_{500}^{4} (b) $R_{500}^{2/500}$, $W_{500}^{1;0.1}$ and S_{500}^{1}

Fig. 9.16 Solar Spot Number time series, 5 epoch, $m = 500$

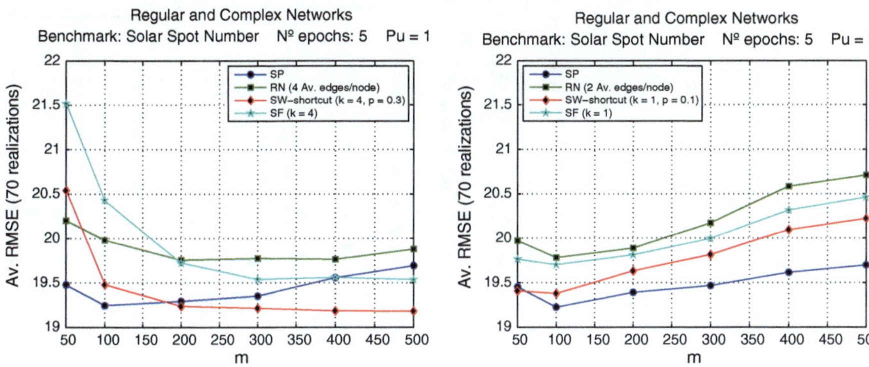

(a) $R_m^{4/m}$, $W_m^{4;0.3}$ and S_m^4 (b) $R_m^{2/m}$, $W_m^{1;0.1}$ and S_m^1

Fig. 9.17 Solar Spot Number time series, variation of m

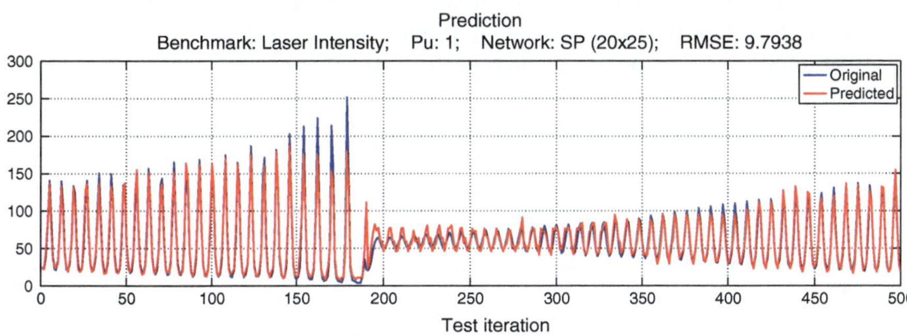

Fig. 9.18 Laser intensity with SP network, and 1 epoch and $m = 500$

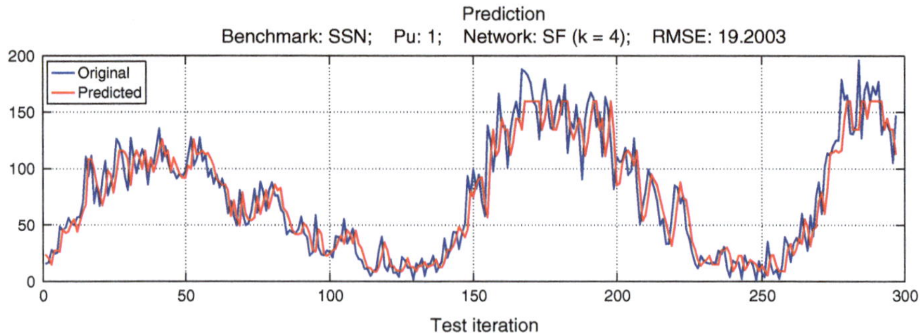

Fig. 9.19 Solar Spot Number with S^4_{500} network and 5 epochs

Fig. 9.20 Performance behavior when m increases in the MGTS benchmark

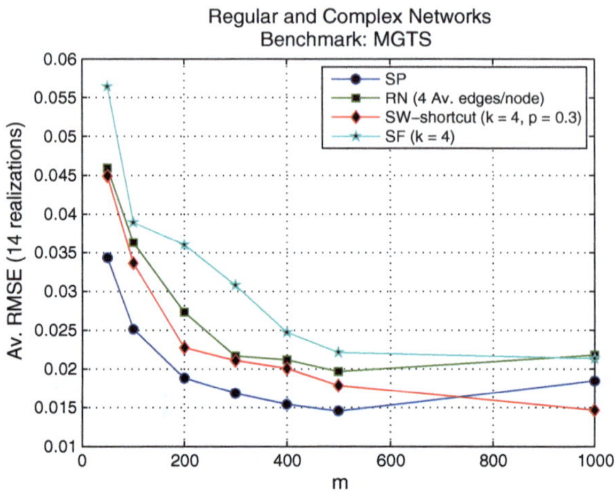

results about the minimum RMSE. We show the error in prediction depending on P_u for all the topologies, and two cases: $m = 50 \ll m_h$ and $m = 500 \gg m_h \approx 250$ in Figs. 9.15 and 9.16 respectively. What it is interesting here, is that SW and SF with parameters $k = 4$ and $p = 0.3$ are able to predict as good as the regular topology. In fact, SW with $k = 4$ and $p = 0.3$, can beat the best SP performance, for a reasonable number of neurons (Fig. 9.16a).

We also show in Fig. 9.17 the simulation results, when m is the parameter of variation, for a fixed $P_u = 1$. Similarly to the case of the laser intensity data in Fig. 9.14, SF and SW, with $k = 4$, constantly reduce the error as m increases, as opposed to the SP and RN cases. The best result for the regular topology is obtained for a low value of $m \approx 100 < m_h$, whereas the lowest error in general can be found for $W_m^{4,0.3}$, when $m > m_h$.

Besides the ratio $\frac{m}{k}$, Figs. 9.14 and 9.17 also suggest a strong dependence of the length of the training set (Fig.9.18). Thus, having few values of a time series also makes the turning point to appear sooner (≈100 in Fig. 9.17). In previous Sect. 9.5.2, we did not observe any performance decreasing in the spatial network, for a high number of neurons $m = 1000$, just because m_h was also significant large (Fig. 9.19). As a matter of fact, Fig. 9.20, where a MGTS with 7000 training samples ($m_h \approx 420$) is analyzed, show similar results to the case of laser intensity and SSN time series, as discussed.

In the *laser intensity* series and the *solar spot number* there is a potential benefit of SF and SW nets when $m > m_h$ and the number of available samples is low.

9.8 Coalitions and Complex Networks for SOM

The work introduced in this section continues some ideas from the previous Chap. 8, as it also has its origins in the work presented in [8], that describes cellular Evolutionary Algorithms (cEAs) with Coalitions (EACO), and takes advantage of both cellular and island population models. After such work, [5] analyzes the strong similarities between cEAs and SOM models, and argues about the strong potential for using similar approaches as a way to define dynamic neighborhoods by means of coalitions, and apply them to SOM models using an algorithm similar to EACO.[3] Finally this work was successfully developed and applied in [6], where neighborhoods also considered coalitions, that can be understood as small panmictic subpopulations (i.e., islands) that emerge in the cellular topology providing dynamic areas of BMU influence.

In this and the next sections we will extend and refine the work presented in [6], concerning the algorithm CASOM that uses coalitions and SOM for Time Series Prediction. First, we introduce the basics of the algorithm, then a more formal definition, and finally we present a set of results that enhance even more the ones obtained in the previous sections.

9.8.1 Introducing a General Coalitional Algorithm for SOM

An algorithm for training the SOM network with coalitions could be defined as:

For each input pattern $x(t)$:

1. Find the closest neuron with index b in the map, that becomes the BMU: $b(x(t)) = argmin||x(t) - w_i(t)||, \quad \forall i \in \{1 \ldots m\}$
2. Extend the neighborhood of the BMU with *similar* neurons in the grid: $c_b(t+1) = \{j \mid j \in c_b(t) \bigcup InfectNeighbors(b,t)\}, j = 1 \ldots m$
3. Update each neuron in the *new* neighborhood of the BMU according to the rule: $w_i(t+1) = w_i(t) + \alpha(t) Hnc_{bi}(t)[x(t) - w_i(t)], \quad i = 1 \ldots m$
4. Repeat the process until a certain stop condition is met.

[3] In fact, a traditional and very successful combination of Cellular Automata and Artificial Neural Networks (similar to SOM) are Cellular Neural Network models [7].

A main difference with the classical SOM algorithm is the coalition creation process, introduced in item 2, where a BMU neuron b tries to link neurons with index j, in order to spread its neighborhood. This dynamic coalition creation is the way that every BMU uses to extend its influence, i.e., by dynamically infecting/joining[4] other neurons to its neighborhood.

Another difference with the classical SOM algorithm is the new neighborhood function $Hnc_{bi}(t)$ in step 3 that takes into account if the neurons b and i belong to the same coalition or not. Even we can define multiple neighborhood functions to deal with coalitions, but a basic proposal could be:

$$Hnc_{bi}(t) = \begin{cases} if((c_i(t) = \emptyset)\&||b(t) - i(t)||) \leq r(t)) \Rightarrow hn_{bi}(t) \\ elseif \quad (c_b(t) = c_i(t)) \Rightarrow hc_{bi}(t) \\ else \Rightarrow 0 \end{cases} \tag{9.13}$$

where if the BMU neuron is independent, then those neurons close to it are in the neighborhood, and we apply a neighborhood function $hn_{bi}(t)$. Otherwise, if a BMU has a coalition, those neurons in the coalition are also its mates, and we apply a coalition function $hc_{bi}(t)$.

9.8.2 CASOM: A Coalitional Algorithm for SOM

Now, we proceed to define in more detail the particular coalitional training algorithm used in this article. Table 9.1 provides an enumeration of the different variables, symbols and parameters described along the text, in order to simplify the task of understanding the algorithms presented in this section.

The main cycle is presented in Procedure 6. Initially all parameters are initialized according to Table 9.1. Then, we iterate the main cycle over all the training samples, and for a certain set of epochs, in order to configure adequately the neuron weights. The rest of this algorithm is intuitive, and it can be summarized as: the new input pattern and its corresponding output value are obtained as inputs, then the BMU is determined. In line 6 the BMU infects other neurons in the map (described in Procedure 7). In line 7 the BMU updates its weights, and the weights of the neurons related with it (described in Procedure 9). Finally, in line 9 the learning rate α is decreased. The variation $\Delta\alpha$ has been defined in a simple way to reach zero linearly at the end of the training phase (see Table 9.1), being constant during each epoch, but other definitions can be done.[5]

After defining the main cycle of the algorithm, now we describe the behavior of the procedures instantiated by it. Procedure 7 describes the actions done by the

---

[4]Infection means here that coalitions try to join nearby neurons. We will also see an alternative *joining* method, where neurons are the ones that request to join coalitions.

[5]Note that in the first part of this chapter, related to SOM, it was constantly decreased after every sample, i.e., $\Delta\alpha = \frac{1}{T_s.T_e}$.

Procedure 6: CASOM Algorithm

```
1 Set parameters according to Table 9.1;
2 while Not EndEpoch do
3     x^in = getInputPattern();
4     x^out = getOutputValue ();
5     b = getBMU (x^in);
6     b.infectNeurons ();
7     b.updateCoalitionNeurons (x^in, x^out);
8 end
9 α = α - Δα;
```

BMU to infect other neurons depending on its role: a coalition leader (former BMU), a coalition member or an independent neuron. This procedure implements the game theory model introduced in [5] and mentioned before, i.e., each BMU tries to extend its influence as far as possible over the rest of the neurons.

In line 1 of the procedure we check if the BMU is a coalition leader (i.e., it was a previous BMU), then it infects its direct neighbors (that could belong to another coalition), and also the neighbors of its coalition members. In line 4, it checks if it is a coalition member, and as it became a BMU, then the first thing it does is to get independence, and then infect its direct neighborhood. Finally, if the neuron is independent, then it infects its direct neighbors as in the classical SOM.

Procedure 8 describes the procedure for infecting neurons in a particular set (coalition neuron neighbors or own neighborhood). Basically, for every neuron in the set,

Table 9.1 Relevant definitions

Identifier	Description	Initial value
b	Best Matching Unit	–
x^{in}	New input value	–
x^{out}	New output Value	–
w_i^{in}	Input pattern stored in neuron i	–
w_i^{out}	Output pattern stored in neuron i	–
P_u	Probability of updating a neuron	[0, 1]
P_{inf}	Probability of infecting a neuron	0.05
h_{bi}	Neighborhood function value	0.5, –
hc_{bi}	Coalitional neighborhood function value	0.5, –
T_s	Number of training samples	–
T_e	Number of training epochs	20
α	Learning rate	0.99
$\Delta\alpha$	Learning rate decrement	$\frac{1}{T_e}$
StopCondition	End of the training phase	$T_s.T_e$

Procedure 7: b.infectNeurons ()

```
if isCoaLeader() then
    infect (Neighbors);
    infect (CoaNeighbors);
else if isCoaMember() then
    getIndependence ();
    infect (Neighbors);
else                                        // b is independent
    infect (Neighbors);
end
```

it considers if such neuron is not a leader (avoiding to modify other BMUs), if it is already a coalition member; and if it has not been a BMU during this training epoch.[6] Then, if a random value is lower than the learning rate multiplied by the probability of infection, the neuron *i* is infected and becomes part or this BMU's coalition. Note that this is done to produce coalitions with bigger size at the beginning of the training process, but to reduce their effect when the network should be relatively stable, i.e., near the end of the training process. As soon as the simulation gets closer to its end it becomes much more difficult to infect other neurons, and coalitions break down as their members get independence when becoming BMUs. The reason for this behavior is also related with the results obtained with P_u and SOM in Sect. 9.5.2, i.e., to reduce the amount of neurons that become updated as the simulation goes by.

Procedure 8: b.infect (*neuronSet*)

```
foreach i ∈ neuronSet do
    if (not isLeader (i)) & (b ≠ getLeader (i)) & (timesBMUinEpoch(i) == 0) then
        if (rand() < α.P_inf) then
            infect (i);
        end
    end
end
```

Finally, Procedure 9 describes the set of actions taken to modify the weight of the neurons that are in the BMU's coalition and its direct neighbors, as happens in the SOM case. The BMU updates those neurons using the neighborhood function for coalitions hc_{bi}. Finally the input and output weights of the BMU are also updated in the way proposed in the VQTAM model.

[6] Note that all coalition leaders have been BMUs during a training epoch, but not all BMUs are coalition leaders as their coalition members could have been stolen by other BMUs.

Procedure 9: b.updateCoalitionNeurons (x^{in}, x^{out})

```
updateNeuronsInCoalition (hc_bi);
w^in = w^in + α.(x^in − w^in);
w^out = w^out + α.(x^out − w^out);
```

We point out that Procedure 9 only updates neurons in BMU's coalition, i.e., it do not take into account the direct neighborhoods of a neuron. Therefore, we only use the hc_{bi} function, from Eq. 9.13, that affects only the neurons in the coalition, except the BMU. Besides, we do not need the P_u parameter used for SOM, as we have an equivalent one, which is P_{inf} to model how easily a BMU infects other neurons to join them to its coalition.

Summarizing the conditions for creating, joining or leaving a coalition, we find:

- A certain neuron can only belong to one coalition.
- Neurons compete to extend their coalitions, and they will try to extend their influence linking other neurons from other coalitions.
- Neurons from a given coalition are updated by the BMU, so the coalition behaves like a panmictic island that extends the influence of a certain BMU.
- Coalition neurons from a BMU can be infected (stolen) by another BMU.
- There is a parameter, denoted as probability of infection ($P_{inf} \in [0, 1]$), to model how easily a BMU can infect other neurons.
- All the learning process is governed by the learning rate parameter ($\alpha \in [0, 1]$), that decreases along the training phase; and allows more changes at the beginning, but keeps the network stable near the end.

9.9 Experimental Results Obtained with CASOM

In this section we present the experimental results obtained using the CASOM algorithm. First, we analyze how coalitions and results evolve depending on the probability of infection P_{inf}. Then, we consider an alternative method for joining coalitions, instead of the infecting procedure. Afterwards, we present the results obtained by CASOM over several training sets. Then, we compare these results with the ones obtained by SOM. Finally, we consider the possibility of reconnecting the neurons of the network in real time, i.e., use dynamic networks trying to improve the network connectivity.

In the next figures, the error will be measured through the average RMSE (Root Mean Square Error) over 70 simulation runs,[7] 20 epochs per run and a spatial network with a neighborhood of four neurons.

[7]Note that this is a difference with [6], where the average MAE (Mean Absolute Error) was considered.

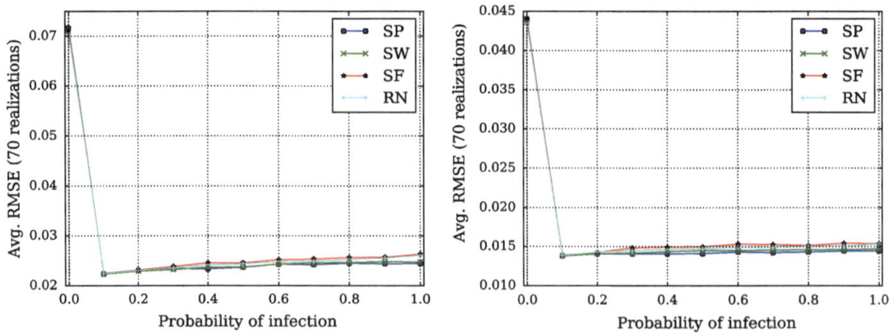

(a) Results for SCSTS (b) Results for MGTS

Fig. 9.21 CASOM results varying P_{inf} and using 144 neurons

9.9.1 Influence of the Infection Parameter

In order to clarify the effect of P_{inf} over the different types of networks, Fig. 9.21 presents the results obtained by CASOM over SCSTS and MGTS, considering the four different topologies. Observing the figure, we can realize that values around ($P_{inf} = 0.1$) provide the best results for almost all type of networks; considering the different topologies and time series used in this chapter. This is coherent with the results obtained by low P_u values for SOM, indicating that big modifications on the network cause the deterioration of the weights already learnt. Therefore, ($P_{inf} = 0.1$) and a spatial topology with a basic von Neumann neighborhood will be our configuration by defect, applied in most of the simulations presented next.

Figure 9.22 presents a snapshot of the grid in the CellNet simulator, after 2000 samples of the BRTS benchmark, with 178 independent cells, 33 BMUs and 113 cells in BMUs coalitions; meaning a total number of 324 neurons in the grid. The figure shows the independent cells in black color, while the BMUs coalitions appear in other colors.

Figure 9.23 describes how the number of coalitions, their members and the number of independent cells evolve in the MGTS and BRTS benchmarks. After the first 300 samples these graphics remain almost stable, but slowly decreasing to zero as the simulation approaches to its end. As the reader may observe, while the coalition number (CN) is similar in both figures, the number of cells in coalitions (CC) is higher in the BRTS case, meaning that in average the coalition size is bigger in BRTS (around 1000 samples); so the BMUs influence a higher number of neurons in the BRTS case. This is a key property of using dynamic coalitions, as they can adapt to the particular conditions of time series.

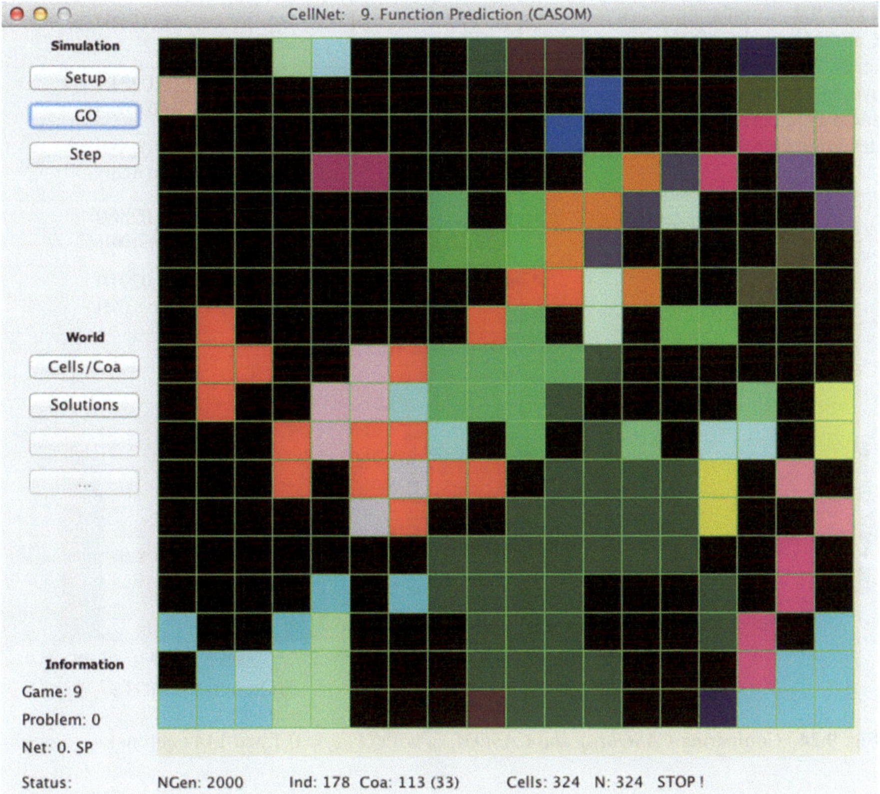

Fig. 9.22 CellNet snapshot with coalitions over the grid

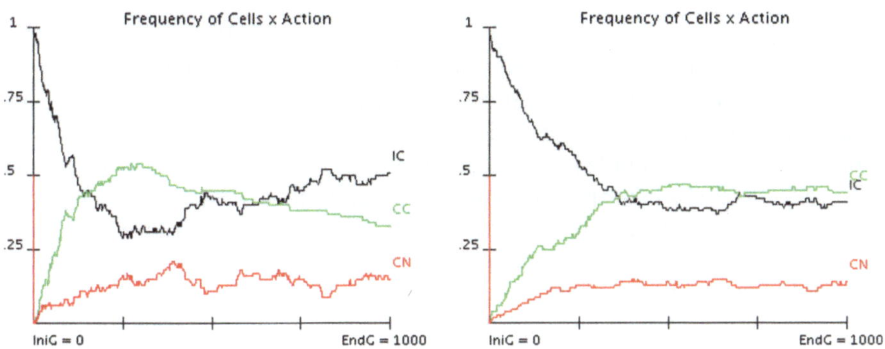

(a) Type of cells in MGTS (144 neurons)

(b) Type of cells in BRTS (324 neurons)

Fig. 9.23 Independent cells (IC), coalition cells (CC) and number of coalitions (CN) with a MGTS and BRTS training sets over a spatial network ($P_{inf} = 0.1$)

Table 9.2 Comparing $CASOM_{inf}$ and $CASOM_{join}$ for SCSTS. We show the average RMSE over 70 runs, and include the standard deviation below

Topology / Algorithm	$CASOM_{inf}$	$CASOM_{join}$
Spatial	0.02245 ±0,0.00101	0.02515 ±0,0.00145
Small-world	0.02221 ±0,0.00082	0.02550 ±0,0.00152
Scale-free	0.02242 ±0,0.00103	0.02860 ±0,0.00188
Random	0.02248 ±0,0.00092	0.02910 ±0,0.00210

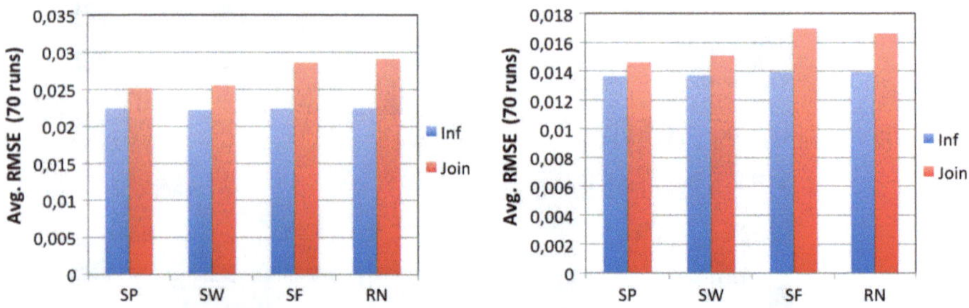

(a) Results for SCSTS (b) Results for MGTS

Fig. 9.24 Comparing $CASOM_{inf}$ and $CASOM_{join}$ with ($P_{inf} = 0.1$ and 144 neurons)

9.9.2 Infection Versus Joining

There are two main alternatives about how coalitions, i.e., BMUs expand their influence:(i) weak neurons, that never became BMUs or were updated, could ask to join strong *nearby* coalitions, or (ii) frequent BMUs expand their influence among nearby neurons through the infection procedure. The first mechanism was proposed in [5], following the ideas from [8]. There, a neuron *joins* a new coalition if he has not been very *successful* in the past becoming a BMU. Alternatively, in [6], the *infection* mechanism was proposed. Here we revise both alternatives, and see how infecting is better than joining, from a practical point of view; and besides it is computationally simpler to implement and much more efficient.

A comparison of both mechanisms can be found in Table 9.2 for the SCSTS, where below the average value appears the standard deviation. Figure 9.24 shows graphically the values of the table, and also includes the ones obtained for the MGTS case. The horizontal axis shows the four topologies considered here: SP, SW, SF, and RN (from left to right). As we can deduce from the table and see in the figure, $CASOM_{inf}$ is better for all the topologies; but the big advantage of $CASOM_{inf}$ relays in the computational efficiency. On the one hand, $CASOM_{join}$ is about five times slower for the network used in the figure, as it has to consider all the neurons in

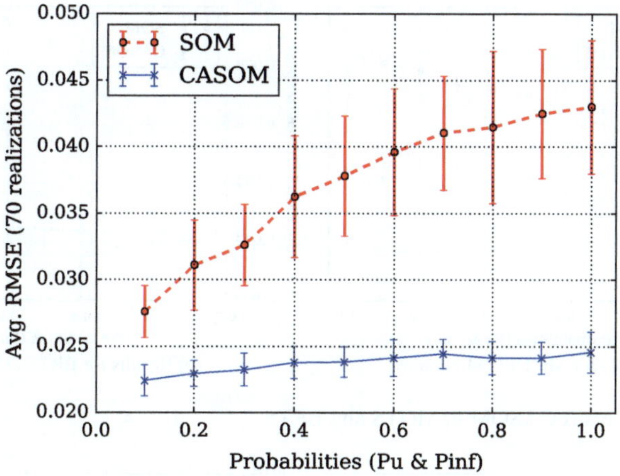

Fig. 9.25 SOM versus CASOM in SCSTS (144 neurons)

the network to allow them to decide if they join a coalition, get independence or change to another one; depending on each neuron context. On the other hand, with $CASOM_{inf}$ each BMU just performs the infection procedure after every input sample, and over a reduced set of neurons (its direct neighborhood plus the neighborhoods of its coalition neurons). Therefore, we have selected $CASOM_{inf}$ in all the results presented next.

9.9.3 SOM Versus CASOM

Now, we are going to compare the two algorithms SOM and CASOM over different time series using the spatial topology, and changing the values of P_u for the SOM case and P_{inf} for the CASOM case in the interval [0.1, 1] to see the evolution of the average RMSE depending on these relevant parameters.

Influence of P_u and P_{inf}

Figure 9.25 shows the resulting average RMSE, with P_u or P_{inf} in the horizontal axis depending if we use SOM or CASOM, respectively. The figure also includes the standard deviations that, as can be seem, are negligible for CASOM, meaning it is more stable concerning the results.

Figure 9.26 shows the results for the MGTS and BRTS cases. Watching the figures two observations are clear: the evolution of the curves depends stronger on the probability of updating the neurons P_u, in the SOM case, while in CASOM the influence of P_{inf} is much smaller. Besides, we remark that CASOM outperforms SOM in all cases.

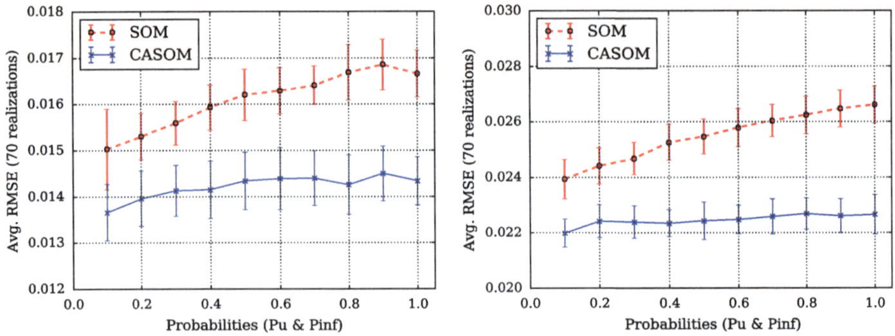

(a) Results for MGTS (144 neurons)

(b) Results for BRTS (324 neurons)

Fig. 9.26 SOM versus CASOM in MGTS and BRTS

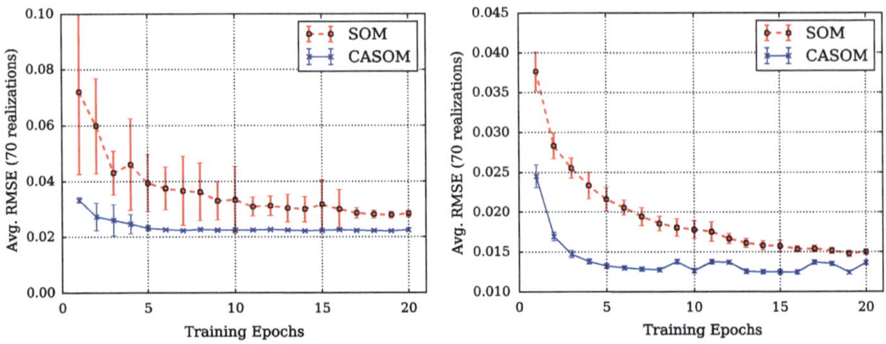

(a) Results for SCSTS

(b) Results for MGTS

Fig. 9.27 Comparing $CASOM_{inf}$ and $CASOM_{join}$ with (P_u, P_{inf} = 0.1 and 144 neurons), and changing the training epochs

We have applied the Wilcoxon unpaired signed rank sum statistical test to assess statistical confidence in the comparison of the two algorithms presented in these figures, and the results are relevant.

Influence of the Training Epochs

Figure 9.27 shows the evolution of the results for SOM and CASOM, when changing the training epochs. It becomes clear the fastest convergence of CASOM, that with more than 3 epochs already converge to its final results, while the convergence of SOM is much slower; as it needs more than 15 epochs to obtain the best results. Due to this results, in the next section we are going to use only 5 epochs to compare both algorithms using dynamic networks and real time series.

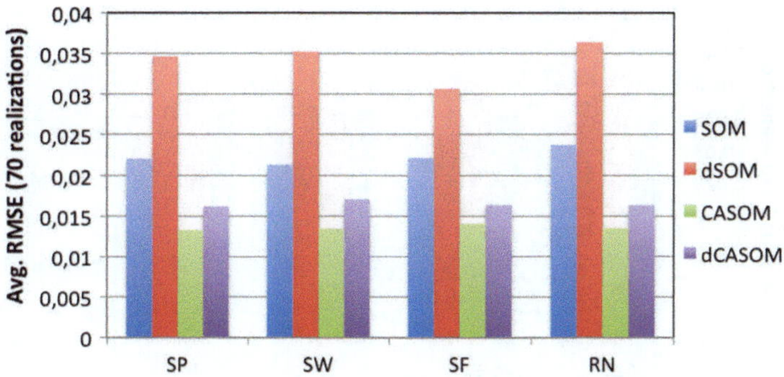

Fig. 9.28 Average RMSE over MGTS ($P_u = 0.1$, $P_{inf} = 0.1$ and 144 neurons)

9.10 Dynamic Networks

In this last section we will consider the use of dynamic networks, i.e., networks that change the links between their neurons during the execution. This is know in the scientific literature as partner switching or rewiring, and means that starting from a certain topology we can end in other different one. The idea is to allow the neurons to change their neighborhood along the training process in order to find better neighbors with closer values.

We have considered two approaches to change the neighborhood of a cell, when becoming a BMU:

1. *Best neuron*: a BMU neuron b_i leaves the neighbor n_f which is far away from its own weights, and joins the closer neighbor of a neighbor nn_c if it is also closer than the deleted link, i.e., $dist(b_i, nn_c) < dist(b_i, n_f)$.
2. *Random neuron*: in this case the BMU neuron b_i also leaves the faraway neighbor n_f, but connects with any other random neuron n_r if $dist(b_i, n_r) < dist(b_i, n_f)$.

Now, we are going to present the results obtained using dynamic networks, starting from an initial spatial network, with four neighbors; and we will use the notation dSOM (dynamic SOM) to refer to a SOM applied over an initial network that dynamically changes the neighborhoods of its neurons, and becomes a different type of network. For the same reasons, we use the notation dCASOM.

In our experiments, we did not find significative differences when using any of the two approaches mentioned before (best or random neuron), as in both cases the results do not improve. Nevertheless, in the next figures we present the results for the *best neuron* case. We also have used 5 epochs for training the neural networks, given it is enough for CASOM to achieve stability.

Figure 9.28 shows the average RMSE obtained starting with the different topology types (in the horizontal axis) over 70 runs.[8] It becomes clear that dSOM (second bar

[8]Remember that for each run a new network is generated.

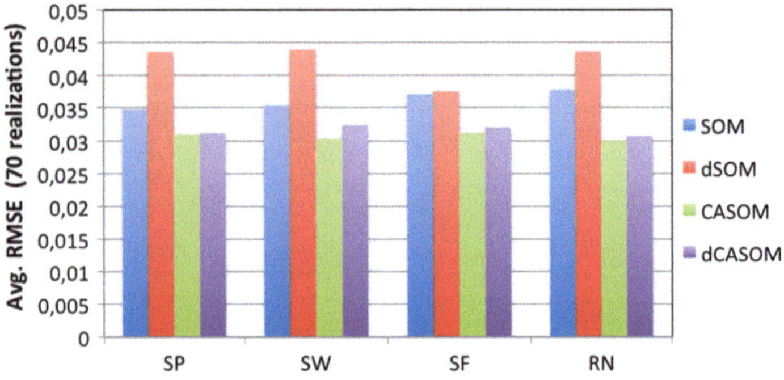

Fig. 9.29 Average RMSE over laser intensity ($P_u = 0.1$, $P_{inf} = 0.1$ and 256 neurons)

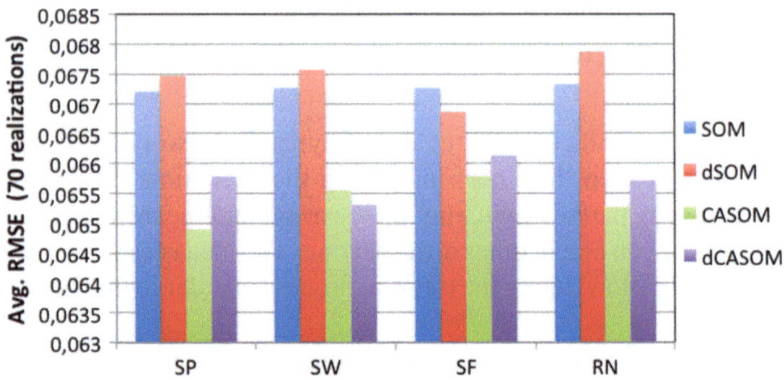

Fig. 9.30 Average RMSE over SSN ($P_u = 0.1$, $P_{inf} = 0.1$ and 441 neurons)

in red) is particularly affected by the rewiring effect in all the topologies considered. On the other side, the CASOM algorithm (3rd bar in green) provides the best results in all the cases, but the interesting result is that it is not too much affected by changing the links in the network in the dCASOM case. This means that it can produce good results even when some links in the network are broken or reassigned.

Finally, we are going to explore all the algorithms with the rewiring option over the real time series introduced in Sect. 9.7. These are the *laser intensity* time series (LITS) and the *solar spot number* time series (SSNTS). The training and testing conditions are the same reported in the previous section for the SOM case, except that here we use five epochs, and that the series are normalized, so the RMSE is bounded to [0, 1].

Figure 9.29 provides the results for the laser intensity time series. Again we observe how dSOM is the algorithm most affected by changes in the network connections. In this case, and given the strong difficulty to predict this complex real time series, the improvement of CASOM over SOM is not so significative, but it is still a bit better than SOM, even when rewiring is applied (dCASOM).

Figure 9.30 shows the results for the Solar Spot Number time series, which are slightly better for CASOM compared to SOM.

To summarize the results in the comparisons provided using these two real time series, we point out that even the results obtained by CASOM are slightly better than the ones obtained by SOM; the key point here are the advantages shown by CASOM along these last sections, and we can summarize them as:

1. First, CASOM does not need any predefined neighborhood, as it adapts to the series under analysis along the training process.
2. CASOM is only slightly affected by the infection probability parameter P_{inf}, that seems to be better at lower values, but even values closer to 1 provide better results than SOM.
3. CASOM is only slightly affected by changes in the network links, as its coalitions can adapt to changing neighborhoods; still providing reasonable results.
4. Finally, CASOM needs much less epochs than SOM to be trained appropriately, and the results are better than the ones obtained by SOM trained over a higher number of epochs.

9.11 Conclusions

In this chapter we consider the Time Series Prediction problem (TSP), and the use of Self-organizing Maps (SOM) as our basic neural network model to apply. We have selected the VQTAM model from the recent literature as a particular case of simple SOM application in univariate TSP scenarios. We have conducted a topology-based TSP performance analysis, based on extensive numerical simulations, taking into account the classical rectangular 2-D lattice; and also different complex networks, such as small-world, scale-free and random networks, for connecting the neurons within the SOM. We also have considered the use of coalitions, and we have redesigned our Coalitional Algorithm for SOM (CASOM), which adapts the neuron neighborhood to the time series under analysis by means of dynamic coalitions.

For the case of the regular 2-D lattice, we have shown that the probability of neighbor update, when a von Neumann neighborhood is considered, highly depends on the number of neurons available in the topology. Mainly, low influence of the BMU is desired when the SOM has a relatively low number of neurons, and viceversa. In addition, we have studied good practices for building the training set, suggesting to foster a uniform sampling of the input space, and a proportional training of the sampled space. We have also found that the prediction results can be enhanced by choosing non-linear functions for the learning rate process. Additionally, on the one hand we have found that complex topologies with standard parameters generally performs worse than a regular lattice. However, by appropriately tuning the parameters, the performance in prediction can significantly improve, when the SOM has relatively few neurons with respect to the length of the training set.

Finally, we have introduced a redefined version of CASOM, as a way to create dynamic neighborhoods, by means of coalitions. First we have studied the influence of the infection parameter, used by CASOM to create the coalitions; and we have realized that the influence is minimal, even lower values are preferred, and this makes CASOM a parameter free algorithm. The results obtained by CASOM have been better than the ones provided by SOM in all the topologies and time series considered in this chapter, even when the network structure changes along the training process (for instance by means of rewiring techniques). Besides, and more remarkable, the number of training epochs, needed by CASOM to stabilize the weights of its neurons (around five), is three times lower than SOM (more than fifteen).

Future work should consider the dynamic adaptation of the neural network size to the input series. In this chapter we have used an heuristic rule for determining the lattice size, but an adaptive approach should be much more convenient and efficient. Another interesting extension of the work presented here would be its evaluation with multivariate datasets.

Notes

This chapter is based in previous work published in [5, 6, 10]. Here we provide a combined introduction to the context, a reformulation of the CASOM algorithm that becomes an improved and parameter free version of the previous algorithm. We also provide an extensive set of simulations and the analysis of CASOM under multiple scenarios, including static and dynamic topologies.

Acknowledgements This work was partially supported by the European Regional Development Fund (ERDF) together with the Galician Regional Government under agreement for funding the Atlantic Research Center for Information and Communication Technologies (AtlantTIC).

References

1. Abinader, F.M., de Queiroz, A.C.S., Honda, D.W.: Self-organized hierarchical methods for time series forecasting. In: 23rd IEEE International Conference on Tools with Artificial Intelligence (ICTAI), pp. 1057–1062 (2011)
2. Alahakoon, D., Halgamuge, S.K.: Dynamic self-organizing maps with controlled growth for knowledge discovery. IEEE Trans. Neural Netw. **11**(3), 601–614 (2000)
3. Barreto, G.A.: Time series prediction with the self-organizing map: a review. Perspect. Neural-Symb. Integr. Stud. Comput. Intell. **77**, 135–158 (2007)
4. Barreto, G.A., Araújo, A.F.R.: Identification and control of dynamical systems using the self-organizing map. IEEE Trans. Neural Netw. **15**(5), 1244–1259 (2004)
5. Burguillo, J.C.: Playing with complexity: From cellular evolutionary algorithms with coalitions to self-organizing maps. Comput. Math. Appl. **66**, 201–212 (2013)
6. Burguillo, J.C.: Using self-organizing maps with complex network topologies and coalitions for time series prediction. J. Soft Comput. **18**(4), 695–705 (2014)
7. Chua, L., Yang, L.: Cellular neural networks: theory. IEEE Trans. Circuits Syst. **35**(10), 1257–1272 (1988)
8. Dorronsoro, B., Burguillo, J.C., Peleteiro, A., Bouvry, P.: Evolutionary Algorithms based on Game Theory and Cellular Automata with Coalitions. In: Handbook of Optimization. Intelligent Systems, vol. 38, pp. 481–503. Springer, Berlin (2013)
9. Furao, S., Ogurab, T., Hasegawa, O.: An enhanced self-organizing incremental neural network for online unsupervised learning. Neural Netw. **20**, 893–903 (2007)

10. Garca-Rois, J., Burguillo, J.C.: Topology-based analysis of self-organizing maps for time series prediction. Soft Comput. 1–18. (2015)
11. Higham, D.J., Higham, N.J.: MATLAB Guide. Society for Industrial and Applied Mathematics, Philadelphia, PA (2000)
12. Jiang, F., Berry, H., Schoenauer, M.: The Impact Of Network Topology On Self-organizing Maps. GEC09, Shanghai, China (2009)
13. Kohonen, T.: Self-Organizing Maps, 3rd edn., Springer, Berlin (2001)
14. Kohonen, T.: Self-organized formation of topologically correct feature maps. Biol. Cybern. **43**(1), 59–69 (1982)
15. Kohonen, T.: Essentials of the self-organizing map. Neural Netw. **37**, 52–65 (2013)
16. Mackey, M.C., Glass, J.: Oscillation and chaos in physiological control systems. Science **197**, 287 (1977)
17. MIT toolbox for Network Analysis. Strategic Engineering Research Group (SERG). Massachusetts, U.S.A. [Online]. http://strategic.mit.edu/downloads.php?page=matlab_networks
18. Newman, M.E.J., Moore, C., Watts, D.J.: Mean-field solution of the small-world network model. Phys. Rev. Lett. **84**, 3201–3204 (2000)
19. Ni, H., Yin, H.: A self-organising mixture autoregressive network for FX time series modelling and prediction. Neurocomputing **72**, 3529–3537 (2009)
20. Ouyang, Y., Yin, H.: A neural gas mixture autoregressive network for modelling and forecasting FX time series. Neurocomputing **135**, 171–179 (2014)
21. Palit, A.K., Popovic, D.: Computational Intelligence in Time Series Forecasting: Theory and Engineering Applications, 1st edn. Springer, Berlin (2005)
22. Taylor, A., Higham D.: CONTEST: A Controllable Test Matrix Toolbox for MATLAB. Department of Mathematics, University of Strathclyde, Glasgow, G1 1XH, Scotland, UK. [Online]. http://www.mathstat.strath.ac.uk/outreach/contest/toolbox.html (2008)
23. The Santa Fe Time Series Competition Data. [Online]. http://www-psych.stanford.edu/~andreas/Time-Series/SantaFe.html
24. Turing, A.: The chemical basis of morphogenesis. Phil. Trans. R. Soc. **237**, 5–72 (1952)
25. Ultsch, A.: Emergence in self-organizing feature maps. In Ritter, H., Haschke, R.(eds.) Proceedings of the 6th International Workshop on Self-Organizing Maps (WSOM '07). Bielefeld, Germany.(2007)
26. Von der Malsburg, C.: Self-organization of orientation sensitive cells in the striate cortex. Kybernetik **14**, 85–100 (1973)
27. Watts, D.J., Strogatz, S.H.: Collective dynamics of small-world networks. Nature **393**, 440–442 (1998)
28. Weigend A., Gershefeld. N.: Time Series Prediction: Forecasting the Future and Understanding the Past. Addison-Wesley. (1993)
29. Wikimedia Commons by Agor153 - Own work, CC BY-SA 3.0. https://commons.wikimedia.org/wiki/File:SOMsPCA.PNG
30. Wikimedia Commons by Mcld - Own work, CC BY-SA 3.0. https://commons.wikimedia.org/w/index.php?curid=10373592
31. Yang, S., Luo, S., Li, J.: An extended model on self-organizing map. In: Proceedings of the 13 International Conference on Neural Information Processing (ICONIP'06). (2006)
32. Yin, H. The self-organizing maps: background, theories, extensions and applications. In Computational Intelligence: A Compendium. (2008)
33. Yin, H., Allinson, N.M.: Self-organizing mixture networks for probability density estimation. IEEE Trans. Neural Netw. **12**(2), 405–411 (2001)

Chapter 10
Coalitions of Electric Vehicles in Smart Grids

Gabriel de O Ramos, Juan C. Burguillo and Ana L.C. Bazzan

Abstract In this chapter, we introduce the use of self-organised coalitions in smart grid scenarios for finding a coalition structure that maximises the systems' utility. The complexity of such a task is exponential with the number of agents, and optimal coalition formation has been considered impractical. Several heuristic alternatives have been proposed in the research literature to handle such a problem. However, most existing methods approach coalition formation neglecting important aspects like maximising the total revenue or ensuring stability. Nonetheless, these points are fundamental in the context of smart grids, especially when we refer to virtual power plants (VPPs) of plug-in electric vehicles (PEVs), which have very limited energy capacity and small profits. In this chapter, we present two classes of constraints: (i) geographic-based, where the geographic position of PEVs is considered to avoid overloading the energy distribution network; and (ii) user-based, where the preferences of the PEV-users (owners) are taken into account to promote lasting coalitions. We also propose three methods for addressing coalition formation within such constrained scenarios: (i) DCCF, where agents invite neighbours to join their coalitions; (ii) SACF, where agents ask to join their neighbours' coalitions; and (iii) SACF^{+}, which is a natural evolution of SACF, where agents can change their coalitions, thus making the process much more dynamic. In all cases, agents negotiate the formation of coalitions among themselves, each on behalf a single PEV. The presented approaches were evaluated in closed and open world scenarios. Regarding the results, all three methods run in a few milliseconds regardless of the number of agents, achieving near-optimal solutions. In all tested cases, results were above 90% of optimum, on average. In comparison, despite delivering optimal solutions, traditional approaches took several hours and run for up to 20 agents, which

G. de. O. Ramos · A. L. C. Bazzan
Instituto de Informática, Universidade Federal do Rio Grande do Sul, Porto Alegre 15064, Brazil
e-mail: goramos@inf.ufrgs.br

A. L. C. Bazzan
e-mail: bazzan@inf.ufrgs.br

J. C. Burguillo (✉)
Department of Telematics Engineering, School of Telecommunications Engineering,
University of Vigo, 36310 Vigo, Spain
e-mail: J.C.Burguillo@uvigo.es

J. C. Burguillo, *Self-organizing Coalitions for Managing Complexity*, Emergence, Complexity and Computation 29, https://doi.org/10.1007/978-3-319-69898-4_10

represents a small and unrealistic scenario for smart grids. Thus, the proposed approaches show that providing approximate solutions for the coalition formation problem is attainable in smart grids scenarios.

10.1 Introduction

Essential services, including telecommunications, transportation, and industrial activities, strongly depend on electricity. Such a fundamental resource is provided through electricity networks (grids). The grid is the infrastructure responsible for connecting energy producers and consumers by means of transmission (from producers to urban substations) and distribution (from substations to consumers) lines. Conversely, grids have evolved very little despite their importance.

The concept of *smart* grid emerges in this scenario: a fully automated electricity network that intensively monitors and controls its elements, being able to supply energy in an efficient, reliable way [9, 32]. Such a concept is illustrated in Fig. 10.1, where the energy is *transmitted* (green lines) from producers (large power plants) to substations (between green and blue lines) and then *distributed* (blue lines) to the consumers (e.g., homes, buildings, factories). Observe that, different from traditional grids, the energy is produced by a range of devices (from large facilities to home devices).

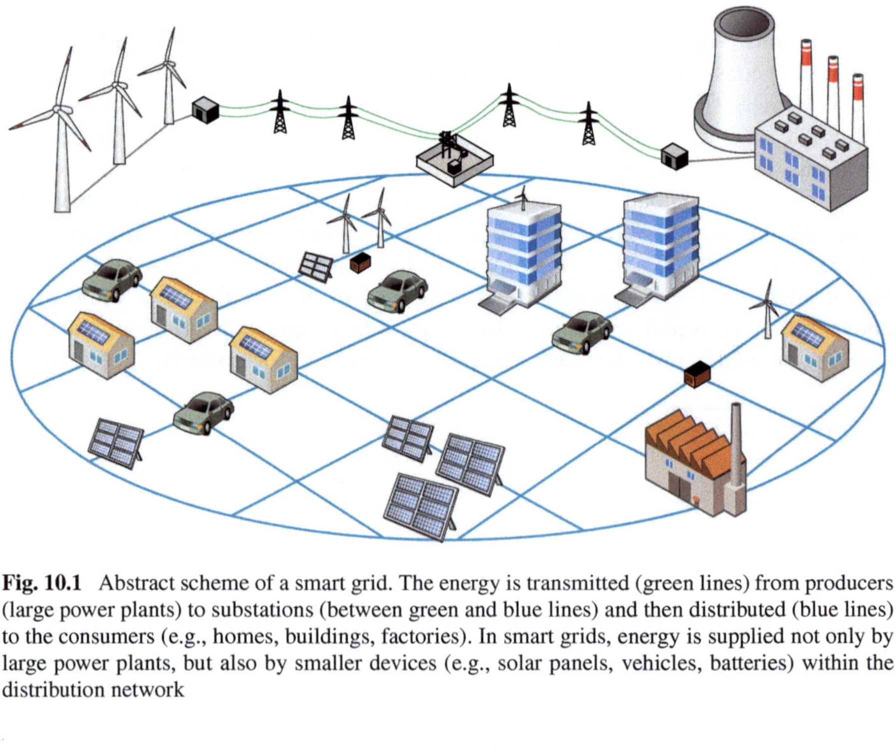

Fig. 10.1 Abstract scheme of a smart grid. The energy is transmitted (green lines) from producers (large power plants) to substations (between green and blue lines) and then distributed (blue lines) to the consumers (e.g., homes, buildings, factories). In smart grids, energy is supplied not only by large power plants, but also by smaller devices (e.g., solar panels, vehicles, batteries) within the distribution network

One of the key features of smart grids refers to their bidirectional flow of energy and communication. Such a feature is also depicted in Fig. 10.1, where some consumers with production capabilities can provide part of their surplus energy to the grid. In other words, its elements can both supply and consume energy. Such a feature paved the way to the so-called distributed energy resources (DER), which are generation/storage devices (including residential ones) connected directly to the distribution network. The deployment of DERs is considered one of the priority areas towards smart grids [9]. An interesting concept here is the so-called Vehicle-To-Grid (V2G), through which plug-in electric vehicles (PEVs) can provide part of the energy available in their batteries to the grid [11, 12]. The V2G mechanism is particularly useful when the grid relies on intermittent, renewable energy sources, such as wind and solar. In such situations, PEVs represent a suitable auxiliary, ready-to-use power source for complimenting the grid's power capacity in the so-called spinning reserve and regulation markets [11].

The participation of PEVs on V2G sessions in a cost-effective way is far from trivial. According to Pudjianto et al. [16], when acting in V2G sessions, PEVs must commit to providing a given amount of energy. However, single PEVs may not manage to fulfil their commitment to the grid. Therefore, many studies have proposed grouping PEVs in the so-called virtual power plants (VPPs), through which PEVs can provide energy in a more predictable, reliable way. Coalition formation has shown to be a suitable alternative to this end [3, 6, 22, 25–28, 36].

In this chapter, we present a review of some representative approaches employing self-organized coalition formation techniques to deliver VPPs of PEVs. Given the focus on coalitions of PEVs, we do not go deeper into the smart grids concept, except when strictly required. The interested reader is referred to [9, 32] for a more comprehensive description of smart grids and related concepts. We mainly consider two classes of constraints: (i) geographic-based, where the geographic position of PEVs is considered to avoid overloading the energy distribution network; and (ii) user-based, where the preferences of the PEV-users (owners) are taken into account to promote lasting coalitions. We also propose three self-organized methods for addressing coalition formation within such constrained scenarios: (i) DCCF, where agents invite neighbours to join their coalitions; (ii) SACF, where agents ask to join their neighbours' coalitions; and (iii) SACF$^+$, which is a natural evolution of SACF, where agents can change their coalitions, thus making the process even more dynamic. In all cases, agents negotiate the formation of coalitions among themselves, each on behalf a single PEV. The presented approaches were evaluated in closed and open world scenarios, and the results show that all three methods achieve near-optimal solutions. Moreover, and regardless of the number of agents, these self-organized approaches run in a few milliseconds; while traditional approaches need several hours to run for up to 20 agents, which represents a small and unrealistic scenario for smart grids. Altogether, we consider that the three self-organized approaches show that providing approximate solutions for the coalition formation problem is attainable in real time for complex scenarios like smart grids.

The present chapter is organised as follows. In Sect. 10.2, we start with the background on coalition formation and some representative applications to smart grids. Section 10.3 formalises the problem of PEVs supplying energy on V2G sessions and presents a formal modelling for it. The presented modelling embodies domain's constraints on the coalition formation process. To this regard, two types of constraints are presented: geographic-based (Sect. 10.4) and user-based (Sect. 10.5). Methods for these modellings are presented and results discussed in their respective sections. A final discussion and future research directions are presented in Sects. 10.6 and 10.7.

10.2 Coalitions and Smart Grids

Coalition formation has shown a particularly suitable way for forming virtual power plants [22]. In this section, we present a brief overview of coalitions in smart grids. We start discussing the background on coalition formation. We then present some representative coalition formation approaches applied to smart grid problems. Finally, we propose the draft of a taxonomy regarding existing works, which is used to delineate the rest of this chapter.

10.2.1 Coalition Formation Background

A coalition is a group of agents that decide to cooperate in order to achieve a common goal, aiming at improving their performance [31]. Given a set of agents $\mathscr{A} = \{1, 2, \dots, n\}$, a coalition is a subset $C \subseteq \mathscr{A}$. A coalition structure is a partition $CS = \{C_1, C_2, \dots, C_{|CS|}\}$ of the set of agents into disjoint (i.e., $\forall C_i, C_j \in CS, C_i \neq C_j, C_i \cap C_j = \emptyset$) and exhaustive (i.e., $\bigcup_{C \in CS} C = \mathscr{A}$) coalitions.

Coalition formation is usually studied in the form of characteristic function games (CFGs). In CFGs, a characteristic function $v : 2^n \rightarrow \mathbb{R}$ assigns a value $v(C)$ to each possible coalition $C \subseteq \mathscr{A}$. The value of a coalition structure CS can be obtained by summing up the values of the coalitions that compose it, i.e., $V(CS) = \sum_{C \in CS} v(C)$.

According to Sandholm et al. [30], the coalition formation process comprises three activities: coalition structure generation (CSG), solving the optimization problem of each coalition, and dividing of the obtained payoff among the agents. In this chapter, we focus on the CSG activity, which regards the formation of coalitions *per se*. In CSG, one aims at finding the optimal coalition structure $CS^* = \arg\max_{CS \in \mathscr{CS}} V(CS)$, i.e., the one with the highest value. However, there is a scalability issue: the number of possible coalitions is $2^n - 1$ and of coalition structures is asymptotically in the order of $O(n^n)$ and $\omega(n^{\frac{n}{2}})$. Furthermore, this problem has proven to be NP-complete [30].

A number of approaches have been proposed to solve the CSG problem. We refer the reader to [19] for a complete survey on the field. One of the most outstanding methods was proposed by Rahwan et al. [21], which is based on integer-partition, and thus known as IP. In their approach, an efficient solution space representation was proposed, where coalitions are grouped by their sizes and coalition structures are grouped by the size of coalitions they have. Based on such representation, the IP algorithm makes use of the branch-and-bound paradigm to provide anytime approximate solutions. Although relatively efficient, the time complexity is high: in the worst case, $O(n^n)$ coalition structures have to be searched. As a result, IP is not scalable for more than a few agents (e.g., finding the optimal solution for 20 agents takes many hours).

Another important approach was proposed by Rahwan and Jennings [17], which makes use of the dynamic programming paradigm. The Improved Dynamic Programming (IDP) algorithm is based on the dynamic programming algorithm of Yeh [39] for the set partitioning problem. Basically, IDP works by deciding for each coalition whether it is better to split it into two small coalitions or to keep it as it is. To this end, the algorithm needs to compute and store the solution and value of all possible coalitions. The IDP's worst case computational complexity is $O(3^n)$, which is much better than IP. However, IDP has a memory requirement of $O(2^n)$ and is not anytime.

Clearly, despite being optimal and generic, both IP and IDP are too slow to be considered for realistic applications. In this sense, many studies have considered exploring the problem's structure.

In a previous work, we have proposed an improved pre-processing phase for IP, where domain information is used in order to prune infeasible coalitions from the search space [24]. In the worst case, however, the time complexity remains $O(n^n)$. In [18], a framework for a particular class of constrained coalition formation (CCF) problems is proposed. In CCF, constraints are imposed on the CSG process through propositional logic statements, and the problem is solved through a divide-and-conquer algorithm. Their approach, however, cannot model the kind of constraints used here. Compact representations of the characteristic function were studied by Ueda et al. [33] and Rahwan et al. [20]. Such approaches exploit the problem's structure in order to make the problem representation more compact. However, they usually provide lower quality solutions.

Coalition formation in synergy graphs was studied by Voice et al. [37]. In a synergy graph, nodes represent agents, edges represent a kind of relation among them (e.g., trust), and a coalition to be feasible must be a connected graph. In this sense, an IDP-based algorithm was proposed to solve the CSG problem. Although faster than IDP (under certain conditions), their approach does not scale well and is not fast enough for real situations. An improvement was proposed by Bistaffa et al. [2], in which the concept of edge contraction was included. Although their approach scales for a higher number of agents, it remains slow for coping with dynamic, real scenarios.

Other approaches include the work of Farinelli et al. [8], which addresses the CSG problem through a hierarchical clustering algorithm. However, although it may run for many agents, the solution's quality is not as good as other approaches. The formation of coalitions in task-oriented domains was studied in the work of Ramchurn et al. [23]. Their approach, nonetheless, is more concerned with solving the tasks (not finding the best coalitions), and is suitable for up to 20 agents. Chalkiadakis and Boutilier [5] have proposed a Bayesian model-based reinforcement learning framework for repeated coalition formation under uncertainty. Such approach, however, is more concerned with agents' learning and decision making, and does not address the CSG problem.

As can be seen, coalition structure generation approaches lack scalability. As such, applying CSG to complex systems may seem infeasible. However, as discussed in the next section, interesting possibilities emerge when the smart grids' structure is taken into account.

10.2.2 Coalition Formation in Smart Grids

This section presents a non-exhaustive review on representative works on the field. These works cover a reasonable range of techniques for handling different problems related to smart grids. For a more comprehensive overview, the reader is referred to [3, 22, 28].

The use of coalitions in smart grids has been widely investigated [22]. One of the main interests of the field has been to increase the reliability of renewable energy production. Chalkiadakis et al. [6] propose coalition formation among distributed energy resources (DERs) to form VPPs. DERs are renewable energy sources with small-to-medium energy capacity, like wind turbines and solar panels. Taking into account that renewable energy sources are intermittent due to weather conditions, their approach suggests grouping DERs in order to aggregate their production, thus improving their reliability and efficiency. The proposed mechanism incentivises DERs to provide accurate estimates of their energy production, rewarding good ones. However, this approach has a primary focus on mechanism design rather than on coalition formation, disregarding how far the solution is from the optimal one.

Another approach is the one of Kamboj et al. [10], which investigates the formation of coalitions among PEVs in order to stabilise the grid's energy supply. Such a task is performed by operating in the regulation market, through which the grid can ensure that the supply will meet the demand. The regulation market basically provides power to the grid whenever demand exceeds supply, and store energy whenever supply exceeds demand. To provide energy, the market usually depends on large batteries (can readily store and supply energy, but are very expensive) and generators (can generate energy, but they are very polluting and take some time to start working). Thus, considering that vehicles remain parked 96% of the time [12], the use of EVs' batteries would help to reduce costs and to improve efficiency of the regulation market. However, such approach addresses coalition formation in an ad hoc fashion, disregarding the solution quality.

In the work of Mihailescu et al. [14], formation of coalitions among producers and consumers is proposed. In their approach, producers who have an increased energy availability are probabilistically selected to coordinate coalitions. Such coordinators are responsible for inviting other producers to join their coalitions. Consumers join the coalitions whose energy profile is more similar to theirs, and also based on their proximity. However, their approach neither addresses coalition formation as CFGs, nor cares about the solution quality.

The formation of coalitions among producers and consumers is also addressed in [36], specifically, among wind turbines and EVs, also forming VPPs. The goal here is more specific: solve the problem of intermittent power generation of the wind turbines through the use of EVs' batteries, in order to increase the reliability of this kind of energy. To this end, a payment scheme to incentivise EVs to join a VPP of wind turbines was deployed. However, aspects concerning the coalition formation problem are not taken into account.

Yasir et al. [38] present a distributed coalition formation approach for microgrids. Basically, a microgrid can be seen as a localised, self-sufficient group of energy producers and consumers [13]. In their work, microgrids are assumed to be fully independent on the main electricity grid. Consequently, microgrids' self-sufficiency must be ensured so that energy generation meets the internal demand. To this regard, coalitions are employed as a mean to enhance the capabilities of microgrids towards self-sufficiency. However, again, coalition formation is approached from an ad hoc perspective.

Coalition formation among PEVs is more formally approached by Bazzan and Ramos in [1]. In their work, the idea of more valuable players is introduced, and a heuristic for finding the optimal solution is proposed. More valuable players model the idea of agents who add more value to coalitions than others agents. The proposed heuristic is able to find the optimal solution much faster than traditional approaches. However, its computational requirements make it intractable for real situations.

Therefore, as seen, there is still a lack of approaches for handling proper coalition formation in complex systems. On the one hand, there are algorithms that generate optimal solutions but are too slow for practical scenarios (as [1] and those from Sect. 10.2.1). On the other hand, there are fast algorithms that do not address coalition formation properly (as the ones presented in this section), relying on ad hoc methods and lacking suitable analysis.

10.2.3 Coalitions in Complex Systems

In general, one cannot expect to propose a *general* coalition formation approach that is both efficient (runs in polynomial time in the size of the problem) and exact (provides provably optimal solutions) for every possible scenario [30]. Consequently, from a simplistic standpoint, coalition formation cannot be said attainable in complex systems, like smart grids. However, literature has strongly evidenced that exploring the domains' structure provides a promising way for efficient coalition formation within such complex situations [1, 2, 24–27, 37].

Along this chapter, we focus on *dynamic* coalition formation approaches that exploit domain's *constraints* to deliver efficient methods for complex scenarios, without compromising the coalition formation formalism. Specifically, we discuss how *real* constraints of smart grids can be used to prune the search space of coalition structures. The aim here is not arbitrarily discarding potential coalitions just to speed up the process. Rather, our objective regards providing an efficient way of ignoring *infeasible* coalitions, i.e., removing coalitions that consume computational resources but end up making no contribution to the final result.

We remark, however, that having a central authority defining the best coalition structure is not realistic within complex, multiagent scenarios. Rather, the agents usually take their own decision with respect to their private interests. In other words, the agents must negotiate the formation of coalitions among themselves in order to maximise their payoffs. Therefore, understanding and improving such a *self-organised* behaviour is a fundamental aspect here. In this chapter, we focus not only on finding near-optimal coalition structures but on developing methods for the agents to achieve good results without relying on a central authority.

In the next section, we develop a formal model of the problem and its constraints and provide all tools required for devising efficient coalition formation methods for smart grids. We then describe, in Sects. 10.4 and 10.5, two classes of methods that couple with the modelled constraints to deliver efficient, self-organised coalition formation approaches. We highlight that, although we focus on a specific scenario, the presented methods are suitable for any scenarios in which coalition formation is constrained to some extent.

10.3 Smart Grid Scenario: Coalitions of Electric Vehicles

In this section, we describe the smart grid scenario and present a modelling for it. The modelling presented in this section is used throughout this chapter, except otherwise stated.

10.3.1 The Scenario

Consider a smart grid in which PEV-agents[1] sell their surplus energy in V2G sessions. Without loss of generality, we ignore PEVs that are not willing to participate in V2G sessions. In this scenario, one aims at forming coalitions of PEVs in a decentralised way to enhance the PEVs profit. Roughly: after a PEV is connected to the grid, it starts negotiating with its neighbours to form a coalition that improves its profit. Coalitions, once formed, last until one of their members decides to leave them. Agents, on the other hand, leave their coalitions when their time availability is over (time to unplug).

The three main aspects to regard in this scenario are: how neighbourhoods are defined, how the scenario dynamism is handled, and how coalitions are paid for their supply. Neighbouring relations are defined in terms of the constraints of the electricity grid. As discussed in Sect. 10.2.3, literature has investigated basically two kinds of constraints: geography-based and user-based. More details on these constraints are presented in Sect. 10.3.2.

[1] We use the terms PEV and agent interchangeably.

Considering the dynamic aspect of smart grids, recall that PEVs can enter or leave the grid at any time. This aspect is not just a modelling definition, but a real feature of the domain. The permanency of PEVs in the grid is ruled by their owners' preferences. To this regard, PEVs should ensure a minimum energy reserve in order to meet their owners' demand. Thus, we assume that PEV-agents know when to stop selling energy to the grid. In other words, PEVs are allowed to leave the grid as soon as they deem necessary.

A final aspect of this scenario regards the monetisation of coalitions. The grid incentivises the formation of coalitions among PEVs through a monetary value, which is proportional to the coalition's power rating and duration, up to certain limits. We assume that the grid is always willing to buy the energy offered by PEVs, i.e., whenever a PEV has energy to sell, the grid will buy it.

10.3.2 Constraints

Previously, in Sect. 10.2.3, we discussed that traditional CSG literature has typically addressed abstract scenarios in which every set of agents is considered a feasible coalition. However, many (if not most) real-world problems cannot be modelled without taking its constraints into account. Such constraints may be due to communication or physical infrastructures, social or trust relationships, among others [7, 22, 34, 35]. In fact, many recent works have studied how such constraints might be handled in the CSG process [2, 24–27, 37].

In the specific scenario under consideration (coalitions of PEVs supplying energy to the grid), we employ the constraints to define a neighbourhood structure among the PEVs. Specifically, let $d : \mathscr{A} \times \mathscr{A} \rightarrow \mathbb{R}$ represent the degree $d(i, j)$ of some *constraint* between two given PEVs $i \in \mathscr{A}$ and $j \in \mathscr{A}$. Moreover, let α be a parameter of the electricity grid limiting such a constraint for any pair of PEVs. To this regard, the *neighbourhood* N_i of PEV i can be defined as the set of agents with whom its degree of constraint is below to α, as in Definition 10.1. The set of all agents' neighbourhoods is referred as $\mathscr{N} = \{N_i \mid i \in \mathscr{A}\}$. In this same line, a coalition is said *feasible* if all of its members are neighbours of each other, as in Definition 10.2. The value of parameter α should be defined in order to represent the grid's requirements. A lower value for α is more restrictive than a higher one, meaning that the coalitions are supposed to last longer, at the cost of being smaller.

Definition 10.1 (*neighbourhood*) The neighbourhood $N_i \in \mathscr{N}$ of a given agent $i \in \mathscr{A}$ is composed of all agents $j \in \mathscr{A} \setminus \{i\}$ for whom $d(i, j) \leq \alpha$.

Definition 10.2 (*feasible coalition*) A feasible coalition C is composed of neighbouring agents, that is, $\forall i, j \in C, i \neq j \ (i \in N_j \wedge j \in N_i)$.

Observe that the constraint degree d can represent any kind of constraint of the smart grid. In this chapter, we approach two distinct constraints. In Sect. 10.4, we present *geographic-based* constraints, in which neighbourhoods are defined in terms of the agents' geographic proximity. We also present *user-based* constraints (Sect. 10.5), where the neighbourhood structure is defined regarding the consumption profile of the PEVs.

10.3.3 Communication Layer

Different from traditional CSG approaches, in this work, coalitions are formed by means of local interactions among the agents. In other words, there is not a central authority responsible for regulating the coalition formation process. In this sense, a communication channel among the agents is required. Here we simplify such a requirement through the use of peer-to-peer (P2P) networks. Basically, a P2P network consists of a hub and many leaves connected to it (consider the Gnutella2 protocol, for instance). Such a topology can be seen as analogous to the smart grid scenario considered here: leaves represent PEVs, and hubs are managed by electrical substations (that control the portion of the distribution network where the PEVs are in). The hub is responsible for providing the P2P infrastructure for the PEVs, and also for limiting the PEVs' interactions to their neighbourhoods.

The P2P network provides two services: (i) information sharing, and (ii) communication. Information sharing service is used to provide agents with all information required for deciding what coalition is better for them, which includes their: energy profile, current coalition, and respective value. Based on the shared information, it can be inferred, for instance, whether a coalition remains feasible if a new agent joins it. On the other hand, the P2P network provides a communication service, which is necessary during the coalition negotiation process.

Along these lines, the P2P network can be seen as a suitable way of providing the required services for our approach. However, it should be noted that the communication layer was modelled to be a seamless interface among the agents. Although we have modelled it as a P2P network, the protocol is not an essential part of our approach. Thus, we no longer focus on P2P technicalities henceforth.

10.3.4 Problem Formulation

The problem can be represented by a graph, where the PEV-agents are expressed by nodes, and the neighbouring relations among them are denoted by edges. An example is presented in Fig. 10.2. As can be seen, the neighbours of agent 1 are the agents 7 and 9. In this case, $\{1, 7, 9\}$ is a feasible coalition because the agents 1, 7 and 9 are neighbours of each other.

The neighbour relation formulated in Definition 10.1 is not transitive. In other words, given three agents 1, 2 and 3, if agent 1 is neighbour of agents 2 and 3, it does not mean that agents 2 and 3 are also neighbours. Considering the example of Fig. 10.2, both agents 8 and 9 are neighbours of agent 3 but there is no neighbour relation between them, so the coalition $\{3, 8, 9\}$ is infeasible. Thus, feasible coalitions are always *complete* subgraphs.

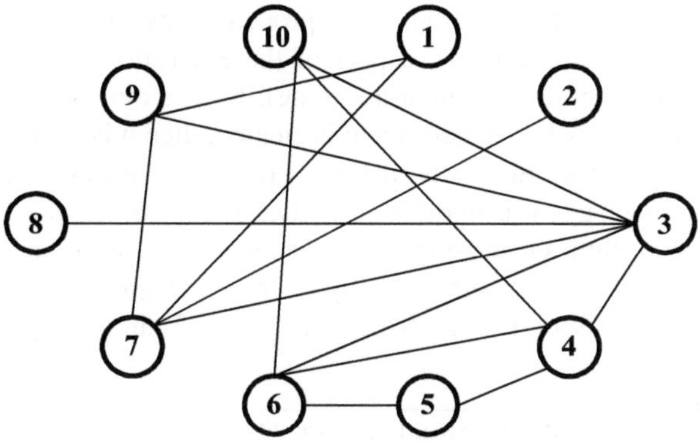

Fig. 10.2 Graph representation of the problem, where nodes represent agents, and each arc represent a neighbour relation between two agents

The problem is formalised in the form of CFGs, and any kind of characteristic function could be used here. Within the present scenario, the characteristic function $v : C \rightarrow \mathbb{R}$ represents how much the grid is willing to pay *beyond* the base price for the amount of energy supplied by a coalition. Specifically, the coalition value $v(C)$ is a function of the coalition C's total power rating W_C and duration (the greater/longer, the better). The total power rating of a coalition can be obtained by $W_C = \sum_{i \in C} w_i$, with w_i denoting the power rating of agent i.

Regarding the payoff division, we start assuming that all PEVs have the same[2] power rating, i.e., $\forall i, j \in \mathscr{A}\ w_i = w_j$. To this respect, we can simplify the payoff division in a way that the value of a coalition is equally divided among its members. Although somewhat naïve, this mechanism is fair. Furthermore, taking the dynamic aspect of the environment into account (coalitions can be created and break at any time), each agent is paid at every time step, considering the amount of energy it has sold to the grid along that specific time step. The time step payoff P_i of agent i is referred as *instantaneous payoff*. Again, observe that the greater the coalition's power rating and availability, the higher the agents' revenue. Thus, the agents are better off forming coalitions.

10.3.5 Simulation

Assuming the problem modelling of the previous subsections, a final aspect to be regarded is the simulation of the smart grid scenario. Usually, deploying a PEV-controlling mechanism in real world is unattainable. To this regard, simulating the desired smart grid scenarios is a fundamental step for validating such mechanisms. Following the proposed modelling, we also present a scheme for controlling and setting up the simulation, as formulated in Procedure 10.

[2]We assume a power rating w_i of 3.3 kW for every agent $i \in \mathscr{A}$. This value is similar to that of some commercial PEVs. However, any other such value could be used here.

Basically, the simulation procedure executes each agent's algorithm and formalises the dynamic aspect of the environment (thus allowing agents to enter or leave the simulation at any time). Each simulation iteration corresponds to a time step.[3] Firstly, if a new agent is created, then it is initialised (as a part of the P2P sign-in process). Secondly, if the set of agents has changed, then the list of neighbours of each agent is updated. Finally, each agent is called to run a single iteration of its algorithm. Observe that the agents' initialisation and execution procedures depend on the method being employed. In this chapter, we present three different methods: DCCF (Sect. 10.4.2), SACF (Sect. 10.4.3) and SACF$^+$ (Sect. 10.5.2).

Procedure 10: Simulation mechanism

```
while simulation is running do
    // if some agent has entered or left
    if set 𝒜 has changed then
        foreach new agent i do
            initialize agent i;
        end
        update all neighbourhoods N_i, ∀i ∈ 𝒜;
    end
    run agent i, ∀i ∈ 𝒜;
end
```

10.4 Geographic-Based Constraints

This section presents coalition formation approaches for scenarios with geographic-based constraints. Under such constraints, the neighbourhood structure is ruled by the geographic distance among the PEVs. The complete description on the modelling of such constraints is presented in Sect. 10.4.1. Following such a modelling, we present two coalition formation approaches: DCCF, in Sect. 10.4.2, and SACF, in Sect. 10.4.3. This section also provides, in Sect. 10.4.4, an empirical evaluation of DCCF and SACF.

10.4.1 Modelling Constraints

In the scenario under consideration, the grid incentivises the PEVs to form coalitions to have a more reliable VPP. However, an important aspect when supplying energy in groups refers to the flow of energy. As discussed in [7, 22, 24, 34, 35], EVs should supply energy only to consumers who are in the same region as them (or just close enough). Such a constraint exists because power lines have a limited energy

[3] We consider that each time step corresponds to 1 min.

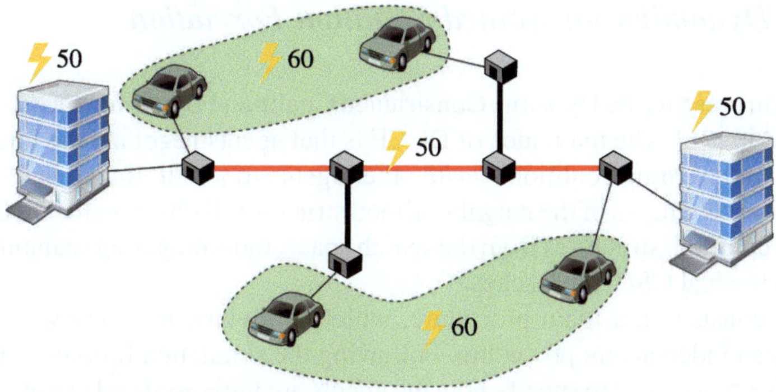

Fig. 10.3 Example of the grid's geographic constraints. Using the coalitions to meet the buildings' demand can potentially overload the red power line

flow capacity. Considering that multiple power lines may be used in order to supply energy to a single consumer, travelling long distances may impose a huge burden on the distribution network. Therefore, the distance among the EVs is a constraint that should be taken into account while the coalitions are being formed. Such a constraint is illustrated in Fig. 10.3, where four PEVs (each with a capacity of 30 units of energy) are arranged into two coalitions (each with 60 units of energy). In the figure, no vehicle is close to each other (they are mainly connected by the red power line). Suppose two building have a demand of 50 units of energy each. The coalitions can clearly meet the buildings' demand. However, the red power line has a limit of 50 units of energy. At this point, supplying the coalitions' energy to the buildings can potentially overload the red power line.

Following the notation from Sect. 10.3.2, let $d(i, j)$ denote the geographic distance[4] between two PEVs i and j. In this context, let α denote the maximum allowed distance between any pair of agents from a given coalition. Importantly, the maximum distance α must be defined by the grid operator in order to better represent the capacity of the grid's power lines. The neighbourhood structure of the agents is then defined considering the degree $d(\cdot, \cdot)$ among the agents and the constraint parameter α of the grid, as in Definition 10.3. To this regard, a PEV can only form a coalition with its peers that are close enough. A coalition that fits into such criterion is said a feasible coalition, as formulated in Definition 10.4.

Definition 10.3 (*geographic-based neighbourhood*) Under geographic-based constraints, function $d(i, j)$ denotes the geographic distance between two agents $i, j \in \mathscr{A}$. Furthermore, the neighbourhood N_i of a given agent $i \in \mathscr{A}$ is composed of all agents $j \in \mathscr{A} \setminus \{i\}$ to which its distance $d(i, j)$ is no greater than α.

Definition 10.4 (*geographic-based feasible coalition*) Given a geographic-based neighbourhood, a coalition is said feasible if and only if all pairs of its members are neighbours.

---

[4]The distance metric does not play an important role in this work. Anyway, geographical distance is a reasonable approximation (in the absence of a better one) for this problem. In real situations, it could be trivially replaced by another one.

10.4.2 Dynamic Constrained Coalition Formation

This section presents the Dynamic Constrained Coalition Formation (DCCF) method, introduced in [26]. The main idea of DCCF is that agents negotiate the formation of coalitions and existing coalitions *invite* other agents to join it. Recall that coalitions are complete subgraphs of the neighbourhood structure. To this regard, DCCF prunes infeasible coalition structures from the search space, thus mitigating scalability issues faced by classical CSG approaches.

DCCF consists in a main procedure, which is divided into several steps, each defined as an independent procedure. Following the simulation formalism presented in Sect. 10.3.5, we need to specify how the agents are initialised and executed. Whenever a DCCF-agent enters the system, it is initialised according to Table 10.1. Then, on every iteration of the simulation, DCCF is executed for each agent through its main procedure. The complete description of DCCF's procedures is presented along this section.

Main Procedure

DCCF's main procedure refers to the agents' execution, and is presented in Procedure 11. Basically, it allows the agents to negotiate among themselves to form feasible coalitions. The negotiation process takes place through information exchange among the agents. Specifically, every agent is able to find feasible coalitions (given its neighbourhood) and negotiate their formation (through message exchange).

The coalition negotiation process is divided into three phases:

- Neighbours invitation (lines 1 to 3 of Procedure 11): singleton agents propose feasible coalitions to their neighbours.
- Invitations processing (lines 4 to 6 of Procedure 11): agents who have received invitations choose, and accept, the best one.

Procedure 11: DCCF – main procedure

required: variables initialised according to Table 10.1;

`// Neighbours invitation phase`
1 **if** $C_i = \{i\}$ *and* $C'_i = \emptyset$ *and* $C'' = \emptyset$ **then**
2 | invite neighbours (Procedure 12);
3 **end**

`// Invitations processing phase`
4 **if** $M_i^{inv} \neq \emptyset$ **then**
5 | process invitations (Procedure 14);
6 **end**

`// Replies processing phase`
7 **if** $M_i^{rpl} \neq \emptyset$ *and* $C_i = \{i\}$ *and* $C'_i \neq \emptyset$ *and* $C'' = \emptyset$ **then**
8 | process replies (Procedure 15);
9 **end**

Table 10.1 DCCF – definition of the procedures' variables

Variable	Description	Initial value
i	id of the agent who is running the procedure	(agent id)
N_i	List of neighbours of agent i	$\{j \in \mathscr{A} \setminus \{i\} : d(i, j) \leq \alpha\}$
C_i	Agent i's coalition	$\{i\}$ (singleton)
C'_i	Coalition proposed by agent i in phase 1	$\emptyset$
C''_i	Coalition accepted by agent i in phase 2	$\emptyset$
M_i^{inv}	List of "invitation messages" received by agent i	$\emptyset$
M_i^{rpl}	List of "reply messages" received by agent i	$\emptyset$

- Replies processing (lines 7 to 9 of Procedure 11): agents who proposed coalitions process the received replies, and form (or not) the coalitions.

An illustrative example of these phases is presented in Fig. 10.4. In the first phase, agent 1 proposes to its neighbours the formation of coalition $\{1, 7, 9\}$. In the second phase, agents 7 and 9 evaluate the invitation and accept it. Finally, in the third phase, agent 1 processes the received replies and forms the initially proposed coalition.

The negotiation phases are described in detail in the following subsections.

Neighbours Invitation Phase

The neighbours invitation phase is performed by every singleton agent i. The rationale behind this phase is simple: the agent computes the coalition that maximises its payoff and then invites the involved neighbours to form it. This phase is structured as in Procedure 12. Firstly, the agent uses Procedure 13 to compute the feasible coalitions within its neighbourhood. Secondly, the agent ignores all coalitions that worse the

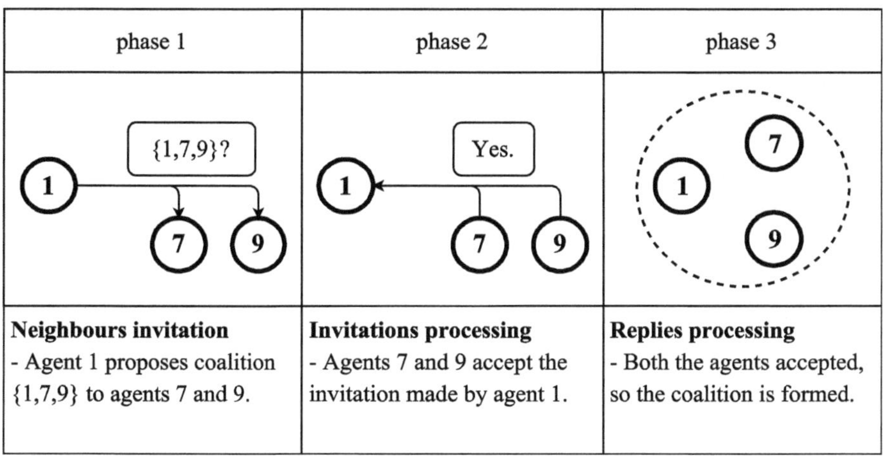

Fig. 10.4 The coalitions negotiation process: in phase 1 the coalition is proposed, in phase 2 the coalition is analysed and accepted, and in phase 3 the replies are processed and the coalition is formed

current payoff of the involved neighbours. Finally, the coalition with highest payoff is selected and proposed by agent i to its neighbours.

Observe that, as the agents are assumed to have the same power rating (as discussed in Sect. 10.3.4), the larger a coalition, the better. To this regard, Procedure 13 ignores feasible coalitions that are subset of others. For instance, suppose two feasible coalitions $\{1, 2\}$ and $\{1, 2, 3\}$ are found. In this case, as the former coalition is a subset of the latter, then only the latter coalition is returned by Procedure 13.

Procedure 12: DCCF phase 1 – inviting neighbours

$\mathscr{F}_i \leftarrow$ Procedure 13 ; `// feasible coalitions`
foreach $C \subset \mathscr{F}_i$ **do**
    foreach $j \in C$ **do**
        if $C_j \neq \{j\} \wedge v(C) \leq v(C_j)$ **then**
            $C \leftarrow C \setminus \{j\}$;
        end
    end
end
$C_i' \leftarrow \arg\max_{C \in \mathscr{F}_i} v(C)$;
`// send an invitation message to agents in` C_i'
update M_j^{inv} with $C_i', \forall j \in C_i'$;

Procedure 13: Finding the largest feasible coalitions within the neighbourhood

$\mathscr{F}_i \leftarrow \emptyset$;
foreach $j \in N_i$ **do**
    $\mathscr{F}_i \leftarrow \mathscr{F}_i \cup \{\{i, j\}\}$;
    foreach $k \in N_j$, $k > j$ *and* $k \in N_i$ **do**
        foreach $C \subset \mathscr{F}_i$ **do**
            if $C \setminus N_k = \emptyset$ **then**
                $C \leftarrow C \cup \{k\}$;
            else
                $C_{new} \leftarrow (C \setminus (C \setminus N_k)) \cup \{k\}$;
                $\mathscr{F}_i \leftarrow \mathscr{F}_i \cup \{C_{new}\}$;
            end
        end
    end
end
return $\mathscr{F}_i$;

Invitations Processing Phase

In the invitations processing phase, agents must reason about the best invitation to accept. The best invitation received is the one that proposes the coalition with highest value. This mechanism is presented in Procedure 14. Roughly speaking, the best invitation can be accepted if the corresponding coalition's value is higher than one of i's current coalition (lines 7 to 9).

One additional case must be handled. If agent i has already proposed a coalition to its neighbours and, coincidentally, the best invitation received is to form the same coalition it has proposed. This case is called a *mutual invitation*, since two (or more) agents are inviting each other for the same coalition. To address this case, all agents reply the invitation only to the agent with lowest ID[5] (as in lines 3 to 6). This way, only the agent with lowest ID will perform the third phase.

Finally, if a coalition has been accepted, then the agent sends a message to the neighbour who has made the invitation, notifying it about its choice (line 12). All non-accepted invitations are rejected.

Procedure 14: DCCF phase 2 – processing invitations

1 $\hat{m} \leftarrow \arg\max_{m \in M_i^{inv}} v(m.coalition)$;
2 $accept? \leftarrow false$;
3 **if** $C_i' \neq \emptyset$ **then**
4   **if** $\hat{m}.coalition = C_i'$ *and* $\hat{m}.sender < i$ **then**
5         $accept? \leftarrow true$;
6   **end**
7 **else if** $v(\hat{m}.coalition) > v(C_i)$ **then**
8     $accept? \leftarrow true$;
9 **end**
10 **if** $accept?$ **then**
11     $C'' \leftarrow \hat{m}.coalition$;
    `// accept invitation` $\hat{m}$
12   update $M^{rpl}_{\hat{m}.sender}$;
    `// reject remaining invitations`
13   update $M^{rpl}_{m.sender}, \forall m \in M_i^{inv} \setminus \{\hat{m}\}$;
14 **else**
    `// reject all invitations`
15   update $M^{rpl}_{m.sender}, \forall m \in M_i^{inv}$;
16 **end**

Replies Processing Phase

Finally, the replies processing phase is performed as in Procedure 15. Essentially, the coalition can be formed only if all replies are positive, i.e., if all agents of the proposed coalition have accepted the invitation. If there is a negative reply, then the proposed coalition does not form.

In the event that the coalition is formed, the members' (P2P-)shared information is updated. In order to simplify the procedure, in this chapter we entrusted the agent who proposed (agent i) the coalition with this task (as lines 4 to 9). Additionally, agents who were already in coalitions must leave them to enter the new one (lines 5

[5]In real scenarios, the ID could be easily replaced by any other comparable code, such as the vehicle's license plate.

to 7). Importantly, agents who receive a *cancellation* or a *coalition finished* message simply return to their initial state, being able to perform phases 1 and 2 again.

Procedure 15: DCCF phase 3 – processing replies

---

if $\exists \hat{m} \in M_i^{rpl}\ (\hat{m}.answer = \text{“}No\text{”})$ **then**
    send a cancellation message to every agent $i \in C_i'$;
else
    foreach $j \in C_i'$ **do**
        if $C_j \neq \emptyset$ **then**
            leave C_j (notify members of C_j);
        end
        $C_j \leftarrow C_i'$;
    end
end

10.4.3 Self-adapting Coalition Formation

In this section we present the Self-adapting Coalition Formation (SACF) method, introduced in [25]. Roughly, SACF-agents negotiate to join existing coalitions in their neighbourhood. As for DCCF (Sect. 10.4.2), coalitions are complete subgraphs of the neighbourhood structure. Consequently, SACF is able to avoid infeasible coalition structures during the search process, thus scaling to more complex scenarios than classical CSG approaches. Observe that SACF and DCCF share some similarities. The key difference between them lies in the negotiation scheme. In DCCF, agents are *invited* to join coalitions. In SACF, on the other hand, agents *ask* to join coalitions. Despite subtle, such a difference makes SACF more flexible than DCCF, as the agents can explore all coalitions in their neighbourhood before deciding. Moreover, SACF can be extended (to SACF$^+$, as shown in Sect. 10.5.2) for allowing the agents to change their coalitions whenever a better one is attainable.

SACF's structure is similar to DCCF: it consists in a main procedure, which is divided into several steps, each defined as an independent procedure. Again, following the simulation formalism from Sect. 10.3.5, one needs to specify how the agents are initialised and executed. Whenever an SACF-agent enters the system, it is initialised according to Table 10.2. After that, for each simulation's iteration, SACF is executed for every agent through its main procedure. This section presents the complete description of SACF's procedures.

Main Procedure

The main procedure performed by each SACF-agent is presented in Procedure 16. The negotiation process is started by agents who are not in coalitions, which are

Table 10.2 SACF – definition of the procedures' variables

Variable	Description	Initial value
i	id of the agent who is running the procedure	(agent id)
N_i	List of neighbours of agent i	$\{j \in \mathscr{A} \setminus i : d(i, j) \leq \alpha\}$
C_i	Agent i's coalition	$\{i\}$ (singleton)
C'_i	Best coalition found by agent i in phase 1	$\emptyset$
C''_i	Coalition accepted by agent i in phase 2	$\emptyset$
M_i^{rqs}	List of "request messages" received by agent i	$\emptyset$
M_i^{rpl}	List of "reply messages" received by agent i	$\emptyset$
M_i^{cnf}	List of "confirmation messages" received by agent i	$\emptyset$
$wait\_4?$	Whether the agent is waiting for the coalition consolidation or not	*false*

called *requesters*.[6] Requester agents can ask their neighbours (i) for permission to join their existing coalitions or (ii) to form a new coalition (if the neighbour is not in a coalition). In the former case, the requester wants to join a coalition that already exists. In the latter case, both the requester and its neighbour are not in coalitions, so a new coalition is going to be created. The agent who has the lowest ID in its coalition is selected as its *leader*.[7]

Procedure 16: SACF – main procedure

required: variables initialised according to Table 10.2;

`// Neighbours' coalitions checking phase`

if $C_i = \{i\}$ *and* $C'_i = \emptyset$ *and* $C''_i = \emptyset$ *and not* $wait\_4?$ **then**
    check neighbours' coalitions (Procedure 17);
end

`// Requests processing phase`

if $M_i^{rqs} \neq \emptyset$ *and* $C'_i = \emptyset$ *and not* $wait\_4?$ **then**
    process requests (Procedure 18);
end

`// Confirmation phase`

if $M_i^{rpl} \neq \emptyset$ *and* $M_i^{rpl}.answer =$ "*Yes*" *and* $C'_i \neq \emptyset$ **then**
    confirmation (Procedure 19);
end

`// Coalition consolidation phase`

if $M_i^{cnf} \neq \emptyset$ *and* $C''_i \neq \emptyset$ *and* $wait\_4?$ **then**
    process confirmations (Procedure 20);
end

[6]It is important to note that, in the beginning, all agents act as requesters.

[7]The leader of a coalition does not have any specific characteristics. Instead, it is just a role assumed by an agent to represent the coalition in negotiation processes, thus avoiding redundant negotiations. When the coalition is broken (i.e., one of its members has left), the leader is also responsible for notifying its members about the event.

The working flow of the agents, which characterizes the negotiation process, is basically divided into four phases:

- Neighbours' coalitions checking (lines 1 to 3): singleton agents (requesters) check the existence of coalitions in their neighbourhoods. The agent sends a join request to the leader of the best coalition found, i.e., the one with highest value. At this point, if there is no coalition in its neighbourhood, the agent can send requests to other singleton agents.
- Requests processing (lines 4 to 6): every agent who has received any request must decide whether the requesters can join its coalition or not. If the agent is a leader then it can directly accept the request. Otherwise, if the agent is singleton, then the negotiation process must advance to the next two phases.[8]
- Confirmation (lines 7 to 9): every requester who was accepted (by a singleton neighbour) must confirm whether it still intends to form the new coalition.
- Coalition consolidation (lines 10 to 12): where new coalitions are created with the requesters who have confirmed their intention to join.

The sequence of these four phases characterizes the coalition creation flow, which is illustrated through two examples in Fig. 10.5. In the case of Fig. 10.5a, the two neighbours of agent 1 are in the same coalition. When agent 1 requests to join the coalition (phase 1), agent 7 accepts it immediately (phase 2), since $\{1, 7, 9\}$ has a greater value than $\{7, 9\}$. In the case of Fig. 10.5b, agent 3 decides to ask agent 8 to join its coalition (phase 1). As agent 8 is also singleton, it does not accept agent 3 immediately (phase 2). Instead, it asks if agent 3 still wants to join it. After agent 3 has confirmed its intention (on phase 3), agent 8 creates the new coalition $\{3, 8\}$ (phase 4).

The four negotiation phases are described in detail in the following subsections.

Neighbours' Coalitions Checking Phase

In the neighbours' coalitions checking phase, agents who are not in coalitions (requesters) verify the existence of coalitions in their neighbourhoods. Such a search behaviour is performed by Procedure 18, where all coalitions in agent i's neighbourhood are checked.[9] For each coalition, the procedure verifies whether it remains feasible if agent i enters it. Among the feasible coalitions identified, the procedure selects (and returns) the one with the highest value.

---

[8] The rationale behind the second case (request received by a singleton agent) is that the requester can also have received requests after having made his request. In this sense, the confirmation is needed to ensure that the requester is still interested in joining the coalition.

[9] If there are no coalitions, then singleton neighbours are considered in this process.

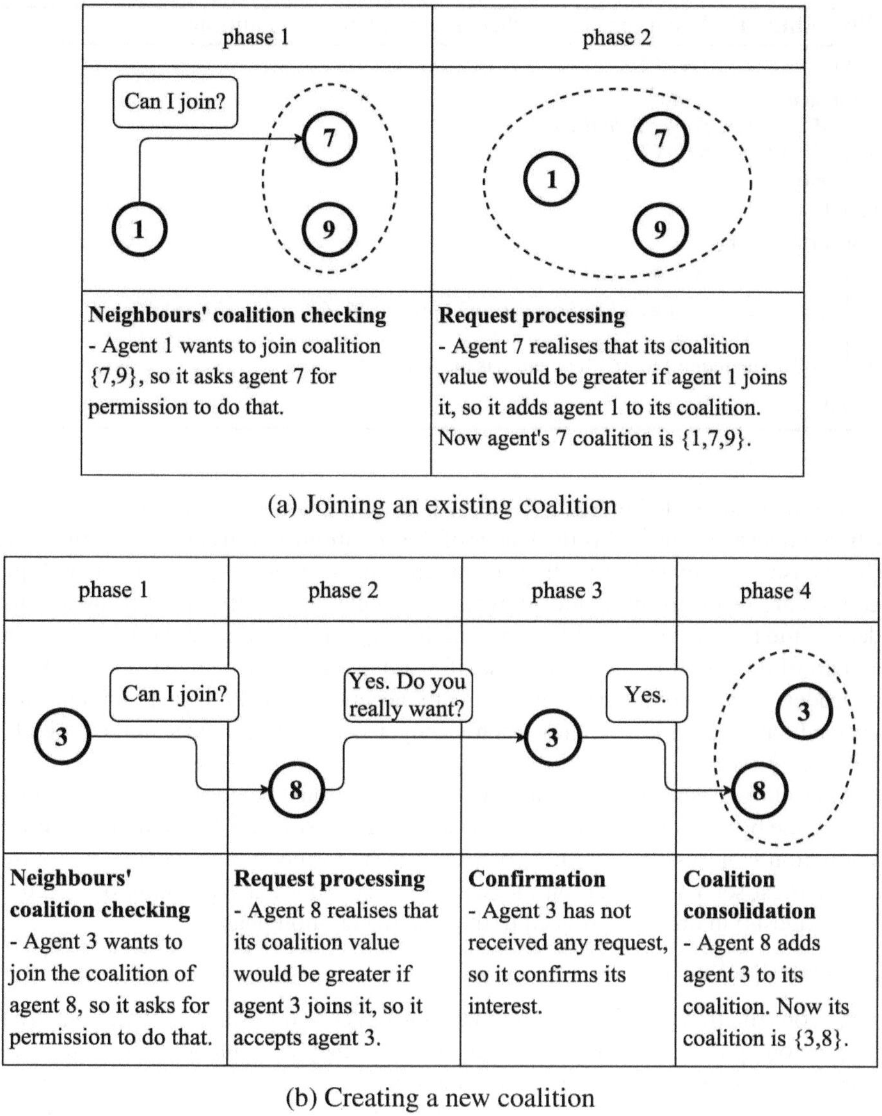

Fig. 10.5 The negotiation process for coalition formation, where the requester can ask its neighbours **a** for permission to join their existing coalitions or **b** to form a new coalition (if also the neighbour is not in a coalition)

Requests Processing Phase

The requests processing phase is performed through Procedure 18. In this phase, agents who have received requests must decide whether or not the requesters can join their coalitions. As previously stated, this procedure can be performed by leader agents and also by singleton agents. This procedure is divided into: requests evaluation (lines 2 to 12) and decisions notification (lines 13 to 25).

Procedure 17: SACF phase 1 – checking neighbours' coalitions

$NC_i \leftarrow \{C_j : \forall j \in N_i\}$;
foreach $C \subset NC_i$ **do**
  if $C \cup \{i\}$ *is not feasible* **then**
    $NC_i \leftarrow NC_i \setminus C$;
  end
end
if $NC_i \neq \emptyset$ **then**
  $C'_i \leftarrow \arg\max_{C \subset NC_i} v(C)$;
  `// request permission for leader of` C'_i `to join its coalition`
  add a join request to M_j^{rqs}, where j is the leader of C'_i;
end

As previously stated in Sect. 10.4.3, two cases are taken into account when evaluating requests: (i) agent i is the leader of the coalition, and (ii) agent i is singleton. These cases are treated according to the amount of requests received. If multiple requests are received (no matter if agent i is singleton or not), then the agent must identify the best feasible coalition among the requesters, and accept only those who are part of it (as in line 11). Otherwise, if just one request is received, and agent i is the leader, then the requester can be directly accepted. Finally, if just one request is received, but agent i is singleton, then it only accepts the requester in the case of a mutual request.[10]

After all requests were evaluated, the requesters must be notified about the decisions made by agent i. If agent i is the leader of an existing coalition, then no further negotiation is necessary, i.e., the coalition can be formed immediately. Otherwise, agent i must send a notification message to every requester (accepting or rejecting their requests) in order to proceed to the next negotiation phases.

Confirmation Phase

In the confirmation phase, requesters who have received positive notifications from the previous phase must confirm whether or not they still intend to form the new coalitions. This phase is performed on the basis of Procedure 19, only by requesters who have asked to form a new coalition with other singleton agents. Through this procedure, the agent is able to withdraw its request if a better one was received.

[10] A request is mutual when two agents propose the same coalition to each other. In this case, only the agent with the lowest ID will accept the request of the other, thus becoming the leader of that coalition.

Procedure 18: SACF phase 2 – processing requests

1 $C_i'' \leftarrow \emptyset$;
2 **if** $|M_i^{rqs}| = 1$ **then**
3   $C^r \leftarrow C_i \cup \{M_i^{rqs}.requester\}$;
4   **if** $C_i \neq \{i\}$ *and* C^r *is feasible* **then**
5     $C_i'' \leftarrow C^r$;
6   **else if** $C_i = \{i\}$ *and* $C_i' = C^r$ *and*
7   $i < M_i^{rqs}.requester$ **then**
8     $C_i'' \leftarrow \{i, M_i^{rqs}.requester\}$;
9   **end**
10 **else**
11   $C_i'' \leftarrow C_i \cup best\_feasible(M_i^{rqs}.requesters)$;
12 **end**
13 **if** $C_i'' \neq \emptyset$ **then**
14   **if** $C_i \neq \{i\}$ **then**
15   **foreach** $j \in C_i''$ **do**
16       $C_j \leftarrow C_i''$;
17   **end**
18   **else**
    `// send an accept to agents in` C_i''
19   add an accept message to M_j^{rpl}, where j are all agents in C_i'';
20     $wait\_4? \leftarrow true$;
21   **end**
  `// reject some requests`
22   add a rejection message to M_j^{rpl}, where j are all agents in $M_i^{rqs}.requesters \setminus C_i''$;
23 **else**
  `// reject all requests`
24   add a rejection message to M_j^{rpl}, where j are all agents in $M_i^{rqs}.requesters$;
25 **end**

Specifically, if agent i has received (in second phase) a request that is better than the one it has proposed (in first phase), then C_i'' will not be empty (as in line 2). In this case, agent i confirms its intention only if the request was mutual[11] (as in lines 3 to 5), otherwise it refuses it. If agent i has not received a better request, then it simply confirms its interest of forming the initially proposed coalition

[11] It is noteworthy that only the agent with the greatest ID operates this phase.

Procedure 19: SACF phase 3 – confirmation

$accept? \leftarrow false$;
if $C_i'' \neq \emptyset$ **then**
  if $C_i'' = M_i^{rpl}.coalition$ *and* $i > M_i^{rpl}.sender$ **then**
    $accept? \leftarrow true$;
  end
else
  $accept? \leftarrow true$;
end
if $accept?$ **then**
  `// send a confirmation message`
  add a confirmation message to M_j^{cnf}, where j is $M_i^{rpl}.sender$;
  $wait\_4? \leftarrow true$;
else
  `// send a cancellation message`
  add a cancellation message to M_j^{cnf}, where j is $M_i^{rpl}.sender$;
end

Coalition Consolidation Phase

Finally, coalition consolidation is performed as in Procedure 20. Here, the singleton agent (who has received the request in the second phase) creates the new coalition with the requesters who have confirmed their interest in forming the coalition with him (in the previous phase).

Procedure 20: SACF phase 4 – processing confirmations

foreach $m \in M_i^{cnf}$ **do**
  if $m.wait\_4?$ **then**
    $C_{m.requester} \leftarrow C_i''$;
  end
end

10.4.4 Empirical Evaluation

Methodology

In this section, we provide an empirical evaluation of DCCF and SACF from different perspectives. The main goals include: (i) how long these methods take to run, (ii) how far the provided solutions are from optimum, and (iii) how well these methods behave in a dynamic scenario, namely a smart grid, regarding PEVs' payoffs, and coalitions' stability. In order to answer these questions, DCCF and SACF are evaluated in two different settings:

Closed world In closed world scenarios, no agent can enter or leave the simulation while it is running. The aim here is to compare SACF against IP and IDP in terms of runtime and solution quality. This evaluation is presented in Sect. 10.4.4.

Open world Here, agents are allowed to enter and leave the system. In this kind of scenario, we aim at evaluating how DCCF and SACF can improve PEVs' performance on V2G sessions. This analysis is presented in Sect. 10.4.4.

The main focus of our approach is on open world settings, which are dynamic and intrinsically more complex. In spite of that, the closed world settings are used to allow a comparison of our DCCF and SACF against IP and IDP, which provide optimal solutions but do not work on dynamic scenarios.

The agents were randomly positioned in a grid-based scenario. Edges were created between pairs of agents whose Euclidean distance is lower than α. In all scenarios tested, both in open and closed world settings, the parameter α was set to 7. The distance here is measured by cells. Indeed, one can imagine that each cell has 10×10 m.

The characteristic function is formulated through Eq. (10.1), where δ is the expected power rating of any coalition, ε defines the maximum value the grid is willing to pay for a given coalition, W_C is the power rating of coalition C, and ρ is the base price for an energy unit. In the equation, the first and second lines assign the values of infeasible and feasible coalitions, respectively.

$$v(C) = \begin{cases} 0, \text{ if } \exists i, j \in C, i \neq j \ (d(i,j) > \alpha) \\ \rho \times \min\left[\left(\frac{W_C}{\delta}\right)^2 \times \varepsilon, \varepsilon\right], \text{ otherwise} \end{cases} \tag{10.1}$$

Regarding the parameters of the characteristic function, the following values were defined: $\varepsilon = 0.9$ (i.e., the grid would pay up to 90% beyond the normal price to a coalition), $\delta = 150$ (i.e., the grid would like to form small VPPs, whose power rating is around 150 kW), $\rho = \$0.5$ (i.e., approximately the energy price per kWh in Brazil, in Brazilian currency). Recall that, as defined in Sect. 10.3.4, the power rating w_i of all PEVs $i \in \mathscr{A}$ was set to 3.3 kW.

Each simulation instance (experiment) is randomly generated based on the above parameters, either for closed and open world settings. All experiments were performed in an Intel(R) Core(TM) i7-2600 3.40 GHz PC, with 16 GB RAM.

Closed World Performance

In closed world settings, DCCF and SACF are compared against IDP [17] and IP [21] (presented in Sect. 10.2) in terms of runtime and solution quality. The latter comparison is relevant because DCCF and SACF find near-optimal solutions, so a comparison against the optimal ones is a useful metric to measure their performance.

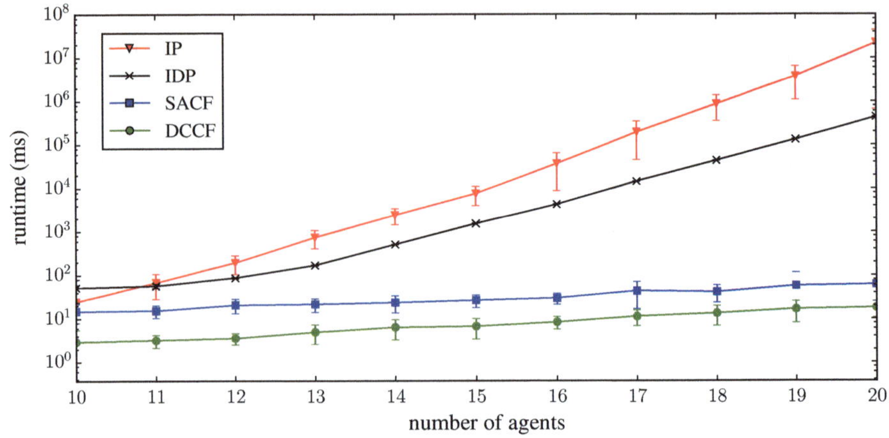

Fig. 10.6 Comparison of DCCF and SACF against IDP and IP showing the runtime (in log scale) for different number of agents (from 10 to 20 agents)

Experiments were run for different number of agents $a = \{10, 11, \ldots, 20\}$. For each number of agents, 30 different scenarios were generated (as described in the methodology of Sect. 10.4.4). In order to accurately compare the approaches, each of them was tested in exactly the same scenarios. Results are presented in Fig. 10.6. In the graph, each point shows the average runtime for the 30 scenarios, with error bars representing the standard deviation.

As can be observed in Fig. 10.6, in terms of runtime, DCCF and SACF outperform the exact algorithms IP and IDP by many orders of magnitude. While the average runtime of DCCF and SACF were lower than 1 s in all sets of experiments, in other algorithms the average runtime increases exponentially with the number of agents. For the sake of comparison, for 20 agents the IP algorithm takes, on average, about 6 h to run. In contrast, DCCF and SACF take around 20 and 60 ms, respectively.

Finally, observe that DCCF is faster than SACF. This can be explained by their negotiation process, which consists in 3 phases for DCCF and 4 phases for SACF. Moreover, recall that DCCF-agents propose coalitions to their neighbours. In this process, the agent is able to check all possible coalitions within its neighbourhood, thus finding the best ones. On the other hand, SACF-agents ask to join the best coalition within its neighbourhood (the coalition already exists). As several non-neighbour agents may ask to join the same coalition, the leader shall choose just one. Consequently, the rejected ones need to restart the process, thus taking longer to find another coalition. Therefore, the negotiation process tends to be much longer for SACF than for DCCF.

Now we analyse how far the solution generated by DCCF and SACF are from the optimum. Results are plotted in Fig. 10.7, where the bars show how far the average solutions of DCCF and SACF are from the optimal ones. Results were normalised (in relation to the optimum) to enhance the visualisation. Error bars plot the standard

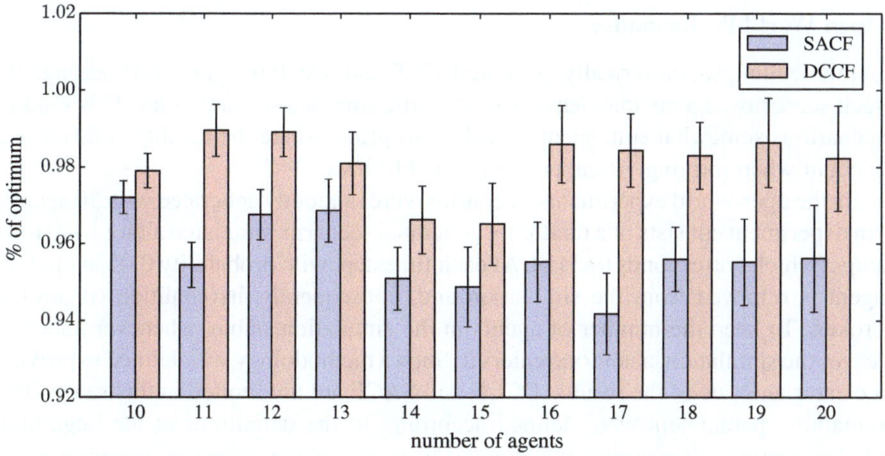

Fig. 10.7 Quality of solution generated by DCCF and SACF, showing how far (normalised) the provided solutions are from the optimal ones, for different number of agents (from 10 to 20 agents)

deviation of each set of experiments. It is important to note that the non-normalised curves behave in an ascending monotonic fashion (as a function of the number of agents).

As shown in Fig. 10.7, the results are very promising. Although DCCF and SACF have taken less than 1 s to run, they were able to find good solutions. In almost all tested cases, the solutions generated by DCCF and SACF achieved more than 94% of the optimal solution. The average quality (over all experiments) achieved was approximately 98.1% for DCCF and 95.7% for SACF. Also, the standard deviation was up to 1.4% in all experiments, for both methods, showing a tendency on producing good solutions.

Comparing DCCF and SACF to each other, DCCF outperformed SACF. Again, such a difference can be explained by negotiation process of these methods. While in DCCF the agents check all possible coalitions before making an invitation, in SACF the agents check only existing coalition within their neighbourhoods. Consequently, the SACF's performance depends on the leader finding a good coalition in the beginning.

Therefore, the advantage of DCCF and SACF over the other algorithms is that they run on a small amount of time, in dynamic, distributed environments, achieving good solutions in all tested cases. Furthermore, DCCF performed slightly better than SACF. However, as shown forward in Sect. 10.5, SACF can be extended for allowing the agents to change their coalitions whenever a better one is attainable. Importantly, however, DCCF and SACF cannot be said to always produce the same results for scenarios other than those experimented here.

Open World Performance

In this section, we empirically evaluate DCCF and SACF in open world settings. In such scenarios, agents may leave and enter the simulation at any time. This kind of scenario is somewhat equivalent to real smart grids, where the locality is taken into account when forming virtual power plants of PEVs.

In the open world experiments, scenarios were randomly generated with 50 agents. An experiment consists of a randomly generated scenario, simulated along 1440 time steps (which corresponds to 24 h). At each timestep, with probability 0.05, a random agent is removed from the simulation and, consequently, its coalition (if any) is broken. To keep the number of agents in the simulation stable, whenever an agent leaves the simulation, a new one enters it. Such a methodology was defined to provide a clearer analysis of the results. DCCF and SACF are not limited in this sense. The remaining parameters were defined according to the definitions at the beginning of this section. Concerning the stochastic nature of the scenarios generation, 30 replications were made in order to provide more precise results. The results shown hereafter represent the average over these replications, except when otherwise stated.

Figure 10.8 presents the average value $V(CS)$ of coalition structures along time. Observe that this value can be seen as the social welfare. As can be seen, the social welfare does not experiment large variations. The largest variations occur when an agent leaves the system (the valleys in the series), accounting for a change no worse than 10% in DCCF and 5% in SACF. Observe that, apparently, the social welfare is becoming better for DCCF and worse for SACF. This is due to the negotiation process of SACF, which relies on the leaders' decisions to deliver good performance. Consequently, SACF is more sensitive to agents entering or leaving the system. However, we highlight that such a deterioration of the social welfare along time is not significant, accounting for less than 10%, on average, throughout the simulation. Therefore, results show that both DCCF and SACF are effective on organising the agents into coalitions.

Another aspect to consider regarding payoffs refers to how much the agents receive along time. This is the so-called agents' instantaneous payoff and is presented in Fig. 10.9. As can be observed, the instantaneous payoff behaves in a similar way to the social welfare: relatively stable along time, but increasing for DCCF and decreasing for SACF. Note that the instantaneous payoff is really low (less than $0.01) as compared to the value of the entire coalition structure. The point is that this is the value received in just 1 min. Moreover, the CFG employed here represents the value the grid pays beyond the base price (as described in Sect. 10.3.4). To this regard, after the 24-h simulation period, the average payoff obtained by each agent was approximately $7.9 for DCCF and $6.3 for SACF. Also, considering the total profit (base price + coalition price) obtained by the agents, the average was of $47.5 for DCCF and $45.9 for SACF. On the other hand, the payoff for singletons was of $39.6. Therefore, agents that were in coalitions received, on average, a profit 19.9% (for DCCF) and 15.9% (for SACF) greater than singleton agents.

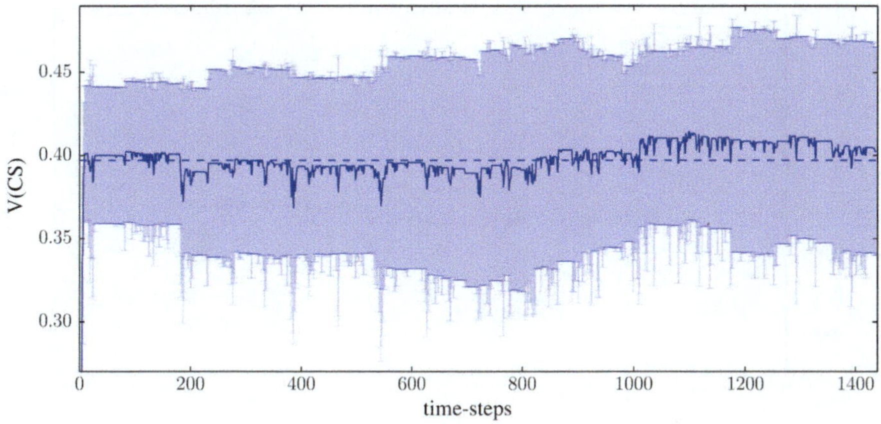

(a) DCCF

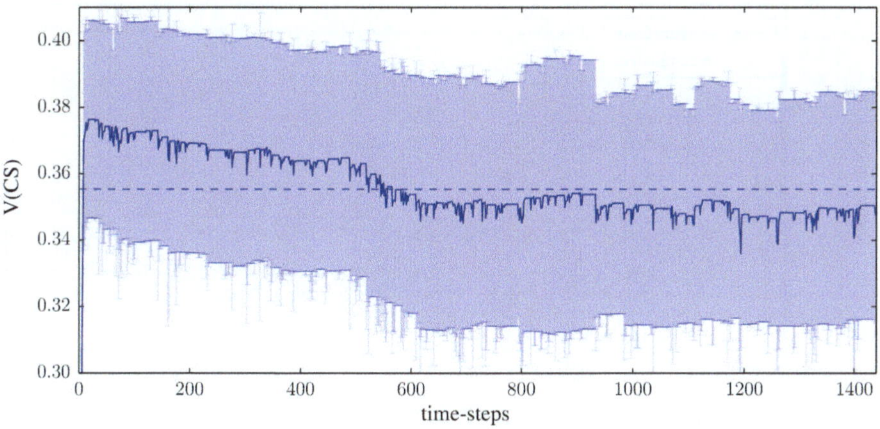

(b) SACF

Fig. 10.8 Average $V(CS)$ along time for DCCF (**a**) and SACF (**b**). The shaded lines present the standard deviation, and the dashed line represents the average over all time steps

Concerning stability, a few aspects must be regarded. Figures 10.10, 10.11 and 10.12 present the average number, size and duration of coalitions, respectively, and Fig. 10.13 shows the average percentage of agents inside coalitions. For all the cases, the expected behaviour is one of stable coalition structures. In fact, as can be observed the coalition structures are reasonably stable from all perspectives presented in the plots. In DCCF, there was an average of 5.8 coalitions with 8.2 agents each, corresponding to 96.6% of the agents within coalitions. In the case of SACF, there were 6.1 coalitions with 8.0 agents each, meaning that 97% of the agents worked

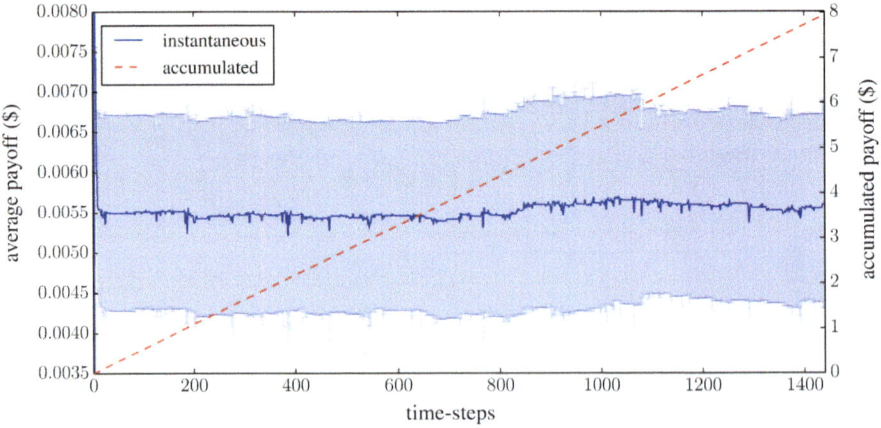

(a) DCCF

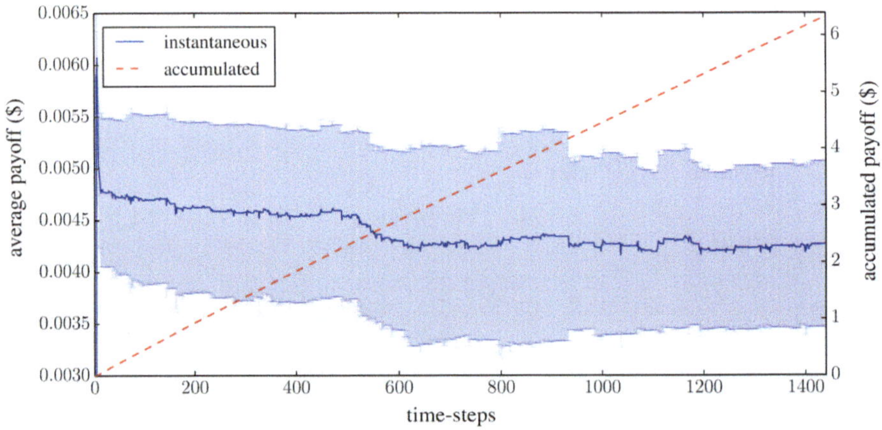

(b) SACF

Fig. 10.9 Average instantaneous (left vertical axis) and accumulated (right vertical axis) agents' payoff along time for DCCF (**a**) and SACF (**b**). The shaded lines present the standard deviation

within coalitions throughout the simulation. Furthermore, the coalitions lasted 377 time-steps for DCCF and 340 time-steps for SACF. These results evidence that the coalitions are: profitable (otherwise, more agents would remain as singletons), stable (they lasted for 25% of the simulation time), and well distributed (around 6 coalitions with 8 agents, each).

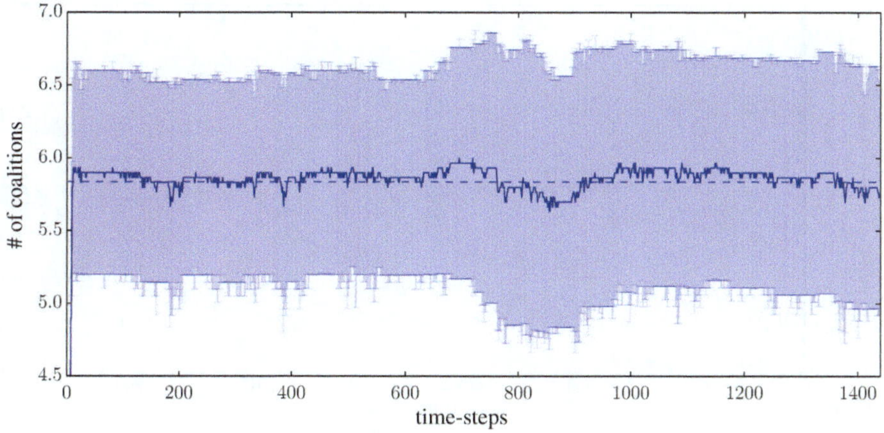

(a) DCCF

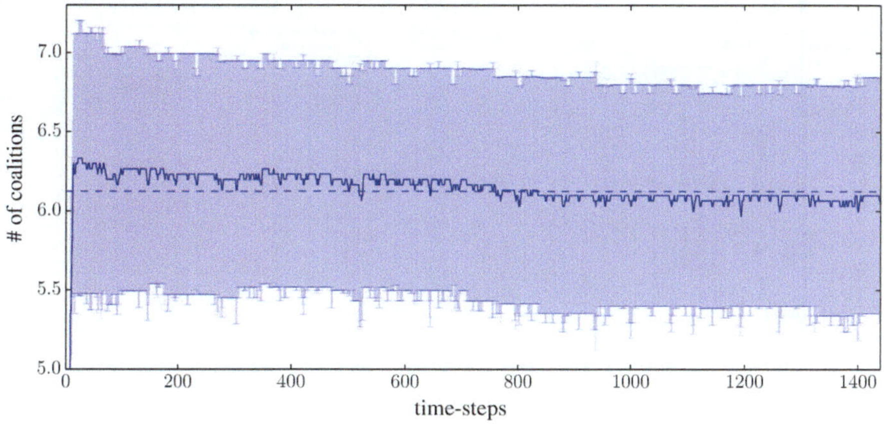

(b) SACF

Fig. 10.10 Average number of coalitions along time for DCCF (**a**) and SACF (**b**). The shaded lines present the standard deviation, and the dashed line represents the average over all time steps

DCCF and SACF presented similar results regarding stability. As shown in Figs. 10.10, 10.11, 10.12, and 10.13, the variance between their results was very small. Along time, no data series (of these particular figures) varied by more than 5%, on average. Interestingly, observe that the coalitions' duration behaves in an ascending fashion (Fig. 10.12). The reason is that coalitions' duration is affected only by agents leaving the system, which is not so frequent. A variation of SACF (SACF$^+$) is able to handle even more dynamic scenario and is presented in Sect. 10.5.

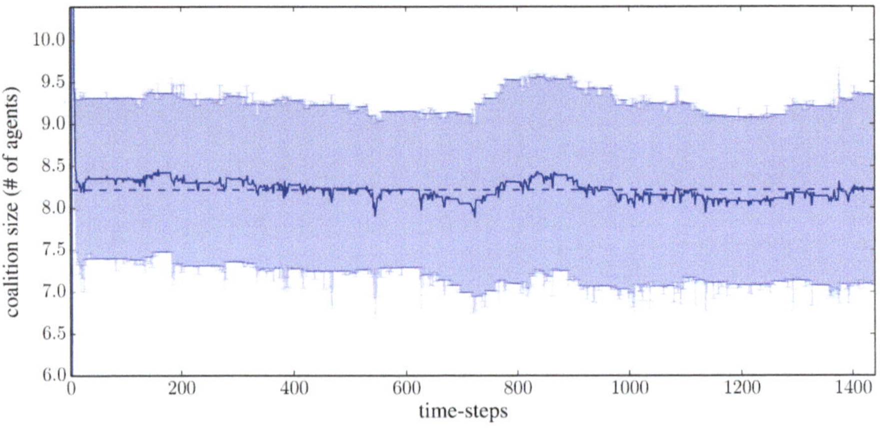

(a) DCCF

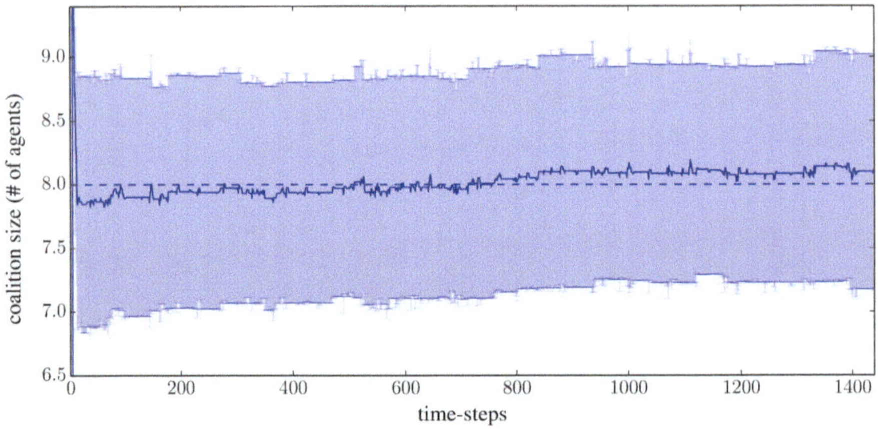

(b) SACF

Fig. 10.11 Average size of coalitions along time for DCCF (**a**) and SACF (**b**). The shaded lines present the standard deviation, and the dashed line represents the average over all time steps

Finally, Fig. 10.14 presents the number of messages exchanged along time. As expected, the amount of messages increases steadily whenever negotiation is taking place, i.e., when agents enter or leave the system. It can be seen that the number of messages exchanged in DCCF is much higher than in SACF. In fact, the communication overhead is the main drawback of DCCF as compared to SACF. The reason is that the negotiation process of DCCF involves each agent inviting all potential neighbours. On the other hand, in SACF, each agent chooses a single coalition (the best within its neighbourhood) and sends a single message to it. The number of messages exchanged throughout the whole simulation was, on average, 5.966 for DCCF and 468 for SACF.

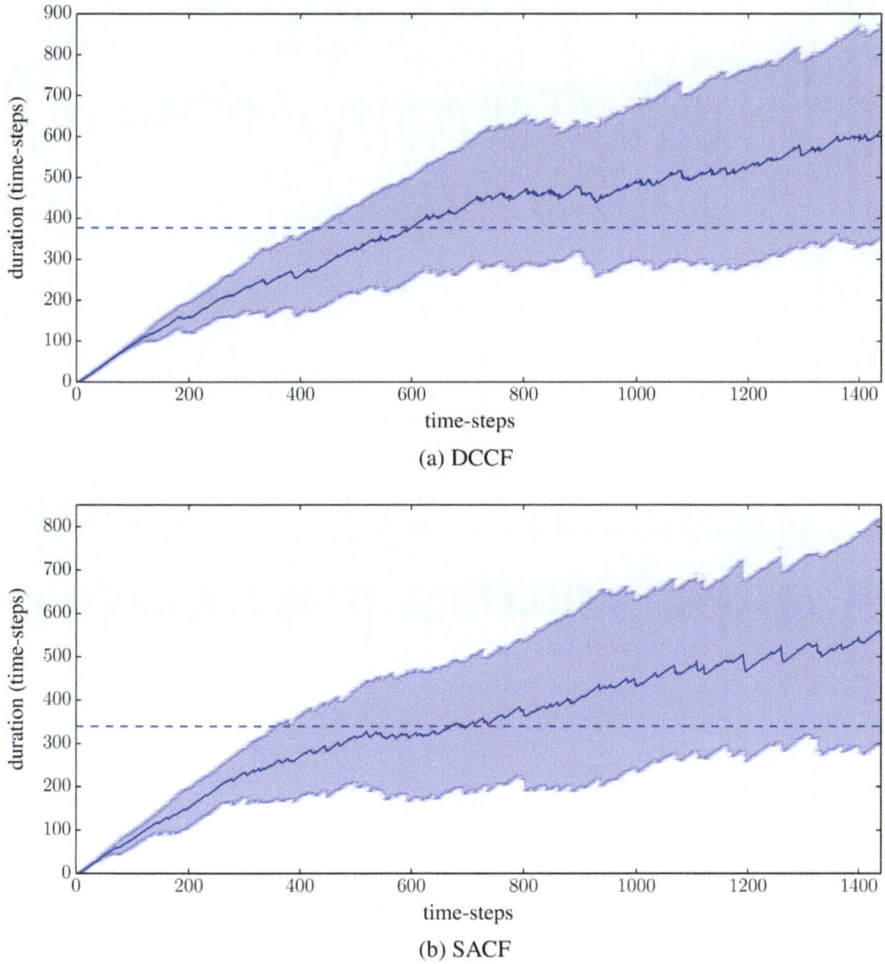

Fig. 10.12 Average duration of coalitions along time for DCCF (**a**) and SACF (**b**). The shaded lines present the standard deviation, and the dashed line represents the average over all time steps

10.4.5 Summary

This section presented DCCF and SACF. The main difference between them lies in the negotiation scheme: in DCCF agents are *invited* to join coalitions, whereas in SACF agents *ask* to join coalitions. To this regard, SACF is more flexible than DCCF, as the agents can explore all coalitions in their neighbourhood before deciding. Moreover, as will be shown in Sect. 10.5.2, SACF can be extended for allowing the agents to change their coalitions whenever a better one is attainable.

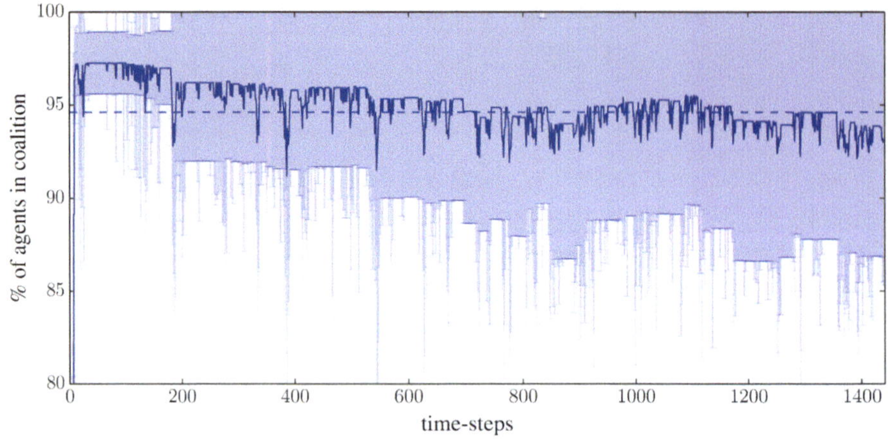

(a) DCCF

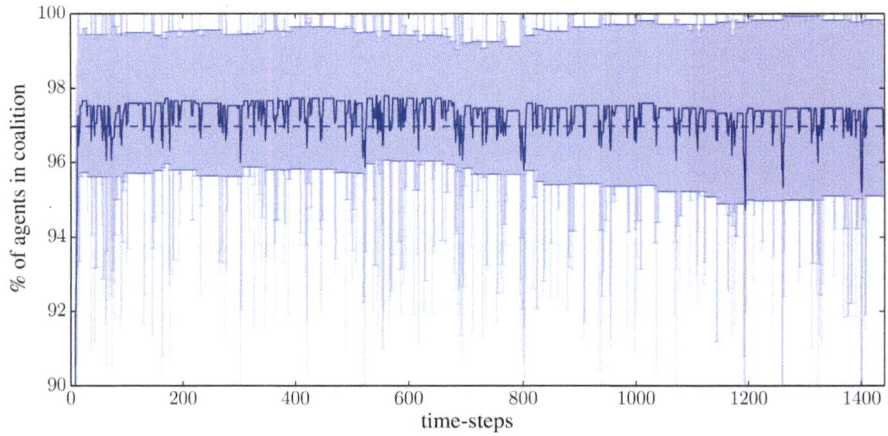

(b) SACF

Fig. 10.13 Average percentage of agents inside coalitions along time for DCCF (**a**) and SACF (**b**). The shaded lines present the standard deviation, and the dashed line represents the average over all time steps

Based on experiments, we showed that both DCCF and SACF are fast (in terms of runtime) and provide reasonably good solutions (close to the optimal). DCCF performed slightly better than SACF in most aspects. However, both approaches have shown reasonably stable and profitable for the agents.

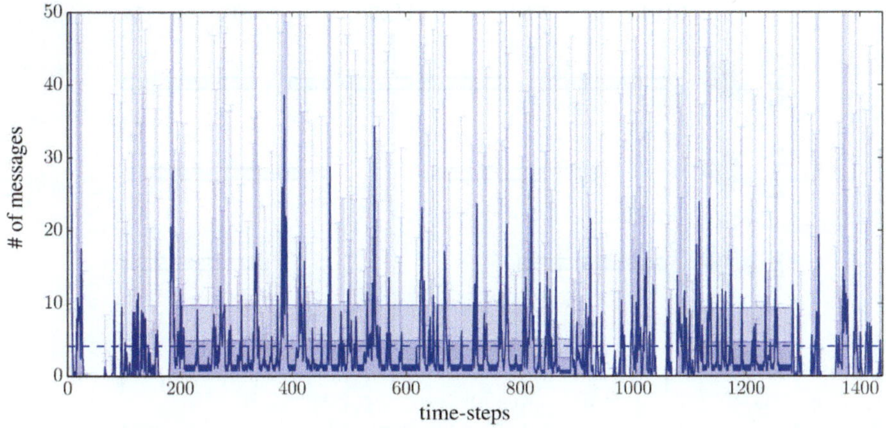

(a) DCCF

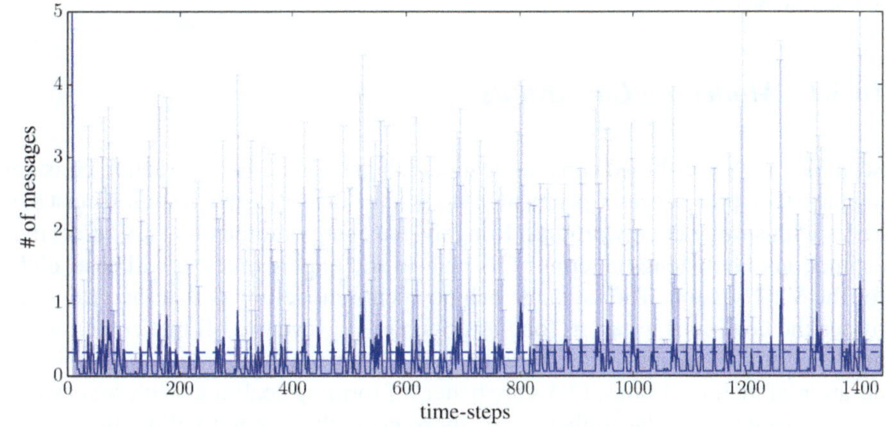

(b) SACF

Fig. 10.14 Average number of messages exchanged along time for DCCF (**a**) and SACF (**b**). The shaded lines present the standard deviation, and the dashed line represents the average over all time steps

10.5 User-Based Constraints

In this section, we present a coalition formation approach for smart grid scenarios with user-based constraints. In such scenarios, the neighbourhood structure is defined considering the PEVs' users usage profiles. The complete description on the modelling of such constraints is presented in Sect. 10.5.1. Following such a modelling, Sect. 10.5.2 presents an extension to SACF (from Sect. 10.4.3). An empirical evaluation of SACF^{+} is presented in Sect. 10.5.3.

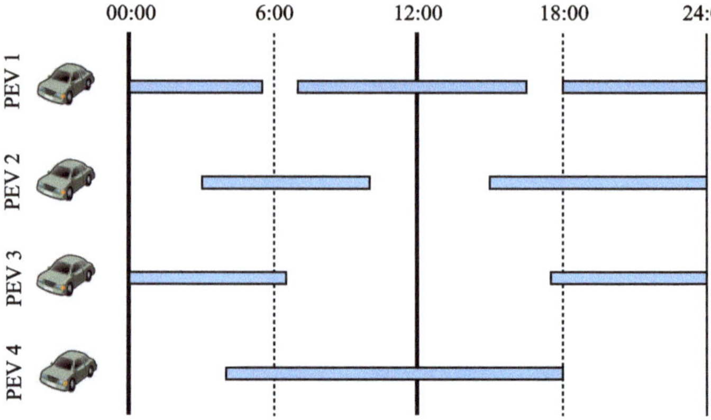

Fig. 10.15 Example of the energy profiles of four PEVs. In the example, the PEVs' energy profiles differ considerably among themselves because of their owners' preferences (e.g., different work/leisure times)

10.5.1 Modelling Constraints

Scenarios with user-based constraints model the neighbourhood structure considering the users' preferences. Specifically, we assume that a feasible coalition must have PEVs with similar energy profiles. Roughly, the energy profile of a PEV refers to an estimation of how long (in terms of its plug-in and unplug times) it will be available for the V2G session. The rationale behind such a modelling is that the grid incentivises the coalitions to last as long as possible. However, recall that coalitions last until any of their members decide to leave it. In this sense, in order to maximise the duration of their coalitions, PEVs are better off forming coalitions with peers whose energy profile is similar to theirs, i.e., those peers that are not willing to leave the grid before them. Grouping agents by their energy profile aims at increasing coalitions' duration, and thus the PEVs profit. From the grid viewpoint, such a modelling translates coalitions into more reliable, stable (coalition negotiation is less frequent) energy sources. Such a constraint is illustrated in Fig. 10.15, where the energy profiles of four PEVs are presented. Observe that, in some cases, the vehicles' energy profiles differ considerably among themselves. This is due to their owners' preferences (e.g., different work/leisure times). Suppose a coalition is being formed at 18:00. Here, any coalition including agent 4 would possibly not be a good deal, since agent 4 will not be available before next day. Thus, at this point, coalition $\{1, 2, 3\}$ would be a better one. Of course, other aspects (e.g., energy availability) must also be taken into account in the coalition formation process. However, the similarity of the agents' energy profiles plays a key role towards more stable, reliable coalitions.

The energy profile of an agent can be simplified as the amount of time[12] it will be available for the V2G session. Let T_i represent the energy profile of agent $i \in \mathscr{A}$. Recall that agents within a coalition should have similar energy profiles. In this sense, given two agents $i, j \in \mathscr{A}$ with $i \neq j$, the similarity $s(T_i, T_j)$ between their energy profiles T_i and T_j is measured by the absolute difference between them, as in Eq. (10.2), where the lower the result, the higher the similarity of the given energy profiles.

$$s(T_i, T_j) = |T_i - T_j| \tag{10.2}$$

The neighbourhood structure is defined considering the degree $d(\cdot, \cdot)$ among the agents and the constraint parameter α of the grid (see Definition 10.3). In the scenarios with user-based constraints, the similarity equation denotes the constraint degree between a pair of PEVs, that is, $d(i, j) = s(T_i, T_j)$. Moreover, we constrain the maximum allowed difference between two energy profiles by means of parameter α. Thus, a PEV can only form a coalition with its peers with similar energy profile, as formulated in Definition 10.5. Again, a coalition that fits into such criterion is said a feasible coalition, as described in Definition 10.6. The value for parameter α represents the grid's requirements. A lower value for α is more restrictive than a higher one, meaning that the coalitions are supposed to last longer, at the cost of being smaller.

Definition 10.5 (*user-based neighbourhood*) Under user-based constraints, function $d(i, j)$ denotes the similarity $s(T_i, T_j)$ between the energy profiles of agents $i, j \in \mathscr{A}$. Furthermore, the neighbourhood N_i of a given agent $i \in \mathscr{A}$ is composed of all agents $j \in \mathscr{A} \setminus \{i\}$ to which its distance $d(i, j)$ is no greater than α.

Definition 10.6 (*user-based feasible coalition*) Given a geographic-based neighbourhood, a coalition is said feasible if and only if all pairs of its members are neighbours.

10.5.2 Self-adapting Coalition Formation with Changing Coalitions

This section presents SACF$^+$ algorithm, introduced in [27]. SACF$^+$ is a variation of the SACF algorithm (from Sect. 10.4.3) that allows the agents to move to another coalition whenever a better payoff is attainable. At first, it may seem that SACF$^+$ is less stable than SACF. However, when the agents can move to better coalitions, SACF$^+$ avoids local optima, thus leading to even stabler coalitions.

[12] Again, we consider each time step corresponds to 1 min.

Table 10.3 SACF$^+$ – definition of the procedures' variables

Variable	Description	Initial value
i	id of the agent who is running the procedure	(agent id)
N_i	List of neighbours of agent i	$\{j \in \mathscr{A} \setminus \{i\} \mid d(i, j) \leq \alpha\}$
C_i	Agent i's coalition	$\{i\}$ (singleton)
C'_i	Best coalition found by agent i in phase 1	$\emptyset$
C''_i	Coalition accepted by agent i in phase 2	$\emptyset$
M_i^{rqs}	List of "request messages" received by agent i	$\emptyset$
M_i^{rpl}	List of "reply messages" received by agent i	$\emptyset$
M_i^{cnf}	List of "confirmation messages" received by agent i	$\emptyset$
$wait\_4?$	Whether the agent is waiting for the coalition consolidation or not	$false$
$chk\_new?$	Whether the agent is trying to join a better coalition	$false$

SACF$^+$ consists of a main procedure, which is divided into several steps, each defined as an independent procedure. Following the simulation scheme of Sect. 10.3.5, the agents' initialisation and procedures must be specified. Whenever an agent using SACF$^+$ enters the system, it is initialised according to Table 10.3. After that, for each simulation's iteration, SACF$^+$ is executed for every agent through its main procedure. This section presents the complete description of SACF$^+$.

Main Procedure

The main procedure is performed by each agent and is presented in Procedure 21. Roughly speaking, the main procedure allows the agents to negotiate within their neighbourhoods in order to form feasible coalitions. The negotiation process is started by singletons, which are called *requesters*.[13] The requesters can ask their neighbours (i) to form a new coalition (in the case of single neighbours), or (ii) for permission to join their existing coalitions. In the former scenario, both the requester and its neighbour are singletons, so a new coalition is going to be created. In the latter scenario, the requester wants to join a coalition that already exists. In this case, the join requests received by a coalition are handled by its *leader*.[14]

[13]As for SACF, in the beginning, all agents act as requesters.

[14]Following the same idea of SACF, the leadership of a coalition is just a role assumed by an agent to represent the coalition in negotiation processes, thus avoiding redundant negotiations. The leader is also responsible for notifying the coalition's members when it is broken (i.e., one of its members has left). Here, the leader of a coalition is the agent with lowest ID (which, in real situations, could be replaced, e.g., by the vehicle's licence plate).

Procedure 21: SACF$^+$ – main procedure

required: variables initialised according to Table 10.3;

`// Neighbours' coalitions checking phase`

if $C_i = \{i\}$ *and* $C_i' = \emptyset$ *and* $C_i'' = \emptyset$ *and not* $wait\_4?$ **then**
  check neighbours' coalitions (Procedure 22);
end

`// Requests processing phase`

if $M_i^{rqs} \neq \emptyset$ *and* $C_i' = \emptyset$ *and not* $wait\_4?$ **then**
  process requests (Procedure 23);
end

`// Confirmation phase`

if $M_i^{rpl} \neq \emptyset$ *and* $M_i^{rpl}.answer =$ *"Yes"* *and* $C_i' \neq \emptyset$ **then**
  confirmation (Procedure 24);
end

`// Coalition consolidation phase`

if $M_i^{cnf} \neq \emptyset$ *and* $C_i'' \neq \emptyset$ *and* $wait\_4?$ **then**
  process confirmations (Procedure 25);
end

The coalition negotiation process is divided into the following four phases:

- Neighbours' coalitions checking: requesters check for coalitions in their neighbourhoods and send a joining request to the leader of the one with the highest value.
- Requests processing: every agent who has received a request must decide whether the requester(s) can join its coalition or not. If the agent is a leader, then it can directly accept one or more request. Otherwise, if the agent is a singleton, then the negotiation process must advance to the two next phases.[15]
- Confirmation: every requester who has received a positive answer (necessarily sent by a singleton) must confirm whether it still intends to form the new coalition.
- Coalition consolidation: where new coalitions are created with the requesters who have confirmed their join intention.

Essentially, as for SACF (in Sect. 10.4.3), the coalition negotiation process takes place in two different occasions. In the first, two singletons negotiate the formation of a new coalition (see the example of Fig. 10.5b). In the second negotiation scenario, on the other hand, the coalition already exists, and the single agent asks to join it (see the example of Fig. 10.5a). In SACF$^+$, however, besides these two situations, agents within coalitions are also allowed to check for better ones. Such a self-adapting

[15] The second phase is important because the requester may also receive requests. Thus, a confirmation is needed for ensuring the requester is still interested in joining the coalition.

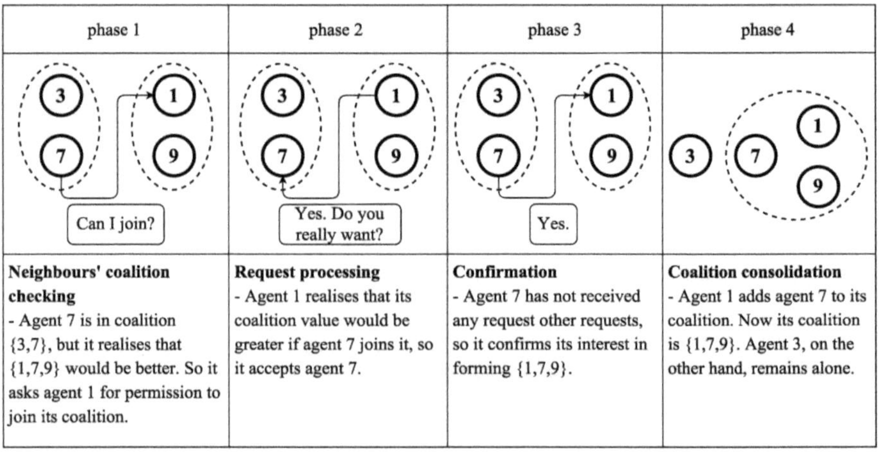

Fig. 10.16 The negotiation process of SACF$^+$ when an agent decides to leave its current coalition to join a better one

mechanism enables agents to improve their revenue whenever possible. In such case, the agent proceeds as it were joining a new coalition. The only difference is that, in the event the agent is admitted to the coalition, it has to cancel its current coalition before joining the new one. The idea behind allowing agents to check for better coalitions is to avoid local optimal solutions. The agents are always allowed to check for better coalitions. An illustrative example is presented in Fig. 10.16. Here we have two coalitions: $\{1, 9\}$ and $\{3, 7\}$. At some point, agent 7 realises that its value would be better if it joins coalition $\{1, 9\}$, and then it asks agent 1 for permission to join its coalition (phase 1). When agent 1 receives the request (phase 2), it accepts it because $\{1, 7, 9\}$ has a higher value than its current coalition $\{1, 9\}$. As soon as agent 7 confirms its intention (on phase 3), agent 1 creates the new coalition $\{1, 7, 9\}$ (phase 4). At this point, agent 3 becomes a singleton.

The negotiation phases are described in detail in the following subsections.

Neighbours' Coalitions Checking Phase

In this phase, agents check for existing coalitions in their neighbourhoods. Such a search behaviour is performed as in Procedure 22, where all coalitions (and even other single agents) in agent i's neighbourhood are analysed. For each coalition, the procedure checks whether it remains feasible if agent i joins it. We emphasize that only feasible coalitions are considered here (according to Definition 10.2). Among the feasible coalitions identified, the best one is selected, i.e., the one where the energy price is higher. Then, given the selected coalition, the agent asks its leader for permission to join it.

Procedure 22: SACF$^+$ phase 1 – checking neighbours' coalitions

foreach $C \subset NC_i$ **do** is preceded by $NC_i \leftarrow \{C_j : \forall j \in N_i\}$;

```
NC_i ← {C_j : ∀j ∈ N_i};
foreach C ⊂ NC_i do
    if C ∪ {i} is not feasible then
        NC_i ← NC_i \ C;
    end
end
if NC_i ≠ ∅ then
    C'_i ← arg max_{C⊂NC_i} v(C)/|C|;
    if v(C'_i)/|C'_i| > v(C_i)/|C_i| then
        // request permission for leader of C'_i to join its
           coalition
        add a join request to M_j^{rqs}, where j is the leader of C'_i;
        if C_i = ∅ then
            // agent is trying to form a new coalition
            chk_new? ← false;
        else
            // agent has found a better coalition to join
            chk_new? ← true;
        end
    end
end
```

It is important to note that, when negotiating a coalition, the agents do not care about the value of such coalition, but about the energy price (per unit) on it. The reasoning here is a quite simple: as the payoff (value) of a coalition is equally divided among its members, the agent's payoff is calculated through the energy price on the coalition. Consider, for instance, that two coalitions have the same value, but different number of agents. In this case, on the basis of the problem definition, the coalition with fewer agents would have a higher energy price. In this sense, during the negotiation process, only the energy price is considered.

This procedure can be also performed by agents that are already in a coalition, as long as the new coalition is better than its current one. In this case, a special flag called *chk_new?* is activated so as to properly handle this process in next phases.

Requests Processing Phase

The requests processing phase is performed as in Procedure 23. In this phase, agents who have received requests must decide whether or not the requesters can join their coalitions. As previously stated, this procedure might be performed either by leader or single agents. This procedure is divided into: requests evaluation (first *if* block of Procedure 23), and decisions notification (second *if* block of Procedure 23).

Procedure 23: SACF$^+$ phase 2 – processing requests

```
C_i'' ← ∅;
if |M_i^rqs| = 1 then
    C^r ← C_i ∪ {M_i^rqs.requester};
    if C_i ≠ {i} and C^r is feasible and v(C^r)/|C^r| > v(C_i)/|C_i| then
        C_i'' ← C^r;
    else if C_i = {i} and C_i' = C^r and i < M_i^rqs.requester then
        C_i'' ← {i, M_i^rqs.requester};
    end
else
    C_i'' ← C_i ∪ best_feasible(M_i^rqs.requesters);
end
if C_i'' ≠ ∅ then
    if C_i ≠ {i} then
        foreach j ∈ C_i'' do
            C_j ← C_i'';
        end
    else
        // send an accept to agents in C_i''
        add an accept message to M_j^rpl, where j are all agents in C_i'';
        wait_4? ← true;
    end
    // reject some requests
    add a rejection message to M_j^rpl, where j are all agents in M_i^rqs.requesters \ C_i'';
else
    // reject all requests
    add a rejection message to M_j^rpl, where j are all agents in M_i^rqs.requesters;
end
```

Many cases are handled in this procedure. The most easily solvable ones occur when just one request is received. In this case, if the agent is the leader of the coalition, it just needs to check if the new coalition (with the requester) is feasible, and if the energy price on it is better than of the current coalition. If so, then the coalition is accepted. Handling the request in this way, the performance of the coalition cannot deteriorate. On the other hand, if the agent is a singleton, then it probably have requested to join a coalition also. This way, the agent only accepts the request if the resulting coalition is the same as the one it has proposed. We call this a mutual request, since two agents are proposing the same coalition to each other. In this case, only the agent with the lowest ID accepts the request of the other, thus becoming the leader of that coalition.

In another scenario, the agent receives multiple requests. In this case, if all requesters are neighbours, then all can be accepted. Otherwise, all permutations of requesters are computed in order to get the subset that most increases the energy price of the coalition. In Procedure 23 we consider that such a computation is performed by the function *best_feasible*, which trivially computes the best subset of requesters to be accepted.

After all requests were evaluated, the requesters are notified about the decisions made by agent i. If agent i is a singleton, then it sends a notification message to every requester (accepting or rejecting their requests) in order to proceed to the next negotiation phases. Otherwise, if agent i is the leader of an existing coalition, then no further negotiation is necessary, and the coalition can be immediately formed.

Confirmation Phase

In the confirmation phase, requesters who have received positive notifications from the previous phase must confirm whether or not they still intend to form the new coalitions. This phase is performed as in Procedure 24, only by requesters who have asked to form a new coalition with other single agents. Through this procedure, the agent is able to withdraw its request if a better one was received. Specifically, if agent i has received (in the second phase) a request that is better than the one it has proposed (in the first phase), then C_i'' will not be empty. In this case, agent i confirms its intention only if the request was mutual,[16] otherwise it refuses it. If agent i has not received a better request, then it simply confirms its interest in forming the initially proposed coalition.

Procedure 24: SACF$^+$ phase 3 – confirmation

```
accept? ← false;
if C_i'' ≠ ∅ then
    if C_i'' = M_i^rpl.coalition and i > M_i^rpl.sender then
        accept? ← true;
    end
else
    accept? ← true;
end
if accept? then
    // send a confirmation message
    add a confirmation message to M_j^cnf, where j is M_i^rpl.sender;
    wait_4? ← true;
    if chk_new? then
        cancel current coalition;
    end
else
    // send a cancellation message
    add a cancellation message to M_j^cnf, where j is M_i^rpl.sender;
end
```

[16]It is noteworthy that only the agent with the greatest ID operates this phase.

In some cases, the current negotiation, which was started by agent i, corresponds to an attempt of moving to a better coalition (if *chk_new*? is flagged as true). In such scenarios, after agent i has confirmed its join intention, it needs to leave its current coalition, informing the leader about the event. The leader, in turn, breaks the coalition, leaving the remaining agents to negotiate new coalitions.

Coalition Consolidation Phase

Finally, the coalition consolidation phase is performed as in Procedure 25. In this phase, the single agent (who has received the request in the second phase) creates the new coalition with the requesters who have confirmed their interest in forming the coalition with him (in the previous phase).

Procedure 25: SACF$^+$ phase 4 – processing confirmations

```
foreach j ∈ M_i^cnf.sender do
    if agent j has accepted then
        C_j ← C_i'';
    end
end
```

10.5.3 Empirical Evaluation

Methodology

In this section, we provide an empirical evaluation of SACF$^+$'s performance from different perspectives. The methodology is similar to one used in Sect. 10.4.4. Again, the main goals include analysing: (i) how long SACF$^+$ takes to run, (ii) how far the provided solutions are from optimum, and (iii) how well SACF$^+$ behaves in a dynamic scenario, namely a smart grid, regarding PEVs' payoffs, and coalitions' stability. In order to answer these questions, our approach is evaluated in two different settings:

Closed world In closed world scenarios, no agent can enter or leave the simulation while it is running. The aim here is to compare SACF$^+$ against IP and IDP in terms of runtime and solution quality.

Open world Here, agents are allowed to enter and leave the system. In this kind of scenario, we aim at evaluating how SACF$^+$ can improve PEVs' performance on V2G sessions. Recall that neither IP nor IDP are suitable for such dynamic scenarios, thus SACF$^+$ is not compared with them in open world settings.

The main focus of SACF$^+$ is on open world settings, which are dynamic and intrinsically more complex. In spite of that, the closed world settings are used to allow a comparison of our approach against IP and IDP, which provide optimal solutions but do not work in dynamic scenarios. It is noteworthy, however, that SACF$^+$ is not compared against DCCF (Sect. 10.4.2) and SACF (Sect. 10.4.3). Indeed, such a comparison would not be fair given that the scenario modelling and characteristic function defined in this section differ from that of previous sections.

The characteristic function is formulated through Eq. (10.3), where δ is the expected power rating of any coalition, ε defines the maximum value the grid is willing to pay for a given coalition, W_C is the power rating of coalition C (Sect. 10.3.4), $T_C = \min_{i \in C} T_i$ is the duration of the coalition, and ρ is the base price for an energy unit. In the equation, the first and second lines assign the values of infeasible and feasible coalitions, respectively.

$$v(C) = \begin{cases} 0, \text{ if } \exists i, j \in C, i \neq j \ (d(i,j) > \alpha) \\ W_C \times \left(\rho \times \min\left[\frac{W_C \times T_C \times \varepsilon}{\gamma}, \varepsilon\right]\right), \text{ otherwise} \end{cases} \tag{10.3}$$

Regarding the parameters of the characteristic function, the following was defined. The normalising constant γ needs to be high enough to normalise the product of a coalition's power rating and energy profile, thus it was set to 10,000. Parameter ε was set to 0.9, i.e., the grid would pay up to 90% beyond the base price to a coalition. The base price ρ of an energy unit was set to \$0.50.

The agents' parameters were defined as follows. The power rating w_i of all PEVs $i \in \mathscr{A}$ was set to 3.3 kW. With respect to the PEVs' energy profile, its time unit was defined as half-hour periods. Recall that an energy profile here represents how long the PEV is going to be connected to the grid. In this sense, for each PEV $i \in \mathscr{A}$, a value T_i is drawn from a normal distribution $\mathscr{N}(\mu, \sigma)$, bounded by t_{min} and t_{max}, with $\mu = 8$ (4 h), $\sigma = 4$ (2 h), $t_{min} = 1$ (0.5 h), and $t_{max} = 16$ (8 h). Both the PEVs' power rating and energy profile are set for each agent at instantiation time. Observe that we do not intend to be realistic with respect to all the parameters' values employed here. In the case of the PEVs' energy profile, however, it can be said that sampling it from a normal distribution is a good approximation. In theory, however, these definitions do not affect the results of our approach.

Each simulation instance (experiment) is randomly generated based on the above parameters, either for closed and open world settings. All experiments were performed in an Intel(R) Core(TM) i7-2600 3.40 GHz PC, with 16 GB RAM.

Closed World Performance

In closed world settings, experiments were run for a different number of agents $n = \{10, 11, \ldots, 20\}$. For each n, 30 different scenarios were generated. In order to provide an accurate comparison, both algorithms were tested in exactly the same scenarios. The results are presented in Fig. 10.17. In the plot, each point represents the average runtime over the 30 scenarios, and error bars represent the associated standard deviation.

As can be observed in Fig. 10.17, our approach outperforms IP and IDP in terms of runtime by many orders of magnitude. The average runtime of SACF$^+$ was almost constant for a varying number of agents. On the other hand, the average runtime of IP

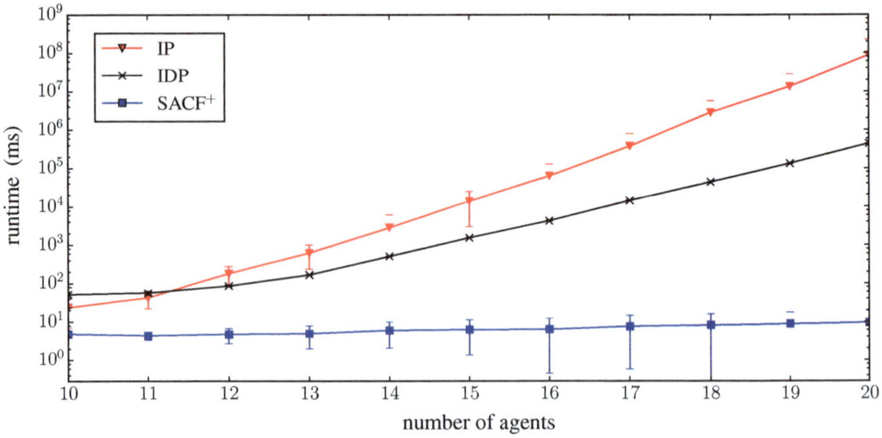

Fig. 10.17 Comparison of SACF$^+$ against IDP and IP showing the runtime (in log scale) for different number of agents (from 10 to 20 agents)

and IDP increases exponentially in the number of agents. For example, the average runtime for 20 agents was greater than 24 h for IP and greater than 7 min for IDP, while SACF$^+$ took less 10 ms, a difference of many orders of magnitude. This difference is due to the fact that, in our approach, agents locally negotiate the formation of feasible coalitions. Consequently, infeasible coalitions are neither considered during the search process. The IP algorithm, in turn, prunes portions of the search space based on estimated upper bounds. As infeasible coalitions have value zero, they do not allow IP to prune the search space. In fact, literature has reported the inability of IP to handle constrained scenarios [37]. Regarding IDP, though it is not sensitive to the distribution of coalition values, it is not able to prune infeasible solutions from the search space.

Now we analyse how far the solutions generated by SACF$^+$ are from the optimum. The results are plotted in Fig. 10.18, where each bar represents the average solution quality for a different number of agents. Results were normalised in order to show the percentage of optimum achieved by SACF$^+$. It is worth noting that the non-normalised bars behave in an ascending monotonic fashion (as a function of the number of agents). Error bars present the standard deviation.

As seen in Fig. 10.18, the results are promising. With a lower running time, SACF$^+$ was able to find good solutions in all experiments. The average solution quality over all experiments is approximately 96.5%, with a standard deviation of 5.5%. In almost 90% of the scenarios, SACF$^+$ has achieved more than 90% of the optimal solution. Furthermore, the optimum was found in almost a half of the scenarios. Such results suggest a trend of SACF$^+$ to produce good solutions. Based on this, it can be said that SACF$^+$ is efficient on providing good and fast results, thus being suitable for the dynamic scenarios considered here. It should be noted, however, that SACF$^+$ cannot be said to always produce good results for scenarios other than those experimented here.

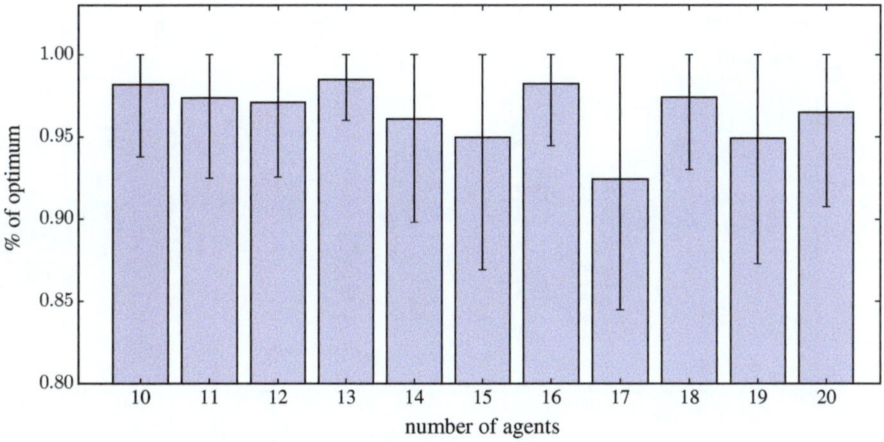

Fig. 10.18 Quality of solution generated by SACF$^+$, showing how far (normalised) the provided solutions are from the optimal ones, for different number of agents (from 10 to 20 agents)

Open World Performance

In this section, SACF$^+$ is empirically evaluated in open world settings. In such dynamic scenarios, at any time agents may leave and enter the simulation. Despite abstract, this kind of scenario is equivalent to a real smart grid, where PEVs may enter or leave V2G sessions according to their energy profiles, which in turn are ruled by their owners' preferences. The idea here is to evaluate how well SACF$^+$ behaves in such settings, trying to provide an idea of the suitability of our approach for real situations.

For the open world experiments, scenarios were randomly generated with 50 agents. An experiment here consists of a random scenario simulated along 1440 time steps, which corresponds to 24 h. To keep the number of agents in the simulation stable, whenever an agent leaves the simulation, a new one enters it. Given that the agents' profiles were modelled in half-hour periods, changes (some agents may leave, and consequently new ones may enter) in the set of agents may occur every half hour. We highlight that such a methodology was defined to provide a clearer analysis of the results, and our approach is not limited in this sense. The remaining parameters were defined according to the definitions at the beginning of this section. Concerning the stochastic nature of the scenarios generation, 30 replications were made in order to provide more precise results. The results shown hereafter represent the average over these replications, except when otherwise stated.

The average value ($V(CS)$) of coalition structures along time is presented in Fig. 10.19. In the plot, long plateaus are followed by short, steep valleys. Each plateau represents a half-hour period, and each valley represents a change in set of agents (i.e., an agent has left or entered the system). Such behaviour is related with the PEVs' energy profiles, whose granularity is of half-hour periods. To this regard, on each half-hour period, some agents may exit the system (and others may enter, to

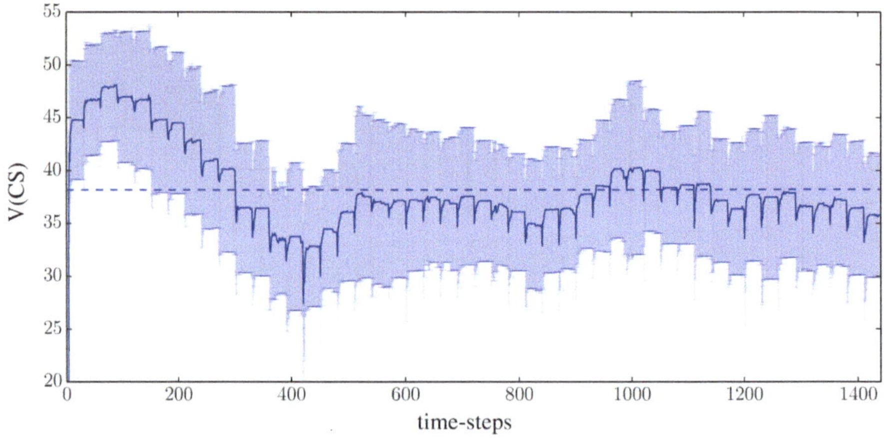

Fig. 10.19 Average $V(CS)$ along time for SACF$^+$. The shaded lines present the standard deviation, and the dashed line represents the average over all time steps

replace them). Whenever a PEV leaves the system, its coalition is broken. In such an event, the remaining PEVs from broken coalitions need to negotiate to form new ones. The new agents also take part of such negotiation. On spite of the valleys, however, the rapid increases that follow them indicate that the negotiation process is fast enough to recover the value of the previous plateau.

Importantly, observe that, in the first ten time-slots, the average value is much higher than in the subsequent ones. This behaviour relates with the distribution of the energy profiles. Recall that such a distribution has a mean 8 and a deviation 4, meaning that, at the 8th slot, approximately half of the agents will have exited the system. Consequently, as can be observed, such oscillations can be seen as a translation of the energy profiles along time into the social welfare.

Another interesting observation is that the valleys after plateaus deteriorate the solution in 6.5%, on average, and in 18.6%, at most. Considering that several agents exit the system on each half-hour period, we can say that the coalition structure value does not vary significantly from one plateau to another. This is due to fact that, by limiting PEVs to form coalitions with peers of similar energy profile, we prevent the coalitions to break early due to agents with shorter availability than others.

The plot in Fig. 10.20 presents the average and accumulated agents' payoff along time. Recall that the agents' payoff is measured by means of Eq. (10.3), which gives the payoff of a given agent i on every time step. As expected, the plateaus-valleys behaviour is also present in this plot. Here, the payoff along time is \$0.0128, and the deterioration between plateaus is below 10%, on average. Thus, not only coalitions, but also the agents' payoff remain relatively stable along time.

Regarding the agents profit, the average payoff accumulated by a PEV after 8 h connected to the grid was \$6.16. Recall that the payoff represents only the value beyond the base price. In this sense, the total value accumulated by an agent after

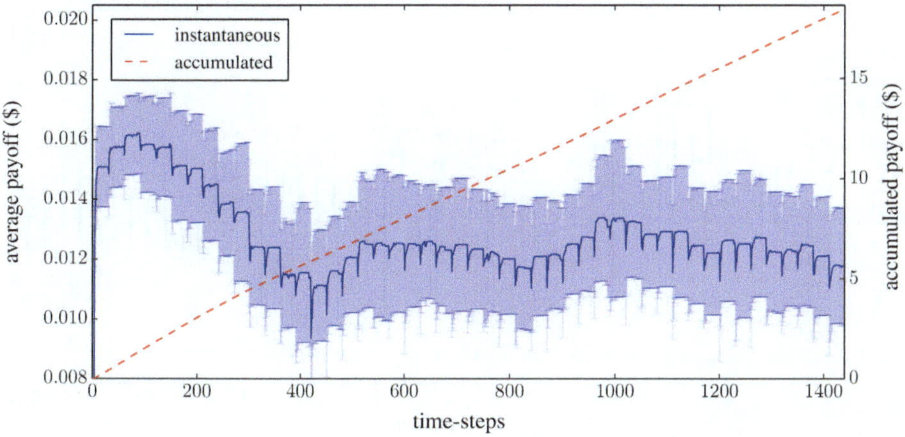

Fig. 10.20 Average instantaneous (left vertical axis) and accumulated (right vertical axis) agents' payoff along time for SACF$^+$. The shaded lines show the standard deviation

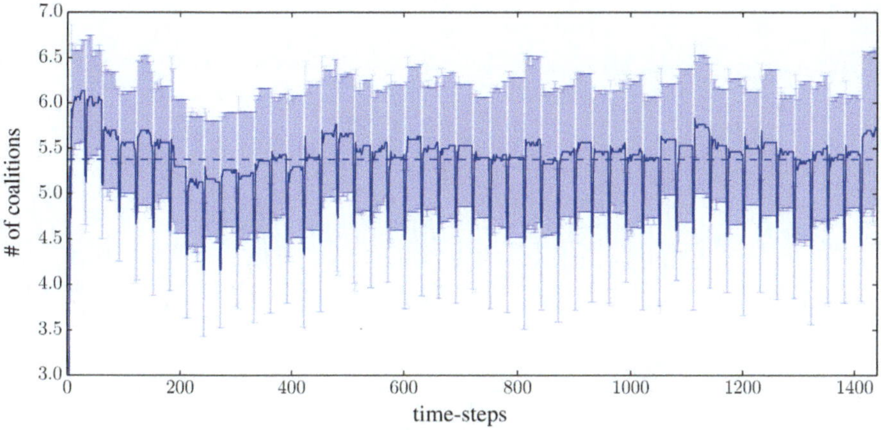

Fig. 10.21 Average number of coalitions along time for SACF$^+$. The shaded lines show the standard deviation, and the dashed line represents the average over all time steps

8 h supplying energy to the grid would be, on average, \$19.36. On the other hand, singletons have a zero payoff, receiving only \$13.2, which refers to the base price. Therefore, agents acting in coalitions received, on average, 46.6% more than singletons. This represents a key advantage of this approach, meaning that agents have a real incentive to form and remain in coalitions.

In terms of coalitions' stability, the following aspects are analysed: average number (Fig. 10.21), size (Fig. 10.22), and duration (Fig. 10.23) of coalitions, and also the average percentage of agents inside coalitions (Fig. 10.24). As expected, the coalition structures were reasonably stable in all cases. Indeed, there was an average of 5.4 coalitions with 8.9 agents each, corresponding to 94.9% of the agents within

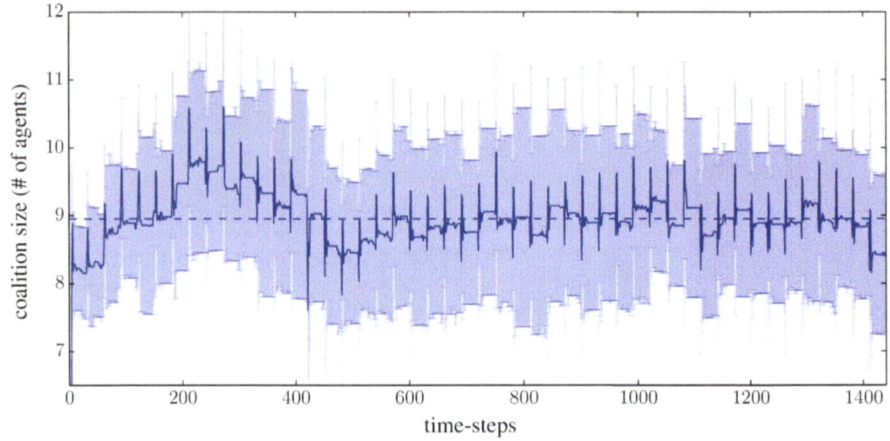

Fig. 10.22 Average size of coalitions along time for $SACF^+$. The shaded lines show the standard deviation, and the dashed line represents the average over all time steps

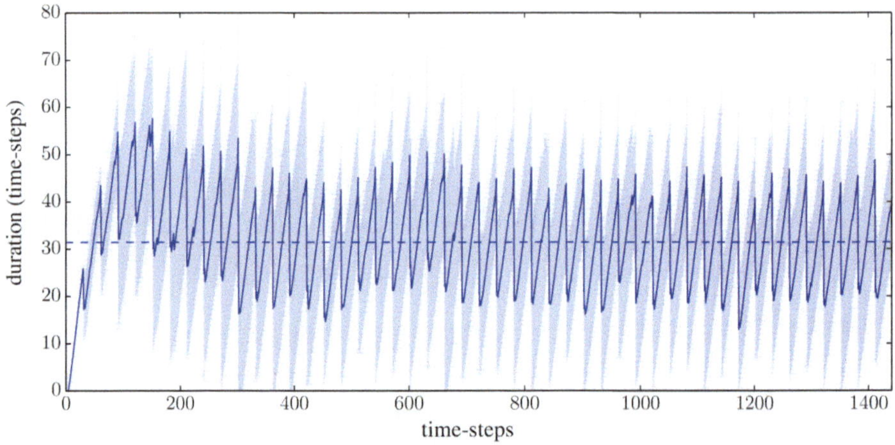

Fig. 10.23 Average duration of coalitions along time for $SACF^+$. The shaded lines show the standard deviation, and the dashed line represents the average over all time steps

coalitions throughout the simulation. However, the coalitions lasted only 31.4 time-steps, on average, which is around 10 times worse than for DCCF and SACF. The point here is that user-based constraints are much more dynamic than geographic-based ones because they are governed by the PEV-users' preferences. Consequently, coalitions here tend to last less than in scenarios with geographic-based constraints. Nonetheless, our results show that the delivered coalitions are profitable (otherwise, more agents would remain as singletons), stable (reasonably small variations from one plateau to another), and well distributed (agents spread in more than 5 coalitions, on average).

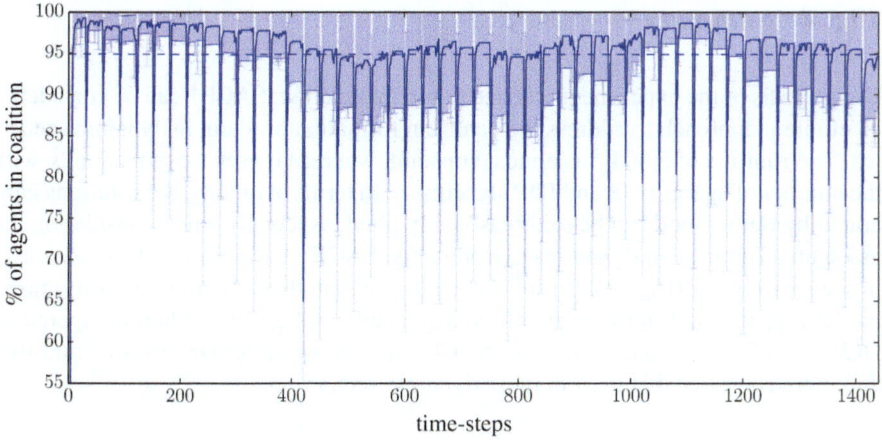

Fig. 10.24 Average number of agents inside coalitions along time for SACF$^+$. The shaded lines show the standard deviation, and the dashed line represents the average over all time steps

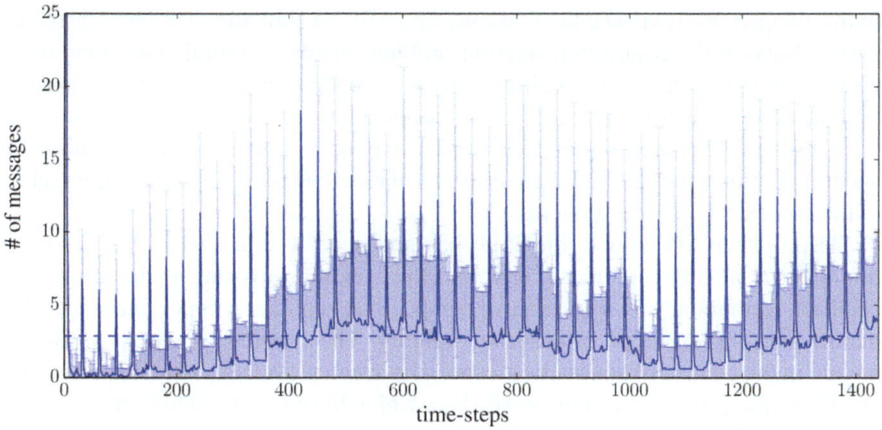

Fig. 10.25 Average number of messages exchanged along time for SACF$^+$. The shaded lines show the standard deviation, and the dashed line represents the average over all time steps

The last aspect to consider is the number of messages exchanged throughout the simulation, as presented in Fig. 10.25. As expected, a peak occurs in the number messages whenever the negotiation process is taking place (i.e., when agents enter or leave the system). On average, 4160 messages were exchanged along each simulation. Recall that the number of messages exchanged by SACF was much lower (Sect. 10.4.4) than SACF$^+$. Obviously, this is due to the additional step of SACF$^+$, which allows the agents to check for better coalitions. However, the number of messages exchanged by SACF$^+$ is still lower than the 5966 by DCCF.

10.5.4 Summary

On the basis of the experiments carried out in this section, SACF$^+$ can be considered an efficient approach to address the coalition formation problem in dynamic scenarios. The runtime of SACF$^+$ is almost constant, whereas it grows exponentially with the number of agents in IP and IDP algorithms. In terms of quality, the solutions generated by our method can be said reasonably good, especially when considering the associated runtime. In dynamic environments, SACF$^+$ has performed satisfactorily in scenarios with 50 agents. Results also indicate that, thanks to the similarity-based modelling, the coalitions were sufficiently stable along time. When compared to DCCF and SACF (from Sect. 10.4), SACF$^+$ has shown faster (in terms of runtime) and more robust (stable even in more dynamic scenarios).

10.6 Discussion

In this chapter, we reviewed the literature on coalition formation for smart grids scenarios. Solving the coalition formation problem involves several tasks. One of the most challenging ones is the coalition structure generation (CSG), which consists in finding a coalition structure (partition of the agents) that maximises the systems' utility. However, the complexity of such a task is exponential in the input size (number of agents). To this regard, optimal coalition formation has been considered impractical in the context of complex systems.

Despite such limitations, literature has proposed several alternatives for coalition formation within complex scenarios. However, most existing methods approach coalition formation from a quite naïve perspective, neglecting important aspects like maximising total revenue or ensuring stability. These points, however, are fundamental in the context of smart grids, especially when we refer to virtual power plants (VPPs) of plug-in electric vehicles (PEVs), which have very limited energy capacity and thus small profits.

Against such a background, we discussed how domain information could be useful when handling coalition formation within complex systems. Specifically, our point is that, by accounting for the constraints existing among PEVs, one can ignore infeasible coalitions from the search process, thus enabling a much more efficient search. As a result, efficient methods for coalition formation become attainable.

In this chapter, we presented two classes of constraints: geographic-based and user-based. In the former, the geographic position of PEVs is considered to avoid overloading the energy distribution network. In the latter, the preferences of the PEV-users (owners) are taken into account to promote lasting coalitions.

We also proposed three methods for addressing coalition formation within such constrained scenarios: DCCF (Sect. 10.4.2), SACF (Sect. 10.4.3), and SACF$^+$ (Sect. 10.5.2). In DCCF, agents invite neighbours to join their coalitions. In SACF, on the other side, agents ask to join their neighbours' coalitions. SACF$^+$ is a natural evolution of SACF, where agents can change their coalitions, thus making the process much more dynamic. In all cases, agents negotiate the formation of coalitions among themselves. The process is fully distributed among the agents, each on behalf a single PEV. DCCF and SACF were employed in the scenario with geographic-based constraints, whereas SACF^{+} was applied in a scenario with user-based constraints. Nonetheless, we highlight that the three methods are not dependent on the domain. In fact, any scenario with some kind of constraint could be addressed using any of the presented methods.

Regarding the results, all three methods run in a few milliseconds regardless of the number of agents, achieving near-optimal solutions. In all tested cases, results were above 90% of optimum, on average. In comparison, despite delivering optimal solutions, traditional approaches took several hours and run for up to 20 agents, which represents a small, unrealistic scenario for smart grids. These results hold for closed world settings (i.e., the set of agents does not change throughout simulation).

The presented approaches were also evaluated in open world settings. Under such scenarios, in the geographic-based model, the average accumulated payoff for agents in coalitions was 15% higher than for singletons. Such an advantage over singletons was even higher for the user-based case, accounting for more than 45%. The user-based approach performed better because the coalitions are created aiming at lasting longer, which improves their profit. However, both cases provide a considerable gain for the agents in coalitions. Therefore, the proposed approaches have shown suitable for the dynamic smart grids scenarios under consideration. Furthermore, it becomes clear that providing approximate solutions for the coalition formation problem is attainable, even in complex systems.

10.7 Research Directions

In this section, we briefly discuss some interesting directions within this line of research presented along this chapter.

Quality guarantees: although the presented methods take into account the main aspects regarding the coalition formation formalism, no formal guarantees were provided on the solutions' quality. To this regard, investigations towards such formal guarantees represent an indispensable future direction. A good starting point here would be employing the general idea discussed in [30] to provide a weak form of guarantee on the coalitions' quality. The concept of synergy graphs introduced in [37] and further developed in [2] could also be useful here. Finally, considering the distributed and local nature of the optimisation process, local search algorithms could also be employed to address our problem.

Multiple constraints: in this chapter, we discussed two approaches for modelling domain constraints in the coalition formation process. Despite the strengths and weaknesses of each approach, both comply with their proposal. An interesting direction would be enhancing the modelling to accommodate not only one constraint, but a few. Such a modelling would be useful when antagonistic constraints are considered. For example, agents could be constrained (considering their neighbourhood relation) regarding their geographic location *and* their energy profiles. More constraints could be added here, like the power rating and the type of DER (see the discussion on heterogeneous coalitions, ahead). The advent of multi-objective optimisation could also be helpful for handling competing constraints (e.g., agents must have similar energy profiles and PEVs must complement the production of wind turbines).

Model uncertainties: as discussed in Sect. 10.1, VPPs of PEVs are useful as auxiliary, ready-to-use power sources. To this regard, such VPPs are compatible with the so-called spinning reserve and regulation markets [11]. However, the PEVs' revenue depends on their amount of energy and availability. The point is that the availability of such devices is conditioned on several factors, ranging from users' preferences to battery conditions. To this regard, modelling such uncertainties is a crucial task towards implementing coalition formation in real-world applications. Therefore, the idea here would be designing mechanisms to penalise PEVs that do not fulfil their commitment to the grid. Related ideas on this topic have been investigated in [5, 6]. Machine learning techniques could be used to allow the agents to predict the preferences of their users without compromising privacy. Finally, such predictions could be coupled with reputation mechanisms to appropriately rank the PEVs based on their predictions.

Heterogeneous coalitions: a further promising direction would be considering heterogeneous DERs in the model. By heterogeneous DERs, we mean wind turbines, solar panels, fuel cells, generators, among other small generation devices. The idea is that such classes of devices have different production patterns, which could complement each other. For instance, while wind turbines depend on weather conditions, generators do not; thus, generators could be used to account for the intermittent production of wind turbines. This idea has been investigated in [14, 36]. Another direction would be including consumers into the coalitions so as to form microgrids. The point here is to form self-sufficient coalitions, where the coalition's total internal demand for energy is met by its internal production. In fact, microgrids represent a priority area towards smart grids [9]. Works on this line include [4, 15, 29, 38].

Notes

This chapter provides an overview of the joint work by Ramos, Burguillo and Bazzan. The methods and results in Sect. 10.4 on DCCF and SACF within scenarios with geographic-based constraints are from [25, 26]. The methods and results in Sect. 10.5.2 on SACF$^+$ within scenarios with user-based constraints are from [27]. We refer the interested reader to this works for more details.

Acknowledgements Ramos and Bazzan are partially supported by CNPq and CAPES grants. This work was also partially supported by the European Regional Development Fund (ERDF) together with the Galician Regional Government under agreement for funding the Atlantic Research Center for Information and Communication Technologies (AtlantTIC). Icons from Openclipart (http://openclipart.org/) were used in Figs. 10.1, 10.3 and 10.15.

References

1. Bazzan, A.L.C., Ramos, G.de.O.: Forming coalitions of electric vehicles in constrained scenarios. In: Bajo, J., Hallenborg, K., Pawlewski, P., Botti, V., Sánchez-Pi, N., Duque Méndez, N.D., Lopes, F., Julian, V. (eds.) Highlights of Practical Applications of Agents, Multi-Agent Systems, and Sustainability - The PAAMS Collection. Communications in Computer and Information Science, vol. 524, pp. 237–248. Springer International Publishing, Berlin (2015)
2. Bistaffa, F., Farinelli, A., Cerquides, J., Rodríguez-Aguilar, J., Ramchurn, S.D.: Anytime coalition structure generation on synergy graphs. In: Proceedings of the 2014 International Conference on Autonomous Agents and Multi-agent Systems, AAMAS '14, Richland, SC, pp. 13–20. International Foundation for Autonomous Agents and Multiagent Systems (2014)
3. Bremer, J., Lehnhoff, S.: Decentralized coalition formation in agent-based smart grid applications. Highlights of Practical Applications of Scalable Multi-Agent Systems. The PAAMS Collection: International Workshops of PAAMS 2016, Sevilla, Spain, June 1–3, 2016. Proceedings, pp. 343–355. Springer International Publishing, Berlin (2016)
4. Chakraborty, S., Nakamura, S., Okabe, T.: Scalable and optimal coalition formation of microgrids in a distribution system. In: IEEE PES Innovative Smart Grid Technologies, Europe, October 2014, pp. 1–6 (2014)
5. Chalkiadakis, G., Boutilier, C.: Sequentially optimal repeated coalition formation under uncertainty. Autonomous Agents and Multi-Agent Systems **24**(3), 441–484 (2012)
6. Chalkiadakis, G., Robu, V., Kota, R., Rogers, A., Jennings, N.R.: Cooperatives of distributed energy resources for efficient virtual power plants. In: Proceedings of 10th International Conference on Autonomous Agents and Multiagent Systems, Taipei, Taiwan, pp. 787–794 (2011)
7. Clement-Nyns, K., Haesen, E., Driesen, J.: The impact of charging plug-in hybrid electric vehicles on a residential distribution grid. IEEE Trans. Power Syst. **25**(1), 371–380 (2010)
8. Farinelli, A., Bicego, M., Ramchurn, S., Zucchelli, M.: C-link: a hierarchical clustering approach to large-scale near-optimal coalition formation. In: Proceedings of the Twenty-Third International Joint Conference on Artificial Intelligence, IJCAI'13, pp. 106–112. AAAI Press (2013)
9. Greer, C., Wollman, D.A., Prochaska, D.E., Boynton, P.A., Mazer, J.A., Nguyen, C.T., FitzPatrick, G.J., Nelson, T.L., Koepke, G.H., Hefner Jr, A.R., Pillitteri, V.Y., Brewer, T.L., Golmie, N.T., Su, D.H., Eustis, A.C., Holmberg, D.G., Bushby, S.T.: NIST framework and roadmap for smart grid interoperability standards, release 3.0. Technical report, National Institute of Standards and Technology, Gaithersburg, MD (2014)
10. Kamboj, S., Kempton, W., Decker, K.S.: Deploying power grid-integrated electric vehicles as a multi-agent system. In: Proceedings of 10th International Conference on Autonomous Agents and Multiagent Systems, Taipei, Taiwan, pp. 13–20 (2011)
11. Kempton, W., Tomić, J.: Vehicle-to-grid power fundamentals: calculating capacity and net revenue. J. Power Sources **144**(1), 268–279 (2005)
12. Kempton, W., Tomić, J.: Vehicle-to-grid power implementation: from stabilizing the grid to supporting large-scale renewable energy. J. Power Sources **144**(1), 280–294 (2005)
13. Lasseter, R.H.: Microgrids. In: Power Engineering Society Winter Meeting, vol. 1, pp. 305–308. IEEE (2002)
14. Mihailescu, R.-C., Vasirani, M., Ossowski, S.: Dynamic coalition adaptation for efficient agent-based virtual power plants. In: Klügl, F., Ossowski, S. (eds.) Multiagent System Technologies. Lecture Notes in Computer Science, vol. 6973, pp. 101–112. Springer, Berlin (2011)
15. Mondal, A., Misra, S.: Dynamic coalition formation in a smart grid: a game theoretic approach. In: 2013 IEEE International Conference on Communications Workshops (ICC), June 2013, pp. 1067–1071 (2013)
16. Pudjianto, D., Ramsay, C., Strbac, G.: Virtual power plant and system integration of distributed energy resources. IET Renew. Power Gener. **1**(1), 10–16 (2007)
17. Rahwan, T., Jennings, N.R.: An improved dynamic programming algorithm for coalition structure generation. In: Proceedings of the Seventh International Conference on Autonomous

Agents and Multiagent Systems (AAMAS'08), Estoril, Portugal, May 2008, pp. 1417–1420 (2008)
18. Rahwan, T., Michalak, T., Elkind, E., Faliszewski, P., Sroka, J., Wooldridge, M., Jennings, N.: Constrained coalition formation. In: The Twenty Fifth Conference on Artificial Intelligence (AAAI), August 2011, pp. 719–725 (2011)
19. Rahwan, T., Michalak, T.P., Wooldridge, M., Jennings, N.R.: Coalition structure generation: a survey. Artif. Intell. **229**, 139–174 (2015)
20. Rahwan, T., Nguyen, T.-D., Michalak, T.P., Polukarov, M., Croitoru, M., Jennings, N.R.: Coalitional games via network flows. In: Rossi, F. (ed.) Proceedings of the Twenty-Third International Joint Conference on Artificial Intelligence, IJCAI'13, pp. 324–331. AAAI Press (2013)
21. Rahwan, T., Ramchurn, S.D., Dang, V.D., Jennings, N.R.: Near-optimal anytime coalition structure generation. In: Proceedings of the International Joint Conference on Artificial Intelligence (IJCAI 07), January 2007, pp. 2365–2371. http://ijcai.org/proceedings07.php (2007)
22. Ramchurn, S., Vytelingum, P., Rogers, A., Jennings, N.: Putting the "smarts" into the smart grid: a grand challenge for artificial intelligence. Commun. ACM **55**(4), 86–97 (2012)
23. Ramchurn, S.D., Polukarov, M., Farinelli, A., Jennings, N., Trong, C.: Coalition formation with spatial and temporal constraints. In: International Joint Conference on Autonomous Agents and Multi-Agent Systems (AAMAS 2010), pp. 1181–1188 (2010)
24. Ramos, G.de.O., Bazzan, A.L.C.: Reduction of coalition structure's search space based on domain information: an application in smart grids. In: 2012 Third Brazilian Workshop on Social Simulation (BWSS), Curitiba, Brasil, October 2012, pp. 112–119 (2012)
25. Ramos, G.de.O., Burguillo, J.C., Bazzan, A.L.C.: Self-adapting coalition formation among electric vehicles in smart grids. In: 2013 IEEE 7th International Conference on Self-Adaptive and Self-Organizing Systems (SASO), Philadelphia, USA, September 2013, pp. 11–20. IEEE (2013)
26. Ramos, G.de.O., Burguillo, J.C., Bazzan, A.L.C.: Dynamic constrained coalition formation among electric vehicles. J. Braz. Comput. Soc. **20**(8), 1–15 (2014)
27. Ramos, G.de.O., Burguillo, J.C., Bazzan, A.L.C.: A self-adapting similarity-based coalition formation approach for plug-in electric vehicles in smart grids. Multiagent Grid Syst. **11**(3), 167–187 (2015)
28. Rigas, E., Ramchurn, S., Bassiliades, N.: Managing electric vehicles in the smart grid using artificial intelligence: a survey. IEEE Trans. Intell. Transp. Syst. **16**(4), 1619–1635 (2015). Aug
29. Saad, W., Han, Z., Poor, H.V.: Coalitional game theory for cooperative micro-grid distribution networks. In: 2011 IEEE International Conference on Communications Workshops (ICC), June 2011, pp. 1–5 (2011)
30. Sandholm, T., Larson, K., Andersson, M., Shehory, O., Tohmé, F.: Coalition structure generation with worst case guarantees. Artif. Intell. **111**(1–2), 209–238 (1999)
31. Shehory, O., Kraus, S.: Methods for task allocation via agent coalition formation. Artif. Intell. **101**(1–2), 165–200 (1998)
32. U. S. Department of Energy. Grid 2030: A national vision for electricity's second 100 years (2003)
33. Ueda, S., Kitaki, M., Iwasaki, A., Yokoo., M.: Concise characteristic function representations in coalitional games based on agent types. In: Walsh, T. (ed.) Proceedings of the Twenty-Second International Joint Conference on Artificial Intelligence IJCAI'11, vol. 1, pp. 393–399. AAAI Press (2011)
34. Vandael, S., Boucké, N., Holvoet, T., Deconinck, G.: Decentralized demand side management of plug-in hybrid vehicles in a smart grid. In: Rogers, A., McArthur, S., Guo, Y. (eds.) Proceedings of the First International Workshop on Agent Technology for Energy Systems (ATES 2010), Toronto, pp. 67–74 (2010)
35. Vandael, S., Boucké, N., Holvoet, T., De Craemer, K., Deconinck, G.: Decentralized coordination of plug-in hybrid vehicles for imbalance reduction in a smart grid. In: Proceedings of 10th International Conference on Autonomous Agents and Multiagent Systems – Innovative Applications Track (AAMAS 2011), May 2011, pp. 803–810. International Foundation for Autonomous Agents and Multiagent Systems (2011)

36. Vasirani, M., Kota, R., Cavalcante, R., Ossowski, S., Jennings, N.: An agent-based approach to virtual power plants of wind power generators and electric vehicles. IEEE Trans. Smart Grid **4**(3), 1314–1322 (2013)
37. Voice, T., Ramchurn, S.D., Jennings, N.R.: On coalition formation with sparse synergies. In: Conitzer, V., Winikoff, M., van der Hoek, W., Padgham, L. (eds.) Proceedings of the 11th International Conference on Autonomous Agents and Multiagent Systems AAMAS '12, Richland, SC, 2012, vol. 1, pp. 223–230. International Foundation for Autonomous Agents and Multiagent Systems (2012)
38. Yasir, M., Purvis, M., Purvis, M., Savarimuthu, B.T.R.: Dynamic coalition formation in energy micro-grids. PRIMA 2015: Principles and Practice of Multi-Agent Systems: 18th International Conference, Bertinoro, Italy, October 26–30, 2015, Proceedings, pp. 152–168. Springer International Publishing, Berlin (2015)
39. Yeh, D.Y.: A dynamic programming approach to the complete set partitioning problem. BIT Numer. Math. **26**(4), 467–474 (1986)

Part III
Evolutionary Games

Chapter 11
Ownership and Trade in Complex Networks

Juan C. Burguillo

Abstract In this chapter we introduce a networked version of the Possessor's and Trader's game, where, among the two basic strategies used in the Hawks-Dove game, hawks (H) and doves (D), it includes two other strategies based on the property of resources: Possession (P), as the right to occupy or possess what one owns; and Trade (T), as the right to buy and sell ownership. The simulations presented in this chapter describe how evolutionary forces, depending on the simulation parameters, allow the emergence of the different type of populations (D, C, P or T) over several complex topologies. The evolution of these populations clearly depends on several parameters as the cost of fighting, the trading values, the network topology, the owner's probability and, under certain conditions, the neighborhood size. We also study the effect of partner switching (a.k.a. rewiring) to discover that, in all the topologies and conditions analyzed, the results are worse than without rewiring, and the global payoff decreases due to the effect of hawks. We also consider the effect of allowing the agents to accumulate payoff during a certain number of rounds, and we have discovered that the global payoff improves significantly when agents accumulate resources during five rounds or more. Finally, we introduce the possibility to create an informal trading social network, where traders seek for other traders, and connect to them; avoiding the hawks in their own neighborhoods. This trading social network is much more successful for traders, and for the global payoff, than the previous all-to-all attempt.

11.1 Introduction

Classical Game Theory [2] provides useful mathematical tools to understand the possible strategies that rationally self-interested agents may follow when choosing a course of action. Evolutionary Game Theory (EGT) [10] is a branch of game theory that models the application of interaction dependent strategies in populations along

J. C. Burguillo (✉)
Department of Telematics Engineering, School of Telecommunications Engineering,
University of Vigo, 36310 Vigo, Spain
e-mail: J.C.Burguillo@uvigo.es

J. C. Burguillo, *Self-organizing Coalitions for Managing Complexity*, Emergence,
Complexity and Computation 29, https://doi.org/10.1007/978-3-319-69898-4_11

generations. EGT differs from classical game theory by focusing on the dynamics of strategy change more than in the properties of strategy equilibrium. In EGT participants do not posses unfailing Bayesian rationality. Instead, they play with limited computing and memory resources. All the requirement is that the players learn by trial and error, incorporate what they learn in future behavior, and die or somehow 'change' if they do not.

Maynard Smith [10], among others, considered how animal behaviors evolved by natural selection of the fittest, and constructed evolutionary game theories modeling those evolutionary processes. In such models, *"deference to possession a dispute resolution strategy based on pre-existing status is evolutionarily preferred over an always-fight strategy, which costs too much, and a never-fight strategy, which yields too little"*.

In The Selfish Gene [5], Dawkins proposes that social ideas, what he calls 'memes', are a non-organic replicator form. Memetics belong to evolutionary games because the evolutionary process is essentially a scenario of replication dynamics based on survival of the fittest [5, 10]. Examples of memes include tunes, catch-phrases, taboos, and fashions among others. In Dawkins' view, the fundamental characteristics of life are replication and evolution. In biological life genes serve as the fundamental replicators in a physical environment; while in human culture memes are the equivalent elements in a brain environment. Both genes and memes evolve by mutation-coated replication and natural selection of the fittest. In the memetics model, less successful individuals and groups within a population imitate the behavior of the more successful ones in order to improve their competence for resources. Accordingly, the 'better' an individual is, the more imitators it has. As a result, the population establishes and self-enforces over time standards of normal behavior. Normal behavior may either be time-independent or it may cycle through a range of behaviors.

As pointed out in on [15], Hirshleifer described the parallelism between evolved animal behaviors (and their evolutionary game theory models) and economically efficient human practices [6–8]. Hirshleifer proposed three metallic norms, in particular: the *Golden Rule of communal sharing*, the *Silver Rule of private rights*, and the *Iron Rule of dominance*. Each, he asserted, has evolved because they have sufficient socioeconomic advantages.

Deference to possession and ownership conventions in human societies range from the simplest intuitive norms, such as wait your turn in a queue, to much more sophisticated norms expressed by the Common Law. Legal property rights are among them [4]. As stated on [15], most ownership rights bundles consist of two primitive strands: (a) Possession, the right to occupy or possess what one owns, and (b) Trade, the right to buy and sell ownership.

This chapter presents a deep study of an evolutionary memetic game model of property ownership and trade, as a way to analyze the conditions for the peaceful resolution of property conflicts. The background of the work presented here takes its roots from [9, 12, 14] and specially from [15]. Here, the work initiated in [3] is reformulated to accomplish hawk-dove games, partner switching and payoff accumulation in different complex network scenarios. Along the chapter we see how the evolution of this populations depends on several parameters as the cost of fighting, the trading values, the network topology, the owner's probability and, under certain conditions, the neighborhood size. We also study the effect of partner switching (a.k.a. rewiring), and we allow agents to accumulate resources during a certain number of rounds to see if the global payoff of the whole population improves significantly. Finally, we introduce an informal trading social network, where traders seek for other traders, and connect to them; avoiding the hawks in their own neighborhoods.

The remainder of the chapter is structured as follows. Section 11.2 introduces the game model, the basic strategies to be performed by the cells, together with the conditions for imitating neighbors actions or to dynamically change the topology. Section 11.3 presents the main results obtained in the simulations for the Possessor's and the Trader's games, which are the two main frameworks explored in this chapter. Finally, Sect. 11.4 draws the main conclusions obtained by this work.

11.2 Game Model

The approach followed in this chapter is a composite game where actions are effectively simultaneous, but every player interacts one-to-one with several neighbors at a time. Considering the action selected by a player A, and the action chosen by its neighbor B, players A and B receive a certain pay-off. The population is structured considering certain network topologies, so the interaction among players are locally restricted to their neighbors, and obtained by iteratively playing one-to-one games. Therefore, each player A interacts only with the n cells of its neighborhood. In evolutionary game theory this is called a m-person game, where $m = n + 1$ in the given case, and the payoff for each player depends on its own actions and the actions taken by the other players in its neighborhood.

11.2.1 Game Basic Strategies

This section introduces the basic strategies available in the game, and the different actions that agents may play. First, the Hawk-Dove game is introduced, and then extended considering the framework from [15], with Possessor and Trader strategies.

The Hawk-Dove Game

We start considering the famous Hawk-Dove game [11], whose payoffs are depicted in left side of Fig. 11.1. In this game, two equally matched parties compete for a resource, valued V by each of them, and we consider only two basic strategies: **Hawk (H)** and **Dove (D)**. Since both parties are equal, in a fight between two hawks, each one has only a one-half chance of winning the asset, so in the immediate payoff we consider the average per each, dividing the whole payoff by two. We also include an expected total cost c for each participant, p.e., to model injuries or the expected

A \ B	Hawk	Dove
Hawk	$\left(\frac{V}{2}-c,\frac{V}{2}-c\right)$	$(V,0)$
Dove	$(0,V)$	$\left(\frac{V}{2},\frac{V}{2}\right)$

A \ B	D	C
D	(P, P)	(T, S)
C	(S, T)	(R, R)

Fig. 11.1 Hawk-Dove game matrix (left) and Prisoner's Dilemma (right)

energy expenditure.[1] Doves retreat when confronted by a Hawk. If two Doves meet, a random one of the two Doves retreats and leaves the other to the spoils; so in the immediate payoff matrix we again divide V by two, but without any extra cost.

In neoclassical game theory, when $(V/2 > c)$ we have the one-shot Prisoner's Dilemma (PD), while when $(V/2 < c)$ we have the Chicken Game. The hawk-dove game has two pure Nash equilibria corresponding to the strategies (Hawk, Dove) and (Dove, Hawk), and has been extensively used to model male contests in biology among other scenarios.

The right side of Fig. 11.1 presents the Prisoner's Dilemma matrix, where the strategies are named defection (D) and cooperation (C). In the payoff matrix, T means temptation, R means reward, P means punishment and S means sucker's payoff. In order to have a PD game we require that $T > R > P > S$. In addition, the iterative version (repeatedly playing) of the PD game (named IPD game) also requires that $2R > T + S$ to prevent alternating cooperation and defection giving a greater reward than mutual cooperation. In any one round (or "one-shot") PD game, choosing defection (D) is a Nash equilibrium, because it rewards the higher payoff for player A whether the opponent chooses defection (D) or cooperation (C). At the same time, the combined payoff for both players A and B is maximized if both cooperate, and here is the dilemma.

In our previous work in [3] we studied the iterated version of the PD, i.e., the famous Iterated Prisoner's Dilemma (IPD) over spatial networks. We used the same values for the IPD matrix as in the interesting work in [14], i.e., $T = 3.5$, $R = 3$, $P = 0.5$, $S = 0$. Note that with those values for T and R we do not fulfill the conditions for the left hawk-dove matrix, as $(T \neq 2.R)$, i.e., certain hawk-dove games are PD games, but not all, and viceversa.

Here we trace back to our main source in [15], to consider hawk-dove games. More information about these games can be found in Sect. 7.7 from Chap. 7.

Introducing Ownership

Finders keepers and *first come, first serve* are not only basic thumb rules in playground citizenship, they are powerful norms that have been recognized by the courts

[1] Usually the payoffs for the (Hawk, Hawk) strategy are denoted as $(V - C)/2$ each, but in this chapter we assume that $c = C/2$; keeping the notation from [15] where $c = h$.

and applied widely in several settings (see [15]). The Possessor strategy models the practice of ownership, and unlike cooperators or defectors, possessors observe convention based on their status; i.e., their behavior depends on whether they are the owner or the intruder of any particular resource.

To model this norm [15] introduces the **Possessor (P)** strategy:

$$P \equiv \begin{cases} D & if\ current\ owner \\ C & if\ current\ intruder \end{cases}$$

To model ownership in our networks, we consider that in any encounter between a player A and any of its neighbors B, there is a probability of ownership P_{owner}, that makes B to consider A as the owner of a certain resource they compete for.

Introducing Trading

A trader (T) is a possessor who is willing to sell or buy a resource when dealing with a fellow trader. In particular, when both owner and intruder of a particular encounter are traders, and the intruder values the resource by V, which is more than the owner's value v, then the intruder purchases the property at a value of x, where $v < x < V$. Yee [15] models this norm introducing the **Trader (T)** strategy:

$$T \equiv \begin{cases} if\ neighbor\ is\ not\ T & behave\ as\ P \\ if\ neighbor\ is\ T & \begin{cases} sell\ for\ x & if\ owner\ and\ v < x < V \\ buy\ for\ x & if\ intruder\ and\ v < x < V \\ behave\ as\ P & otherwise \end{cases} \end{cases}$$

In this case, the payoff for each player is related with the trading process. The payment for the owner will be $(x - v)$, and for the intruder $(V - x)$. As the owner may analyze if the payoff suits its needs, we introduce an additional requisite, with respect to [15]. This additional condition states that the value earned by the owner must be greater than the value it can get behaving as a hawk, i.e., $(x - v) > (V/2 - h)$; otherwise the owner behaves as a possessor.

11.2.2 Network Topologies

If we let every node in the system to interact with the remaining (N-1) nodes, we have a panmictic population, which is the one considered and theoretically analysed in [15]. But, in many real contexts like geography, biology, or telecommunication networks; each node interacts mainly with its neighbors. Therefore, we place agents in a complex network since they provide a realistic model of the topological features found in many nature, social and technological networks (see Chap. 3). In the simulations, we mainly focus on four types of well-known topologies such as spatial (as we did in our previous work in [3]), random, small-world and scale-free networks; since they model the most common networks appearing in Nature and in our human societies. We redirect the interested reader to Sect. 3.5 from Chap. 3 to review these networks and their notations.

11.2.3 Rewiring (Partner Switching)

In most real-world network interactions, relationships are not static, i.e., agents can change the individuals that they are linked to. We denote this option for partner switching as rewiring. Therefore, by means of rewiring agents can modify their neighborhoods in order to contact with new and more promising neighbors.

In our model, we consider that players want to avoid hawks, so if rewiring is allowed; a player may delete any player that has played *hawk* in the previous round, and rewires randomly to another player. We point out that rewiring only happens if the player to be removed does not become isolated, i.e., we do not want disconnected nodes in the network; otherwise the rewiring process is suspended in such round for that agent. This means that the global number of links stay constant in every simulation run.

11.2.4 Memetics Scenario

The model assumes that the payoff obtained by the neighbors of a player is known, so, each player can imitate the action done in the previous round by the most successful peer in its neighborhood, i.e., the one with the highest payoff. This assumption, about public payoff information in a neighborhood, is reasonable since a particular agent knows its payoff, and the payoff obtained by all its neighbors in their one-to-one encounters; so asking the neighbors of a neighbor would allow to recover the whole information. Nevertheless, the idea is not to present here a trading real world model, but to evaluate the strategies in an evolutionary competition; and to see what is the fittest one that becomes the most popular in the population.

11.3 Results

This section presents the results from an extensive set of simulation experiments, done with the game model and the strategies introduced in this chapter; attending to the variation of several simulation parameters. All simulations have been done on a standard Intel dual-core computer with 8 GB of RAM, taking less than a minute per execution (from a hundred to less than a thousand rounds to achieve stability). Except otherwise stated, all the figures presented in this section correspond to an average result over thirty simulation runs, which are really stable (the standard deviation is less than 2% in all cases).

All simulations consider 1600 agents, disposed in a toroidal square lattice of 40×40 in the case of spatial networks. When starting a run, every cell randomly selects its strategy, that may change latter by means of imitation. Each run is composed of a set of iterations in which agents repeatedly play the Possessor's or the Trader's games. The number of iterations varies in each particular run, depending on the simulation convergence and stability. We consider convergence when there are no strategy changes for the whole set of agents during 25 consecutive iterations or when we reach 500 iterations of a particular run.[2] The values considered by defect in the games are: owner's value $v = 1$, intruder's value $V = 5$ and cost $c \in \{1, 2, 3, 4\}$. We also consider the effect of mutation to analyze the stability of the different strategies. For this purpose, in some simulations there is a certain mutation probability (typically with a probability $p_m = 0.01$ or $p_m = 0.005$) that any cell changes its strategy into any other one. Every new round the income of every cell in the lattice is reset to zero, i.e., there is no accumulation of payoffs among multiple rounds. At the end of the chapter we consider what happens if we allow agents to accumulate payoff for several rounds. Finally, unless otherwise stated, the parameters used for building the networks are a von Neumann neighborhood for the spatial network (i.e., four neighbors), $W_{1600}^{4;0.1}$ for the small-world network, $S_{1600}^{4;-2}$ for the scale-free network and $R_{1600}^{0.0025}$ for the random network. These values have been chosen to compare the different topologies in a fair condition trying to have an average of four neighbors per agent.

11.3.1 Possessor's Game

In the Possessor's game, agents have three possible strategies to choose, to be a hawk (H), to be a dove (D) or to be a possessor (P). The last possibility is only available when the agent that is behaving as potential owner has a resource that is desired by a neighbor (modeled by the ownership probability P_{owner}).

Figure 11.2 presents a representative result of the Possessor game for a neighborhood with radio one (i.e., four neighbors). In both figures the horizontal axis presents the probability of being an owner in a confrontation. The Fig. 11.2a graph corresponds to a static network, i.e., agents can change their strategies by imitation but not their neighborhood. The Fig. 11.2b graph corresponds to a dynamic spatial topology, i.e., it starts from a spatial graph, but agents can change their strategies, and also can avoid hawks joining other agents randomly.

Analyzing both graphs in Fig. 11.2 we realize that the dove strategy is not relevant in those two scenarios, as it is closer to 0% for every owner probability. In the static network we also see that as soon as P_{owner} gets closer to 0.4 the percentage of possessors (P) gets higher than the one of hawks (H), and then possessor is the preferred strategy by almost all the agents in the game. In the case that we allow the

[2] Most simulations show relative stability with less than 100 iterations, and this limit was extended five times to provide a higher reliability.

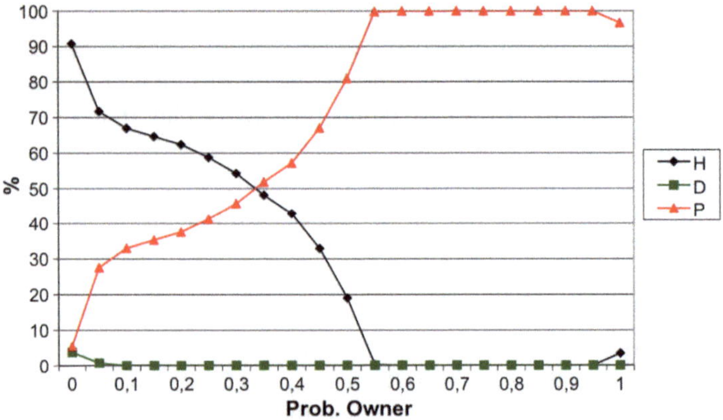

(a) Static spatial topology

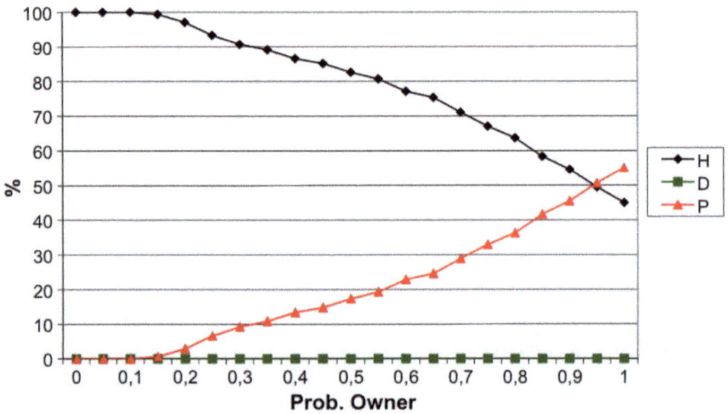

(b) Dynamic spatial topology (rewiring)

Fig. 11.2 Average percentage of strategies obtained changing P_{owner} using 30 spatial networks with radio $r = 1$ and cost $c = 2$

network to dynamically change (figure with rewiring), we see that possessors are less popular than in the static topology, and only when $P_{owner} \geq 0.95$ they become more popular than hawks.

On the one hand, there are not significative changes in the results when introducing mutation. Changing the neighborhood radio also has some limited effect, similar to the one described in [3]. This means that it enhances a bit cooperation with radios 2 (12 neighbors) and 3 (24 neighbors), but after that cooperation is reduced due to the effect of hawks over higher neighborhoods. We will see the effect of the neighborhood radio over spatial networks with more detail in the next section, as the effect over the Trader game is more significative.

On the other hand, the effect of the cost c parameter is relevant. Considering again Fig. 11.1, where we realize that changing c (i.e., the cost of fighting) does not change most of the payoff matrix values, which are $HD = (5, 0)$, $DH = (0, 5)$ and

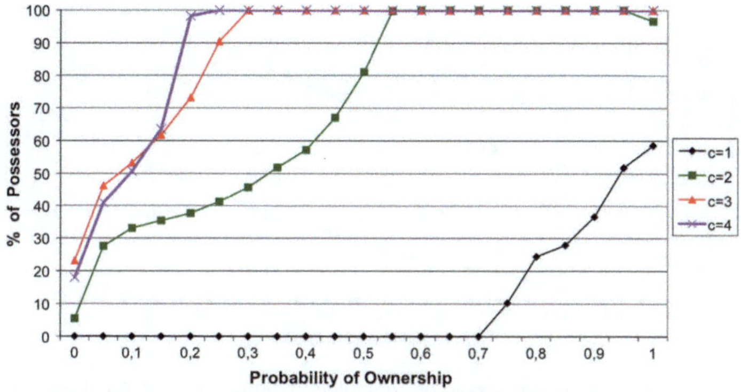

Fig. 11.3 Average percentage of possessors obtained changing P_{owner} for different cost values $c \in \{1, 2, 3, 4\}$ with a spatial topology and radio $r = 1$

$DD = (2.5, 2.5)$. The unique difference appears concerning the pairings between hawks, as $HH = (V/2 - c, V/2 - c)$; which using values of c from one to four means having from $HH_{c=1} = (1.5, 1.5)$ to $HH_{c=4} = (-2.5, -2.5)$. This is shown in Fig. 11.3, where we can observe that in all those simulations, the percentage of doves is nearly zero; thus the figure shows implicitly the percentage of hawks (being the possessors complementary). In the figure we also see how increasing the cost of fighting (c), increases effectively the proportion of possessors for a certain owner probability. Then, when using $c = 1$ and $P_{owner} = 1$ drives a 60% of popularity for the possessor strategy, when using $c = 2$ and $P_{owner} \geq 0.55$ produces that the possessor strategy is the only one used. Increasing the cost to 3 and 4 reduces the ownership probability, needed to obtain a full number of possessors, to $P_{owner} \geq 0.3$ and $P_{owner} \geq 0.2$, respectively. We see that the most interesting scenarios[3] obtained happen with $HH_{c=1} = (0.5, 0.5)$ to $HH_{c=2} = (-0.5, -0.5)$, that corresponds to a Prisoner's Dilemma game and a Chicken game, respectively.

But, what happens when we introduce rewiring into such cost analysis? Will the results of Fig. 11.2b appear again? Figure 11.4 presents the results of such study, and as we can see rewiring is not beneficial for the Possessor game. Therefore, there is a need to keep a stable topological structure in each neighborhood, that is broken when agents join new neighbors, leaving former hawks. The reason is that hawks also can escape from other hawks and join new neighborhoods, to get benefit from them, destroying possessor or cooperating structures. We may think that we could avoid such effect if, besides leaving hawks, we could introduce another rule like "former hawks are not welcome as new neighbors", i.e., allowing agents to accept or not a new neighborhood petition. Curiously, the result is almost the same, i.e., rewiring still produces poorer cooperating results in such case. This confirms the need for

[3]Using $c = 0$ we only have hawks, and it is not represented in the figure. Besides, after $c \geq 4$ the possessors curve remains equal.

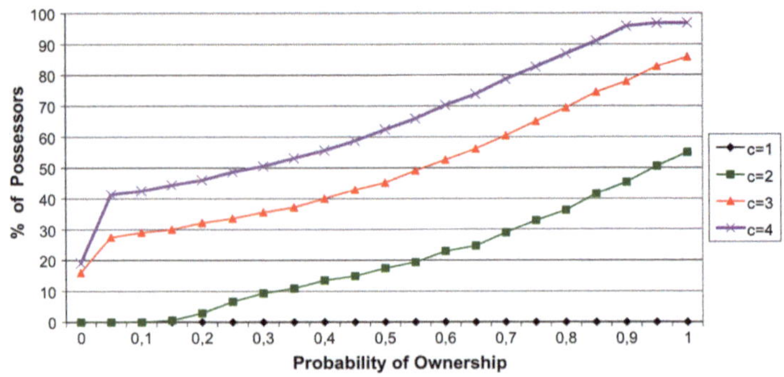

Fig. 11.4 Average percentage of possessors obtained changing P_{owner} for different cost values $c \in \{1, 2, 3, 4\}$ with a spatial topology and radio $r = 1$ and rewiring

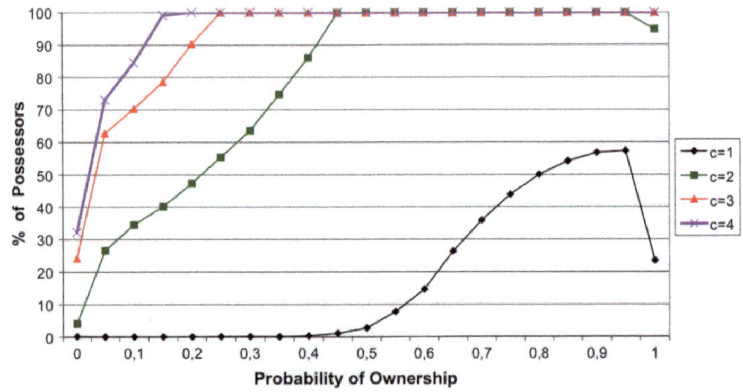

Fig. 11.5 Average percentage of possessors obtained changing P_{owner} for different cost values $c \in \{1, 2, 3, 4\}$ using a small-world topology

getting relative stable structures to avoid hawk behaviors and allowing cooperation or ownership to emerge.

Summarizing, for spatial networks, we have not seen a big influence of mutation, but a limited effect of changing the size of the neighborhood. However, as expected, we have seen that increasing the cost of fighting produces a significative increase in the popularity of the possessor strategy. We also have seen that allowing partner switching (a.k.a. rewiring) to escape from hawks has not a direct benefit for players in the long term.

Can these results be extended to other types of complex networks? Figure 11.5 shows again the average results obtained after 30 runs, generating a different small world network in each simulation. We see somehow a similar pattern to the spatial topology, without rewiring, but when using $c = 1$ or $c = 2$ the percentage of possessors increases, as usually, with P_{owner}; but curiously decreases when $P_{owner} \geq 0.9$. This effect disappears when increasing c. The results obtained using rewiring in the small-world case (see Fig. 11.6) are almost identical to the ones presented in the

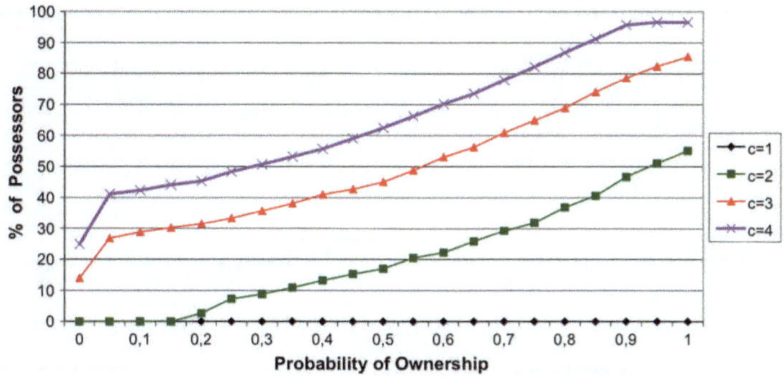

Fig. 11.6 Average percentage of possessors obtained changing P_{owner} for different cost values $c \in \{1, 2, 3, 4\}$ using a small-world topology and rewiring

spatial case (compare both figures). Again, mutation has not a significative impact in the results.

The results obtained in the small-world case are also almost identical to simulating the scenario using a random network or a scale-free network. These include the peculiar effect obtained with $P_{owner} \geq 0.9$ and $c \in \{1, 2\}$, together with the null influence of mutation or the very negative rewiring influence.

11.3.2 Trader's Game

In this section we explore the results obtained when playing the Trader's game with the four possible strategies an agent may chose to play, i.e., to be a hawk (H), to be a dove (D), to be a possessor (P) or to be a trader (T). The last possibility is only available when the agent that is behaving as a potential owner has a resource that is desired by a neighbor (modeled by the ownership probability P_{owner}), when the neighbor agent is also a trader and, besides, when the owner (seller) values the resource less than the intruder (buyer).

Figure 11.7 shows the results of a spatial topology with radio one (four neighbors) for different values of the cost parameter $c \in \{1, 2, 3, 4\}$. Observe that when $c = 1$ (Prissoner's Dilemma game) hawks rule the game for all possible values of the ownership probability, and only when $P_{owner} \geq 0.9$ traders start to become more popular. When $c = 2$ (Chicken's game), the situation starts to change, and hawks are less popular than other strategies as soon as $P_{owner} \geq 0.4$. First possessors become more popular having a coexistence with traders, who become the one and only strategy when $P_{owner} \geq 0.6$. A very similar result, and curve shape happens when $c > 2$, but reducing the ownership probability needed for possessors to appear. Note that the value of P_{owner} needed for traders to appear almost does not change from $c > 1$, and increasing c only affects the initial dominance of the possessors curve. We also can see an alternation of possessors and traders, as happened in our previous

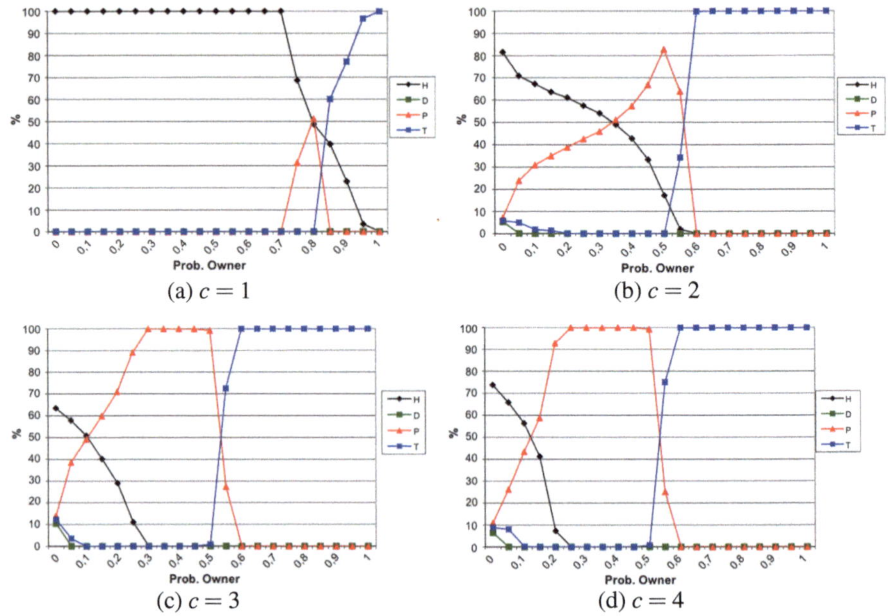

Fig. 11.7 Average percentage of strategies obtained changing P_{owner} using a spatial topology with radio $r = 1$ and cost $c \in \{1, 2, 3, 4\}$

work in [3], but the main difference[4] is that hawks are now more popular for the lowest values of P_{owner}, and that doves do not appear when P_{owner} gets closer to one.

An interesting outcome of the coexistence of all the strategies is the evaluation of the global payoff obtained by the network in each case. Figure 11.8 shows the results obtained, depending on the cost value c. Note that when $c = 4$ we get negative values for the global payoff. However, that is the curve that also produces better global payoff for the lowest values of P_{owner}. In fact, the best results happen when traders appear, and become the unique strategy played by the whole network. This can be seen in Fig. 11.9 that shows in the left side the strategy distribution obtained for two cost values, and on the right side the global payoff produced by the whole network. As we can see, the network payoff increases when possessors appear, and the higher values are obtained when we have traders (see Fig. 11.9d).

As happened with the Possessor's game, mutation here almost does not play any effect over the results; while rewiring has a very negative effect again. Figure 11.10 presents the effect of rewiring starting with the same spatial network topology shown in Fig. 11.7. As we concluded in the previous section, this means that it is needed to have a certain stability in the network topology to allow possession and trading to emerge as successful strategies to be played by the agents.

[4]Note that in [3] we used the typical Prisoner's Dilemma payoff matrix.

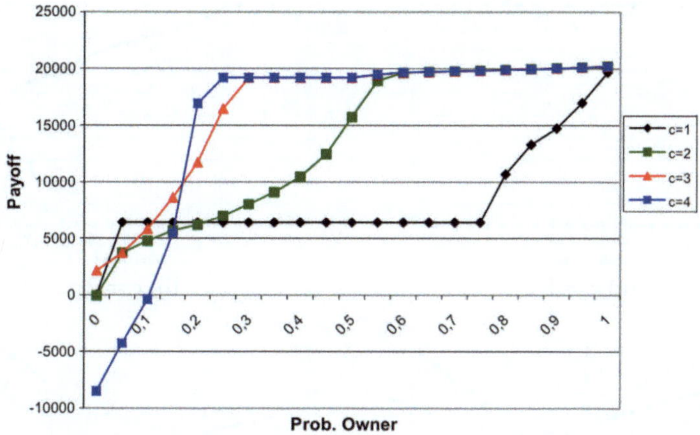

Fig. 11.8 Payoff obtained changing P_{owner} for different cost values $c \in \{1, 2, 3, 4\}$ using a spatial topology with radio one (four neighbors per agent)

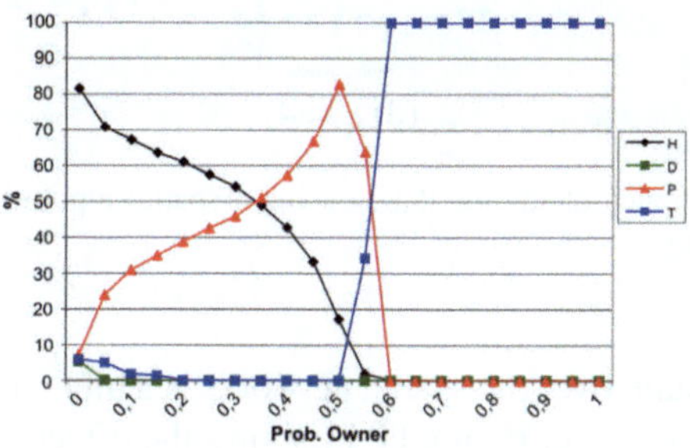

(a) Strategy distribution ($c = 2$)

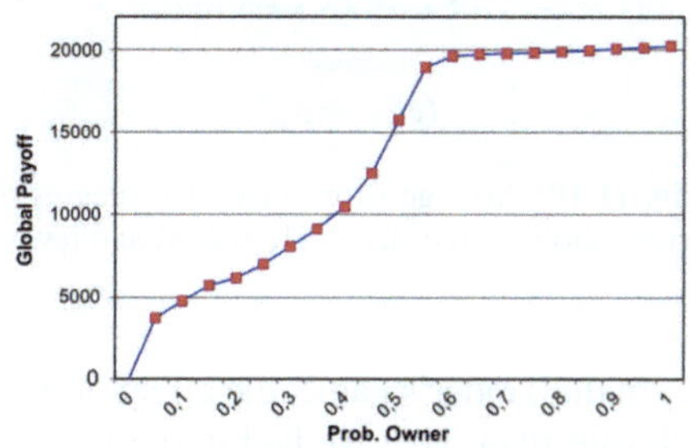

(b) Global Payoff ($c = 2$)

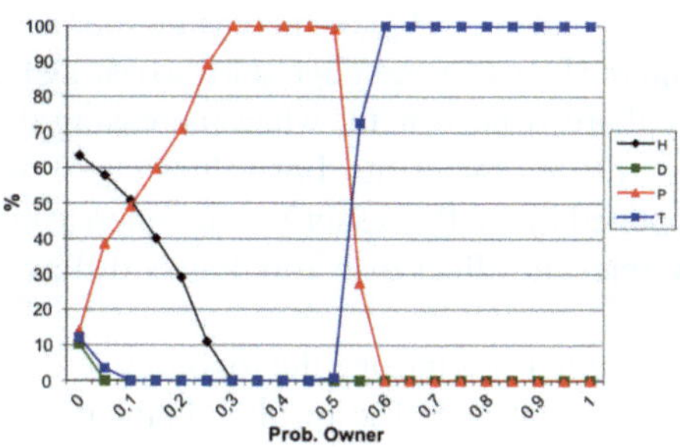

(c) Strategy distribution ($c = 3$)

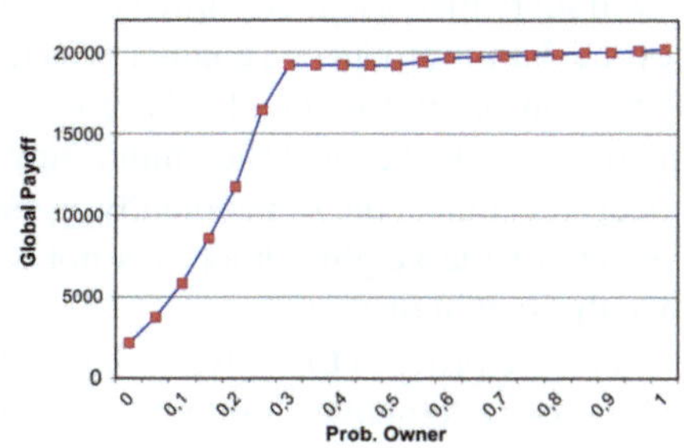

(d) Global Payoff ($c = 3$)

Fig. 11.9 On the left side, Average percentage of strategies obtained changing P_{owner} using a spatial topology with radio $r = 1$, cost $c \in \{2, 3\}$. On the right side, the global payoff obtained by the whole network

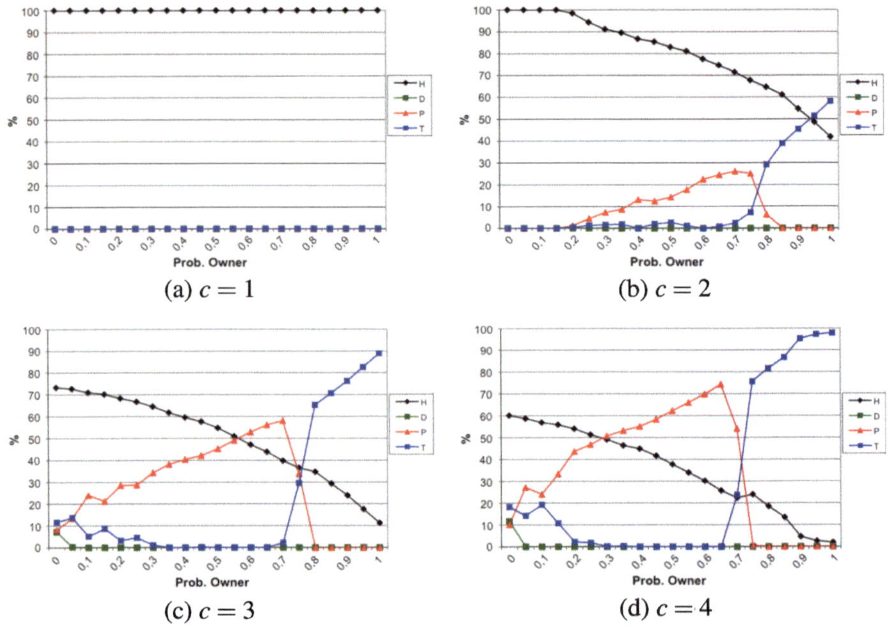

(a) $c = 1$ (b) $c = 2$

(c) $c = 3$ (d) $c = 4$

Fig. 11.10 Average percentage of strategies obtained changing P_{owner} using a spatial topology with radio $r = 1$, cost $c \in \{1, 2, 3, 4\}$ and rewiring

In the Trader's game, the effect of the radio over the spatial networks is a bit more relevant than what we had in the Possessor's one. Figure 11.11 shows the effect of increasing the size of the neighborhood in a spatial network with $c = 2$. We can see how traders increase their influence for lower values of P_{owner}. Increasing the radio beyond 3 units produces a counter effect, and the hawks percentage starts to rise again as we can see in Fig. 11.11d. When $c = 1$ there is no benefit, when increasing the radio, as hawks become the unique strategy played by agents. For values $c \geq 2$ the effects are almost the same. Finally, as happened in the Possessor's game, increasing the size of the neighborhood has not any relevant effect over topologies different than the spatial one.

Now we proceed to analyze what happens when we change the topology to consider other alternatives less regular than the spatial one. The results obtained with small-world topologies are very similar to the spatial case. When $c = 1$, traders become a majority for $P_{owner} = 1$. When $c \geq 2$ we also observe how the results mainly improve for possessors, but not for traders that almost do not change, and become the unique strategy as soon as $P_{owner} > 0.5$. Mutation has no effect, and the use of rewiring is also negative in this topology, as happened with the topologies considered in this study.

Scale-free and random topologies provide the worst results. Figure 11.12 shows the average results for the scale-free topology, and we can notice that although the shape of the figures is similar to the spatial and small-world cases, the results are worse

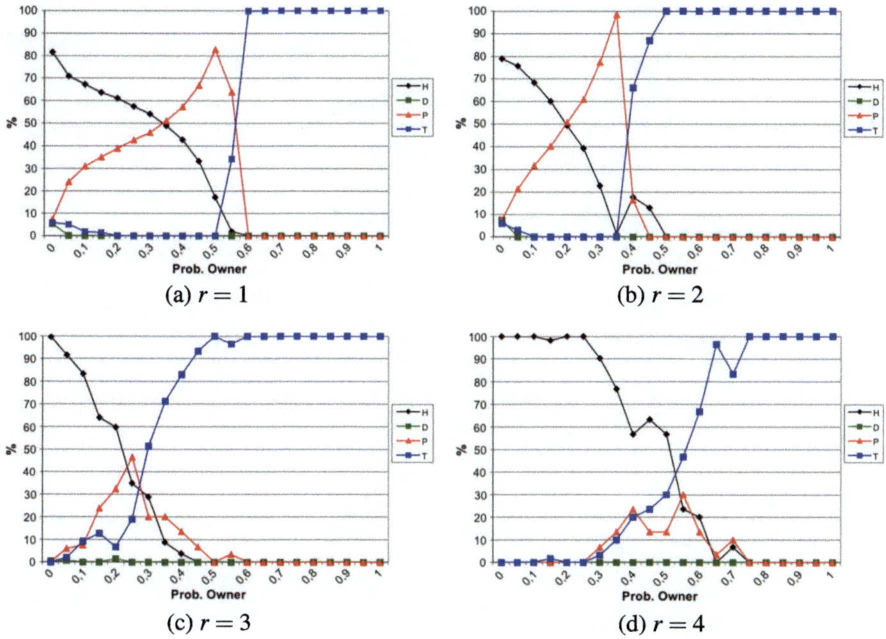

(a) $r = 1$ (b) $r = 2$

(c) $r = 3$ (d) $r = 4$

Fig. 11.11 Average percentage of strategies obtained changing the neighborhood size in a spatial topology with radio $r \in \{1, 2, 3, 4\}$ and cost $c = 2$

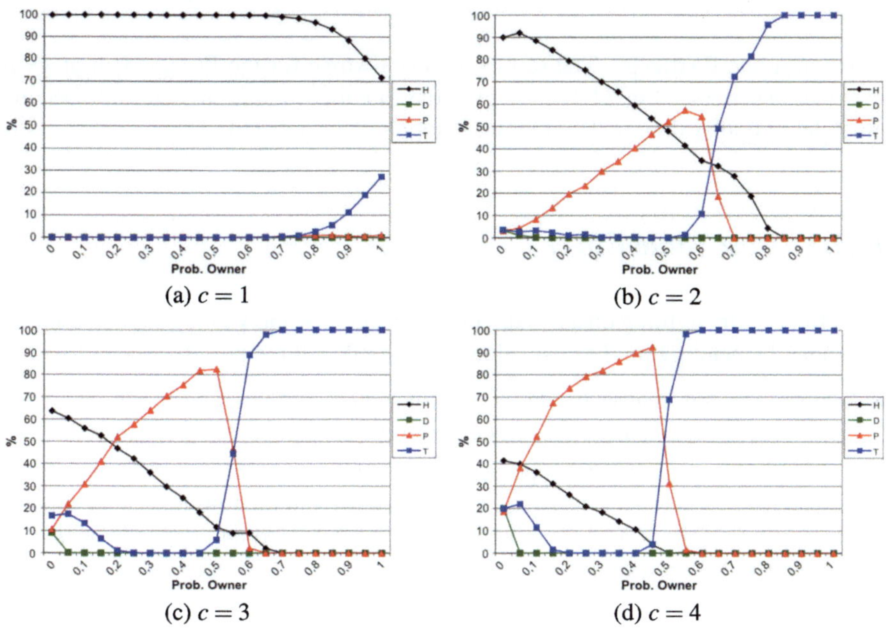

(a) $c = 1$ (b) $c = 2$

(c) $c = 3$ (d) $c = 4$

Fig. 11.12 Average percentage of strategies obtained changing P_{owner} using a scale-free topology and cost $c \in \{1, 2, 3, 4\}$

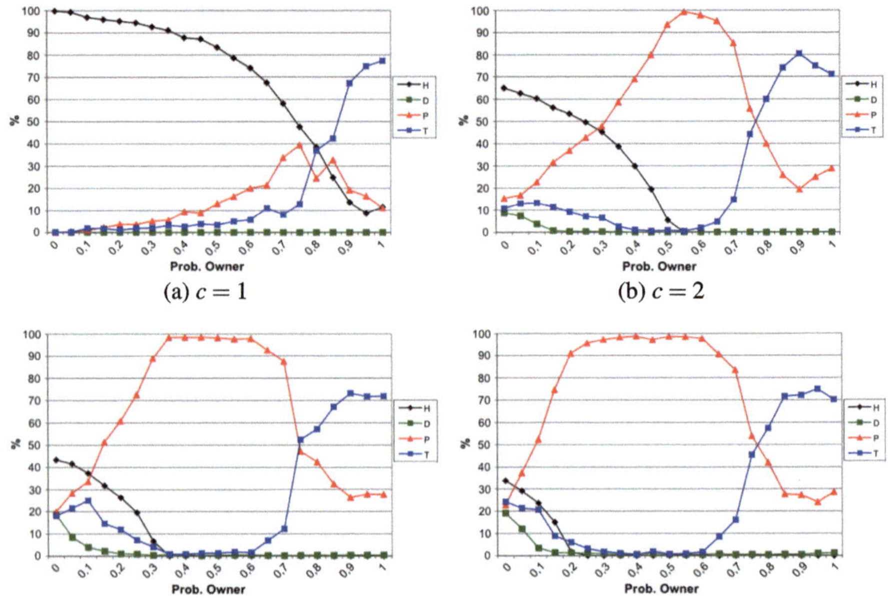

(a) $c = 1$ (b) $c = 2$ (c) $c = 3$ (d) $c = 4$

Fig. 11.13 Average percentage of strategies obtained changing P_{owner} using a random topology and cost $c \in \{1, 2, 3, 4\}$

concerning the P strategy. The results get even worse for the Possessor strategy in the random case (see Fig. 11.13), but even they are the worst for traders, who become much less popular than in the other three topologies analyzed.

11.3.3 Cost Value Effect

The effect of the fighting cost c is directly relevant to what we may expect to find in the game. Depending on the cost value [15] identifies three possible scenarios:

- *Region I*: in this region hawks are the expected strategy, and it happens when $c < \overline{V} - \frac{f}{1+f}\Delta$, where f is the fraction of time an agent behaves as an owner, and $\Delta = \frac{V-v}{2}$ is the incremental expected gain of trading (see [15], p. 18 for details). In our case $\Delta = 2$ and $f = 0.5P_{owner}$ as there are two interactions per round between every two neighbors, and the role of owner is alternated and depends on P_{owner}. Thus, for this region we have:

$$c_{I,II} < 1.25 - \frac{P_{owner}}{1 + 0.5P_{owner}} \tag{11.1}$$

- *Region II*: in this region, traders coexist with other strategies.
- *Region III*: in this last region traders are the expected strategy, and for the cost value we have:

$$c_{II,III} > \overline{V} = \frac{V+v}{4} = 1.25 \tag{11.2}$$

If we look into the simulations presented in this section, we can evaluate how these ideal regions really affect the behavior of agents when, instead of playing with the whole population, they play the game with a limited neighborhood. Along this section we have considered cost values $c \in \{1, 2, 3, 4\}$, so first we see whats the P_{owner} for setting up the frontier between regions I and II. If we set $c = 1$ in Eq. 11.1 we obtain $P_{owner} < 0.285$. Therefore, it means that in a pure competition all against all, with perfect information and rational decisions, hawks will be the ruling strategy when $P_{owner} \in [0, 0.285]$, and then hawks coexist with other strategies for higher P_{owner} values. Besides, traders should rule as soon as $c > 1.25$.

If we analyze Figs. 11.7, 11.12 and 11.13 we see that our results, with a limited interaction among the cells, are rather different to the previous scenario. On the one hand, as soon as we set $c \geq 2$ we have an appearance of traders for $P_{owner} > 0.5$. For $c = 1$ there is a coexistence of strategies, but hawks still dominate for most values of P_{owner} specially in the scale-free and random cases. Moreover, the situation can get even worse in scenarios where rewiring is allowed.

The analytical work described in [15] takes into account an ideal scenario where any agent may play with any other one of the whole population. In our case, when introducing complex networks together with imitation rules, we are restricting agents to play within their neighborhoods (i.e., doing local or mate interactions) that, depending on the network topology, may cause simulations to drift from ideal expected values in a panmictic population.

11.3.4 Accumulating Payoff

In the previous sections we have presented the results obtained when using a classical imitation process where agents evaluate the payoff obtained in the previous round by other neighbor agents and imitate the best agent's strategy. This is a usual approach followed by many evolutionary game approaches in the scientific literature, as actions can be changed in every round and, according to that, every agent evaluates if the neighbors are getting a certain profit or having loses depending on their last action. Nevertheless, we could consider the possibility that agents may accumulate payoff for a certain set of rounds, so imitation can be done considering such accumulated payoff, instead of the classical last round payoff.

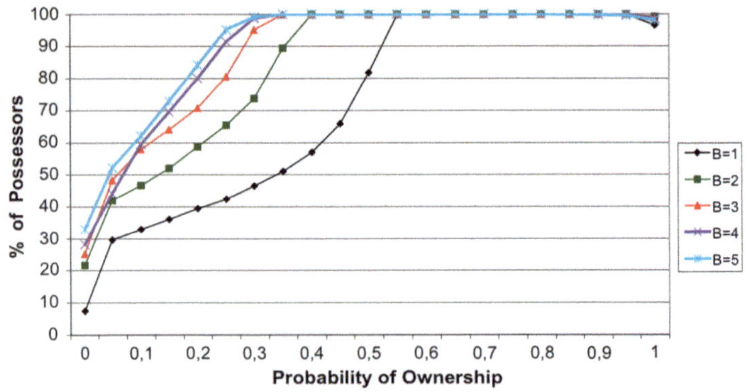

Fig. 11.14 Average percentage of possessors obtained changing the buffer size from one to five to accumulate payoffs, with cost value $c = 2$ using a spatial topology with radio $r = 1$

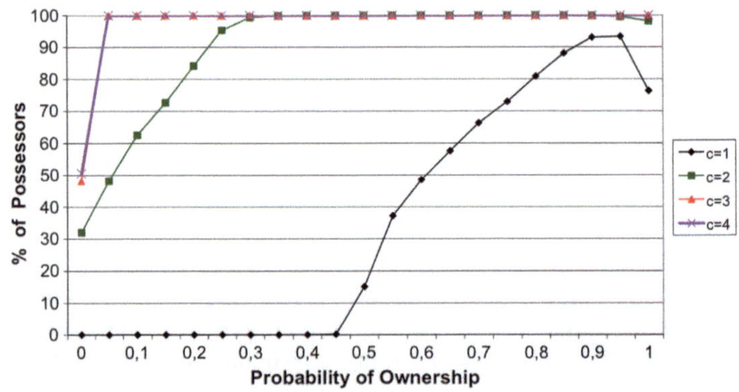

Fig. 11.15 Average percentage of possessors obtained changing P_{owner} for different cost values $c \in \{1, 2, 3, 4\}$ using a spatial topology with radio $r = 1$, accumulating payoff during 5 rounds

Figure 11.14 shows the results obtained by the possessor's strategy in the Possessor's game,[5] considering a cost value $c = 2$ units in a spatial topology of radio one (four neighbors), and accumulating payoff from one to five rounds. We model such accumulation by a FIFO (first in, first out) buffer. We can see how the possessors curve grows to become more popular even for low values of P_{owner}. Increasing the accumulation buffer to more than five rounds does not produce a visible effect, so five will be the value used in the rest of this section to compare with the results achieved using the immediate payoff.

Figure 11.15 presents the results obtained in the Possessor's game, when considering the spatial topology and a buffer of five rounds for payoff accumulation. The reader can compare this figure with the one in Fig. 11.3 to realize that now the results are much better for possessors and, in the end, for the global payoff. Concerning

[5]Remember that in such game, the dove strategy is not used in practice by the agents, so the percentage of hawks is the complementary.

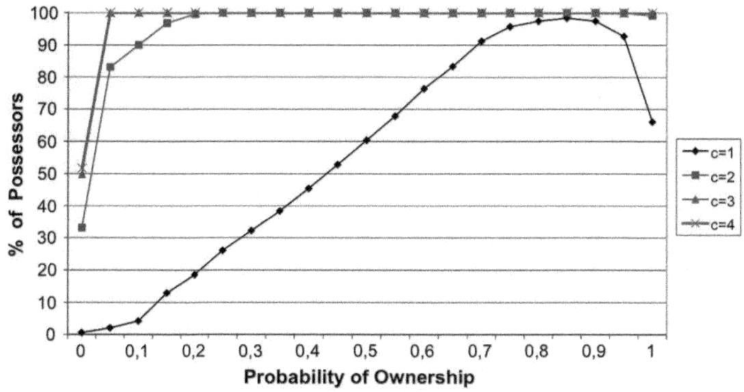

Fig. 11.16 Average percentage of possessors obtained changing P_{owner} for different cost values $c \in \{1, 2, 3, 4\}$ using a small-world network and accumulating payoff during 5 rounds

the small-world topology, Fig. 11.16 shows the equivalent results that suppose also a significative enhancement, compared with the results introduced before in Fig. 11.5 for the instant payoff.

Concerning the Trader's game, the effect of payoff accumulation is a bit more complex than what we have in the Possessor's one. Figure 11.17 shows the different distributions obtained for a spatial topology. If we compare this figure with Fig. 11.7 we can observe how traders dominance appears a bit later than before (now for $P_{owner} > 0.7$), but the dominance of possessors is absolute for values $0.1 < P_{owner} < 0.7$ when $c > 1$.

We also have seen, in the Trader's game results section, that there was a certain benefit increasing the neighborhood size (up to radio 3) in spatial networks, and we may wonder what results provide the combination of increasing the neighborhood size and also the buffer size. Doing that we have found that the neighborhood size effect disappears, and we get the same positive results increasing the payoff accumulation and using the basic neighborhood (four neighbors).

Figures 11.18, 11.19 and 11.20 present the results for small-world, scale-free and random topologies, respectively.[6] The same effect found in the spatial strategy happens in these other topologies, i.e., hawks nearly disappear when $c > 1$ in all cases, being dominated by possessors for values of $P_{owner} < 0.7$. This has a clear advantage as it helps to improve the global payoff per round.

Finally, accumulating payoff produces a significative effect over the global profit obtained by the whole network. Figure 11.21 presents a comparison of the global payoffs obtained by a spatial topology of radio one, with instant payoffs (left side) and accumulating payoff for five rounds (right side). We can see how the emergence of possessors for low values of P_{owner} produces a rising in the global profit obtained by the whole network.

---

[6]We only present values of $c \in \{1, 2\}$ as for higher values of c the strategy distributions almost do not change.

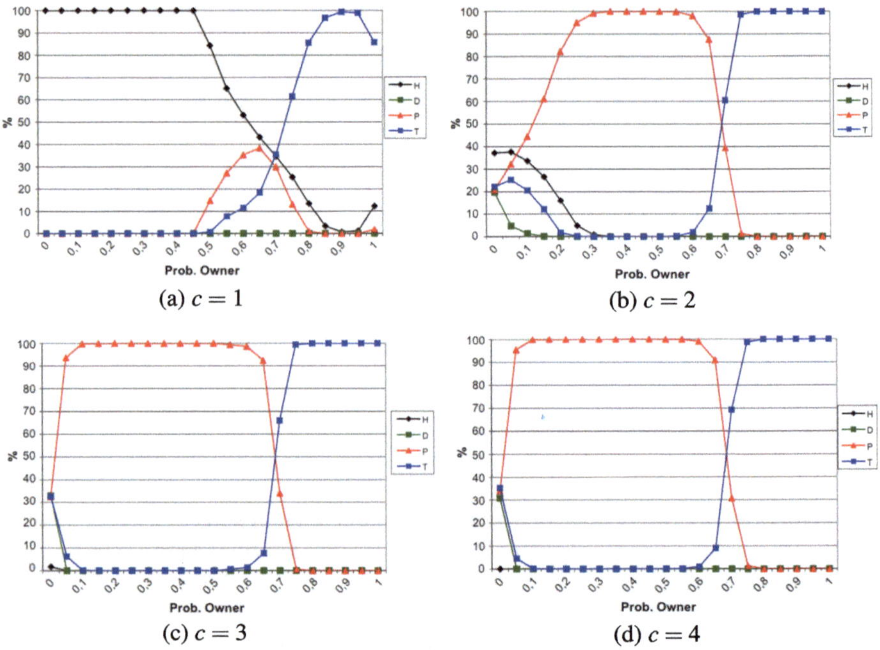

(a) $c = 1$ (b) $c = 2$

(c) $c = 3$ (d) $c = 4$

Fig. 11.17 Average percentage of strategies obtained changing P_{owner} using a spatial topology with radio 1, cost $c \in \{1, 2, 3, 4\}$ and accumulating payoff during 5 rounds

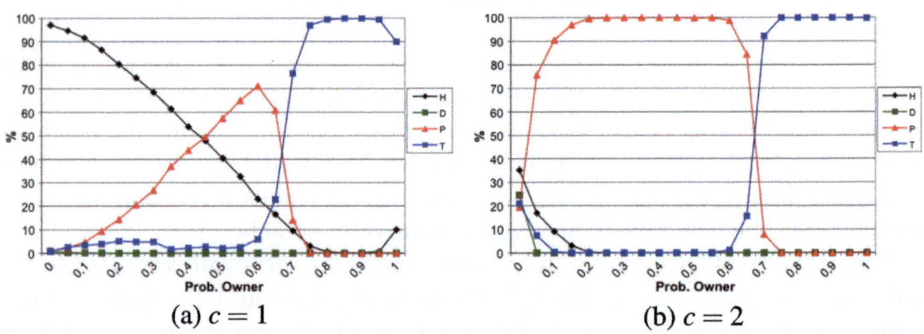

(a) $c = 1$ (b) $c = 2$

Fig. 11.18 Average percentage of strategies obtained changing P_{owner} using a small-world topology with cost $c \in \{1, 2\}$ and accumulating payoff during 5 rounds

11.3.5 A Traders' Coalition

Partner switching looked, a priori, as an interesting alternative to promote traders in order to enhance their opportunities and spread their strategy. However, as we have seen along this section, pure rewiring (just escaping from hawks) did not succeed; either if we allow or not hawks to rewire.

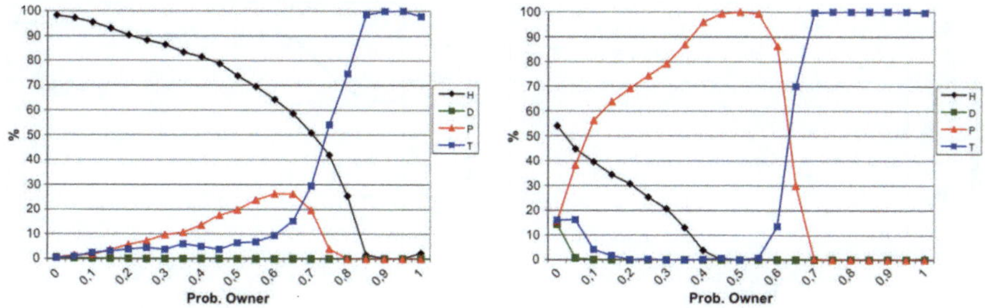

(a) $c = 1$ (b) $c = 2$

Fig. 11.19 Average percentage of strategies obtained changing P_{owner} using a scale-free topology with cost $c \in \{1, 2\}$ and accumulating payoff during 5 rounds

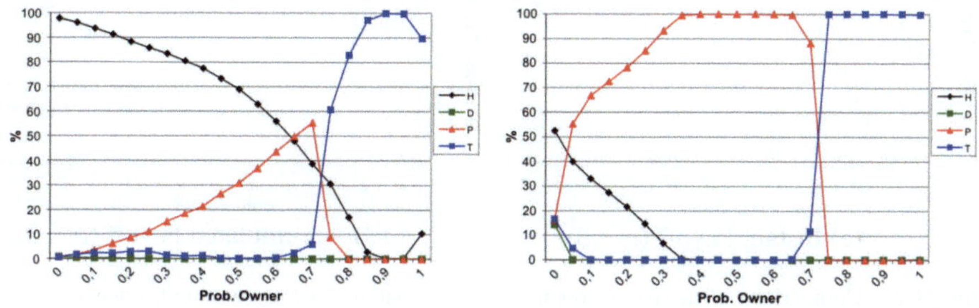

(a) $c = 1$ (b) $c = 2$

Fig. 11.20 Average percentage of strategies obtained changing P_{owner} using a random topology with cost $c \in \{1, 2\}$ and accumulating payoff during 5 rounds

In this last part of the trading game, we study how to create a coalition of traders to spread trading information, and enhance their trading possibilities. In order to do so, we will introduce the possibility of sharing public information among the neighbors. The idea is, from a trader's point of view, to find other traders to rewire and disconnect from hawks present in their neighborhood. To do that, a trader asks its neighbors what traders they know, with the aim of rewiring to them. Using this protocol, if a trader finds a new unknown trader, then it leaves any hawk from its neighbors who has two or more connections (remember that we do not desire disconnected agents). From the point of view of the trader that receives requests to join, it will only accept link requests from other traders; as they all have a common group interest.

Figure 11.23 shows the results for all the topologies, considering intermediate cost values.[7] If the reader compares the results with the previous figures for the topologies without rewiring, we can observe that they are now similar, as a difference to what happened when using rewiring from all to all. In fact, in the random topology, the results are much better than without rewiring (compare with Fig. 11.13). However,

[7] When $c = 0$ hawks are the most popular, even more than without rewiring. When $c = 4$ the results are a bit better for possessor but similar to $c = 3$.

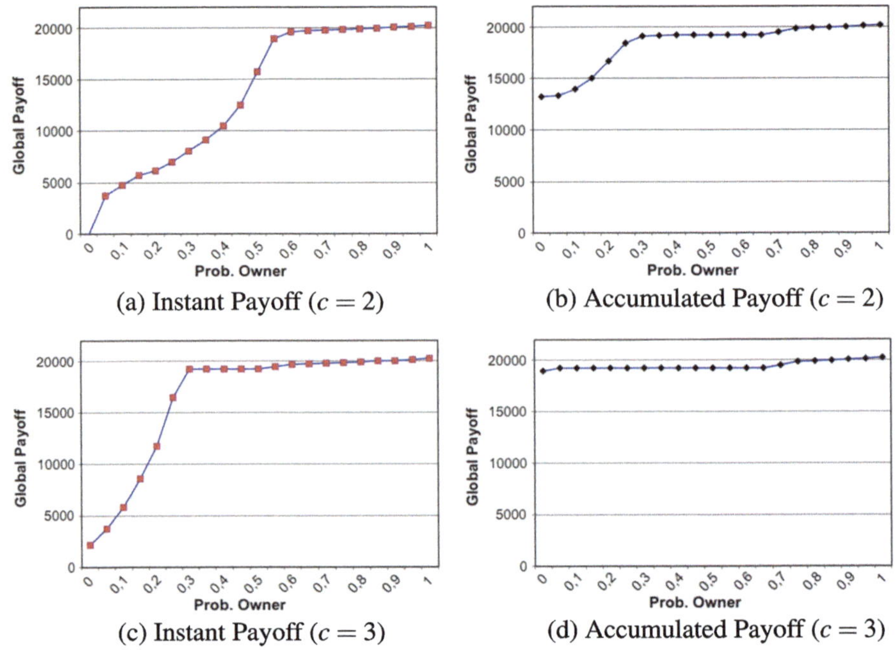

(a) Instant Payoff ($c = 2$)
(b) Accumulated Payoff ($c = 2$)
(c) Instant Payoff ($c = 3$)
(d) Accumulated Payoff ($c = 3$)

Fig. 11.21 Comparison of the payoff obtained with a spatial topology with radio one, cost $c \in \{2, 3\}$. Pictures on the left show the instant payoff, while the ones on the right show the accumulated payoff during five rounds

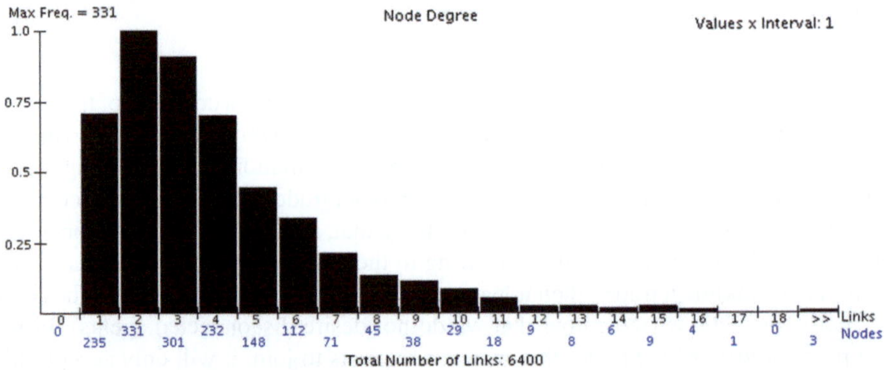

Fig. 11.22 Node degree starting from a spatial network. The horizontal axis shows the number of links per node in black, and the number of nodes having such value in blue. As we have 1600 nodes, in average, there are 4 links per node

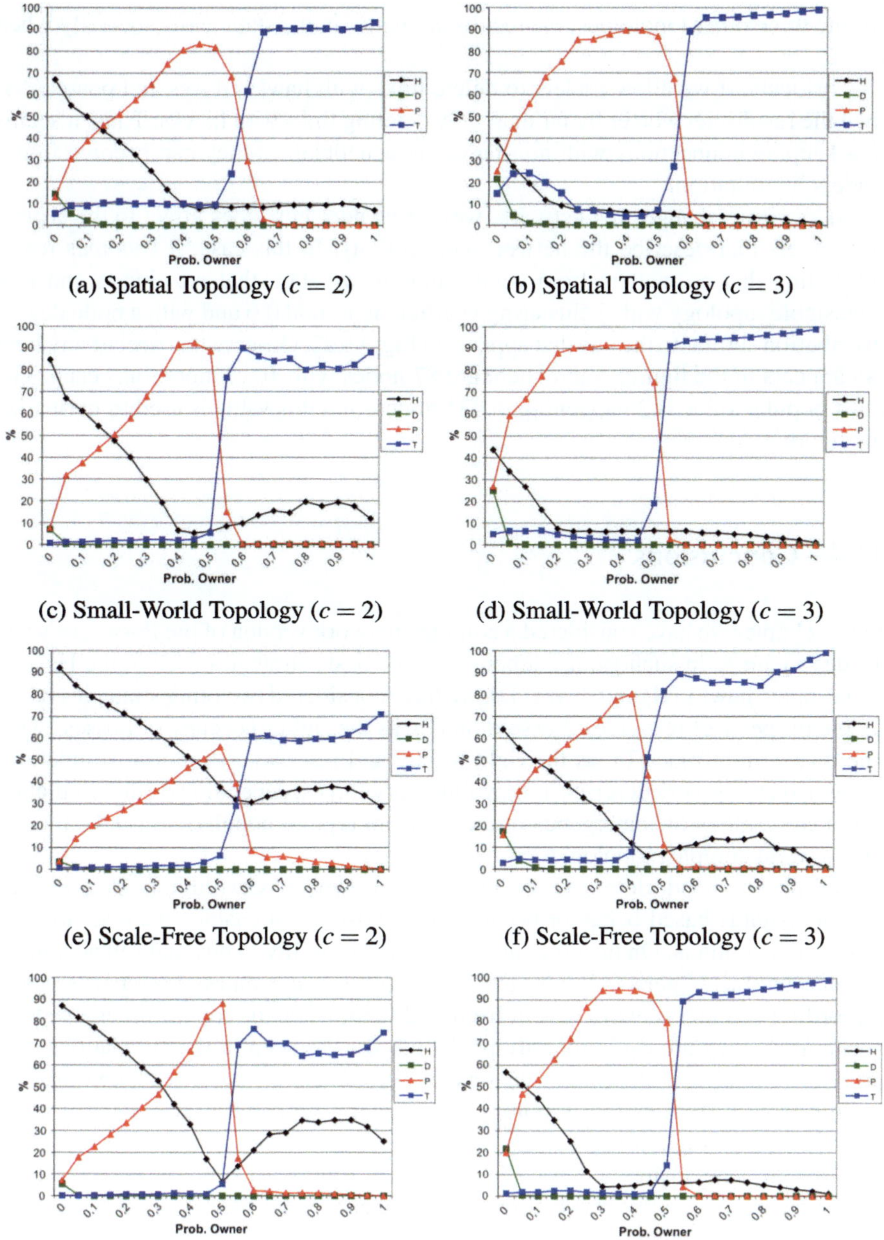

(a) Spatial Topology ($c = 2$) (b) Spatial Topology ($c = 3$)

(c) Small-World Topology ($c = 2$) (d) Small-World Topology ($c = 3$)

(e) Scale-Free Topology ($c = 2$) (f) Scale-Free Topology ($c = 3$)

(g) Random Topology ($c = 2$) (h) Random Topology ($c = 3$)

Fig. 11.23 Average percentage of strategies obtained changing P_{owner} over all the topologies, using instant payoff with cost $c \in \{1, 2, 3, 4\}$, and using a traders' coalition for rewiring for the three other topologies, the results are better in the static case, specially when $c = 0$.

Curiously, if we allow traders to delete links with hawks, doves and possessors, then the results get worser by far than only deleting links with hawks. Thus, it seems that keeping connections with non-hawks is beneficial, as they can become future traders by imitation.

As rewiring changes the network, we can wonder how does affect to the typical parameters that describe the network connectivity. In this case, it becomes really interesting that no matters from what topology we start, that we always end in a metastable topology with a clustering coefficient around 0.6 and with a node degree distribution similar to the one that appears in Fig. 11.22. Observe that a relatively long tail appears in the figure, where there are 87 nodes with 10 or more links each. The most popular value is 2 happening in 331 nodes, and this value is used to normalize the figure.

11.4 Conclusions

In this chapter we have considered a complex network version of the Possessor's and Trader's games. In such games, among the two basic strategies used in the Hawks-Dove game, hawks (H) and doves (D), we have considered two other strategies based on the property of resources: Possession (P), as the right to occupy or possess what one owns; and Trade (T), as the right to buy and sell ownership. The simulations presented in this work describe how evolutionary forces, depending on the simulation parameters, allow the emergence of the different type of populations (D, C, P or T) over several complex topologies.

On the one hand, deference by intruders to owners is evolutionarily preferred over non-status-based behavior because prior possession enables the resolution of possession conflicts. On the other hand, trade is the ability to buy and sell according to what optimizes personal gain; trading does not occur unless both parties gain. Accordingly, traders always benefit from trade and, when the conditions enable their appearance, they are evolutionarily preferred. But, as seen in the simulations, this preference is clearly dependent on several parameters as the cost of fighting, the trading values, the network topology, the owner's probability and, under certain conditions, the neighborhood size.

The main conclusion is that when the cost of fighting (c) and the probability of ownership (P_{owner}) are low, trading cannot appear, being hawks or possessors the most frequent populations, respectively. As soon as the P_{owner} raises, traders can appear, but if the cost of fighting is too low ($c <= 1$) then hawks can succeed and become a majority. When ($c > 1$) possessors or traders can appear, but traders are sensible to have a certain proportion of resources to trade (i.e., $P_{owner} > 0.5$).

We also have considered the effect of mutations, and we have seen that the results do not change significantly (i.e., the strategies are evolutionary stable for the network topologies analyzed). We also studied the effect of partner switching (a.k.a.

rewiring) to discover that, in all the topologies and conditions analyzed, the results are far worse and the global payoff decreases due to the effect of hawks when all-to-all rewiring is considered. We also studied the effect of allowing the agents to accumulate payoff during a certain number of rounds, and we have discovered that results become stable when agents accumulate payoff during five rounds or more. Besides, the decisions taken by the agents under this condition allow the whole network to increase the global payoff, even for low values of P_{owner}, due to the earlier emergence of possessors (P).

Finally, we have introduced the possibility of rewiring to create an informal trading social network. Traders in this network seek for other traders, and connect to them; avoiding the hawks in their own neighborhoods. This trading social network with rewiring has been much more successful for trading, than the previous all-to-all attempt. However, static topologies are more successful in general, except for the case of starting with a random topology where the trading social network since to be more effective.

To sum up, in order to promote the emergence of trading and to reduce conflicts, it is useful to increase the cost of fighting among agents, and to spread ownership among the population. Also, allowing agents to accumulate payoff during several rounds promotes trading, together with the possibility to allow traders to organize into a social trading network. Not surprisingly, many of this conditions appear in human markets.

Future work will consider agent learning their own experiences, i.e., to have adaptive agents that decide their future strategies not only imitating external behaviors; but deducing or learning what to do based on their past experiences. Another interesting approach is to consider the existence of finite resources to buy or sell, how they become distributed along the network and how this distribution depends on network topologies.

Notes

This chapter provides an extension of the work presented in [3], and here we have considered different payoff matrixes, dynamic partner switching and payoff accumulation in different complex network scenarios. In the end, we also introduce an informal trading social network.

Acknowledgements This work was partially supported by the European Regional Development Fund (ERDF) together with the Galician Regional Government under agreement for funding the Atlantic Research Center for Information and Communication Technologies (AtlantTIC).

References

1. Axelrod, R.: The Evolution of Cooperation. Basic Books, New York (1984)
2. Binmore, K.: Game Theory. Mc Graw Hill, Maidenheach (1994)

3. Burguillo, J.C., Peleteiro, A.: Ownership and Trade in Spatial Evolutionary Memetic Games. Lecture Notes in Computer Science **4490**, 455–464 (2010)
4. Calabresi, G., Melamed, A.: Property rules, liability rules, and inalienability: one view of the cathedral. Harv. L. Rev. **1089**, (1972)
5. Dawkins, R.: The Selfish Gene. Oxford University Press, Oxford (1976)
6. Hirshleifer, J.: Economics from a biological viewpoint. J. Law Econ. **20**, 1–52 (1977)
7. Hirshleifer, J.: Privacy: its origin, function and future. J. Leg. Stud. **649**, (1980)
8. Hirshleifer, J.: Evolutionary models in economics and law: co- operation versus conflict. Res. Law Econ. **4**, 1–60 (1982)
9. Langer, P., Nowak, M.A., Hauert, C.: Spatial invasion of cooperation. J. Theor. Biol. **250**, 634–641 (2008)
10. Maynard Smith, J.: Evolution and the Theory of Games. Cambridge University Press, Cambridge (1982)
11. Maynard Smith, J., Price, G.: The logic of animal conflicts. Nature **246**, 15–18 (1973)
12. Nowak, M.A., May, R.M.: Evolutionary games and spatial chaos. Nature **359**, 826–829 (1992)
13. Santos, F.C., Pacheco, J.M., Lenaerts, T.: Cooperation prevails when individuals adjust their social ties. PLoS Comput. Biol. **2**(10), e140 (2006)
14. Schweitzer, F., Behera, L., Mhlenbein, H.: Evolution of cooperation in a spatial prisoner's dilemma. Adv. Complex Syst. **5**(2–3), 269–299 (2002)
15. Yee, K.K.: Ownership and trade from evolutionary games. Int. Rev. Law Econ. **23**(2), 183–197 (2003)

Chapter 12
Promoting Indirect Reciprocity Using Coalitions

Juan C. Burguillo and Ana Peleteiro

Abstract In this chapter we explore an indirect reputation scenario using the donation game, where agents are connected in a social network; and any agent may interact with any other in the population. Agents' neighbors are their close related contacts from which they obtain information, as strategy, payoff and reputation. Besides the agent direct neighborhood, we also model a second set of contacts using coalitions (p.e., mates in a club or in a team), as a way to share information about the environment where agents play. We also include a rewiring mechanism using the neighbors' reputation as a criteria to improve agents' neighborhood, using coalition information as a way to rewire to the best coalition members. Along the chapter, we study in detail how the use of rewiring and mainly coalitions indeed improve cooperation when playing the donation game in a social framework. We also consider other strategy dynamics, using several alternatives for imitating or spreading information about the best decisions to take. Finally, we study the dependency of the model on the initial percentage of defectors, and the results show a strong dependency on the initial conditions, that have a significative influence in the game outcomes.

12.1 Introduction

Cooperation can be achieved in complex real-world encounters that are not limited to direct interactions only, in particular, cooperation can consider prior interactions with other players, i.e., indirect reciprocity. Martin Nowak, in his article from 2006 titled *Five Rules for the Evolution of Cooperation*, put some light to explain the emergence of altruism in human societies from a game theory point of view. Nowak describe five mechanisms by which natural selection in Evolution can lead to cooperation [7]:

J. C. Burguillo (✉)
Department of Telematics Engineering, School of Telecommunications Engineering,
University of Vigo, 36310 Vigo, Spain
e-mail: J.C.Burguillo@uvigo.es

A. Peleteiro
Zalando, Dublin 2, Ireland
e-mail: ana.peleteiro@zalando.ie

J. C. Burguillo, *Self-organizing Coalitions for Managing Complexity*, Emergence,
Complexity and Computation 29, https://doi.org/10.1007/978-3-319-69898-4_12

- *Kin selection*, when natural selection can favor cooperation if the donor and the recipient of an altruistic act are genetic relatives.
- *Direct reciprocity*, happening in games with repeated interactions, like the IPD, where it pays off to cooperate for receiving future cooperations.
- *Indirect reciprocity*, as theoretical and empirical studies show that people who are more helpful are more likely to receive help.
- *Network or social reciprocity*, which relies on geographical or social factors to increase the interactions with nearer neighbors.
- *Group selection*, which determines that groups with a higher percentage of cooperators are more successful as a whole, and grown faster that groups with a high percentage of defectors.

The donation game is an abstract framework used to show how the mechanism of indirect reciprocity operates using players' reputation to promote cooperation [8]. Unlike the case of direct reciprocity, whereby any altruistic act of helping to another player is returned; in indirect reciprocity the altruistic act of helping others is perceived by the community as *helpful*, providing good reputation, and receiving help in return by other players. A donor provides help if the recipient is likely to help others (which often means, if the recipient has helped others in the past). In this case, the cost of an altruistic act for a donor increases the chance to become the recipient of an altruistic act later [8]. Indirect reciprocity is also associated with interactions having short encounters (e.g., one-shot interactions) whereby the effects of direct reciprocity on the interaction outcome are minimized.

Indirect reciprocity has been used by players that compare reputation of potential recipients and cooperate only when the recipient has the same or higher reputation than the donor's strategy. It can be shown that a population of such players can evolve cooperative plays through discriminators that can distinguish players with high reputation (those that have cooperated with other players in past interactions) and cooperate only with such players [8]. Other studies have applied the mechanism of indirect reciprocity in complex interactions where cooperative plays are difficult to evolve. Chong et al. [3] have shown that the mechanism of indirect reciprocity through repeated interactions is less effective in promoting cooperation for interactions with higher number of alternative choices and shorter encounter (e.g., lower number of rounds in a repeated game). However, strategies can evolve to use reputation as a mechanism to estimate behaviors of future partners and to elicit cooperation right from the start of interactions. In such scenarios, cooperation occurs when strategies evolve to maintain high reputation scores.

The notion of coalitions has been studied by the game theory and multi-agent communities for decades. In fact, coalition formation [12, 13] is one of the fundamental approaches for establishing collaborations among self-interested agents. For instance, coalition-based mechanisms as [1, 2, 11] suggest that well-used coalitions indeed facilitate cooperation among self-interested agents.

At the same time, research on games over dynamic topologies has found empirical evidence showing that partner switching leads to cooperative behavior. Along this line, Fu et al. [4] propose a coevolutionary model of the prisoner's dilemma that allows agents to either adjust their strategies or switch their defective partners. Thus, they show that partner switching may help to stabilize cooperation. Also in [5], Fu et al. focus on the effect of reputation on an individual's partner switching problem in a network, showing that using their mechanism, cooperation can prevail. Although in a different framework (the investigation of tag-based coordination), Griffiths et al. [6] show that partner switching can help to increase coordination resilience in the face of malicious behavior.

In this chapter we present a detailed study, based in our model from [9], to analyze the emergence of cooperation among self-interested agents organized in coalitions and placed in a complex network, where agents play the donation game with any other members of the population. The framework described here combines three of the five mechanism described above: indirect reciprocity, social reciprocity and group selection. To achieve such combination our mechanism is based on three main pillars: an indirect reciprocity model, the formation of coalitions among the agents and the use of complex networks with partner switching (i.e., dynamic neighborhoods through rewiring). In [9], we show how cooperation can be improved when using both coalitions and rewiring. Here we study the model in much more detail, considering more scenarios, several strategy dynamics, different rewiring decisions and a more complete analysis over multiple topologies. Altogether, this chapter provides a comprehensive understanding on how cooperation can emerge in networked indirect reciprocity scenarios driven by coalitions.

The rest of the chapter is organized as follows. Section 12.2 introduces the donation game model, that we consider in our framework. Then, Sect. 12.3 introduces our model, based on the donation game, but using coalitions and partner switching (rewiring). Section 12.4 describes the simulation results obtained from our framework under multiple experimental settings. Finally, Sect. 12.5 summarizes the conclusions and open paths for future work.

12.2 Donation Game Rules

Our reputation game is based in the classic donation game published by Nowak and Sigmund [8] involving image scoring strategies, which are a measure of reputation. As described in that paper, the game is composed of several rounds where N agents play a donation game: in each round, a small set of m donor-recipient pairs are chosen. Depending on recipient image score and the donor *generosity* level, the donor provides a certain benefit to the recipient at a certain self-cost. Under such framework, the chance that both players meet again is negligibly small. Thus, direct reciprocity cannot work.

Formally, from each each pair of agents, one is selected as the donor, and the other one as the recipient. Every agent i has a strategy represented by the integer $k_i \in [-5, 6]$ and an image score (reputation) given $s_i \in [-5, 5]$ that depends on its behavior in the past interactions along such round. The donor i has to decide, depending on its strategy (k_i), and its the recipient's j score (s_j), if it donates (cooperates) to the recipient or not. If $k_i \leq s_j$, then agent i donates a benefit b to agent j at a cost c to itself, and increases its image score (s_i) by 1. Otherwise (i.e., $k_i > s_j$), no donation or cost are involved (both obtain zero payoff), but the image score of the donor (s_i) is decremented by 1. Note that the image score of the recipient does not change in any case.

Hence, strategies with $k \leq 0$ are termed cooperative, because individuals with these strategies cooperate with individuals that have not had a previous interaction. Then, we can observe two extreme game-playing strategies, i.e., the strategy with $k_i = -5$ represents always cooperation regardless other agent's score, while the strategy $k_i = 6$ represents always defection in all cases. Other strategies represent various degree of discriminating play, e.g., $k_i \in [-4, 0]$ are discriminators that lean towards cooperation [8].

In our case, after finishing a round, agents imitate the best strategies in their neighborhood, while in [8] agents reproduce themselves, to produce a new population, depending on their obtained payoff. Note that in both cases, depending on the value of m and the random selection, it may happen that there are differences in the amount of times that different agents have played the donation game in a round. However, what is relevant is the evolution of the whole game, and not what happens to a particular agent.

12.3 Model

In this section we introduce our model [9], where we consider a population of N agents where any agent can interact with any other agent (i.e., panmictic interaction) to play the donation game (see Sect. 12.2). However, agents are connected in a complex network, having each of them a set of peers that constitute their neighborhood. We want to model real world interactions over social networks, so agents' neighbors are their close related contacts from which agents obtain information. However, in real world, apart from having a set of direct contacts, people usually belong to several clubs, associations, organizations, or groups in general. We model this second set of contacts with the notion of coalitions, as a way that agents may share some information about the environment where they play. Thus if an agent agrees to become a member of a coalition, it also agrees to share information with the rest of the coalition members. This information sharing helps agents while interacting with the whole population in the panmictic game.

In Procedure 26 we present the basic game behavior, that will be explained detail in the following sections. As a short description, we can see how pairs of agents (line 3) play the donation game during a round (set of encounters), and that any agent has to decide:

- Its action (to donate or not) depending on its own strategy, and the other's image score (line 4). This influences its payoff and image score (line 5).
- To keep independent or join a coalition, and if joining, to which one (line 8).

- Deciding its new strategy for the next round (line 9).
- Changing their neighbors, depending on the image score of the neighborhood (line 10).
- Finally, the payoff and image score are reset for the next round (line 11).

Procedure 26: Game Behavior

1 **Function PlayRound** $(m \leq N)$
2 **for** 1 **to** m **do**
3     $(a_i, a_j) = FindPlayers$ $(i, j \in [1, N]$ and $i \neq j)$;
4     $a_i.DecideAction$ (k_i, s_j);
5     $a_i.ChangeScore()$;
6 **end**
7 **for** *all* a_i **do**
8     $a_i.ChangeCoalition()$;
9     $a_i.ChangeStrategy()$;
10     $a_i.Neighborhood = Rewire()$;
11     $a_i.ResetPayoffScore()$;
12 **end**
13 **EndFunction**

12.3.1 Reputation Sharing

In order to decide their strategy, and to maximize their payoff, agents need to know their opponents' image score. This is a challenging issue, since each of the agents can play with any other in the population. In Nowak's model [8], they use two approaches to solve this problem. First, they consider that image score is public, and that all agents know the image score of any other agent in the population. Second, they also consider that there exist a small percentage of agents (neighbors) that can observe a particular interaction; and only those agents, plus the recipient, update the other agent's image score. The first scenario is an idealistic one, while in the second scenario, each agent has a different perception about the image score of the others.

Here, we consider a different model for reputation, where each agent has a set of neighbors, and this neighborhood represents the direct contacts (friends or mates) that an individual has. We assume that each agent knows the image score of its neighbors. At the same time, we assume that agents may belong to coalitions, that models groups of interest, or organizations, that shares reputation information among its members. Therefore it models a global exchange of information biased by the different coalitions.

Thus, differing to [8] and as in [5], in our model agents are connected to others in a complex network, where each of the agents has a neighborhood. However, as in [8] and differing to [5], each agent may interact with any other agent of the population. We do not consider agents playing only in their neighborhood, since then agents could have a direct reputation from its neighbors. Therefore, as each player may interact with any other in the population, direct reciprocity does not work, as the chances of one player interacting again with the same player are negligibly small [8].

12.3.2 Action Selection

In previous sections we have presented the donation game and how reputation information flows among the agents. Now, in Procedure 27 we proceed to explain how a donor acts in our model when it encounters with a recipient (line 4, Procedure 26).

Once a random pair of agents a_i and a_j has been randomly selected to interact, and their roles are defined, the donor (a_i) checks if the recipient (a_j) belongs either to its neighbors or to its coalition mates (line 2, 27). If it belongs to any of those groups, then we assume that the donor knows the score of the recipient. In the contrary case, as it has no information, it assumes that the image score of a_j is 0 (following [8]). After this, the donor has to decide, depending on its strategy (k_i), if it donates to the recipient, providing a benefit b with a cost c to itself (line 8). This action increases its image score (line 9). On the contrary, if a_i does not donate, both individuals receive zero pay-off, but the image score of the donor is decreased by one (lines 11 and 12).

Procedure 27: Behavior of a donor a_i

```
1  Function ChangePayoffScore (a_i, a_j)
2  if a_i.InCoaOrNeighbor (a_j) then
3  |  s_j = a_j.GetScore();
4  else
5  |  s_j = 0;
6  end
7  if (k_i ≤ s_j) then
8  |  a_i.Donate (a_j, b, c);
9  |  a_i.ChangeScore (+1);
10 else
11 |  a_i.Donate (a_j, 0, 0);
12 |  a_i.ChangeScore (−1);
13 end
14 EndFunction
```

12.3.3 Coalition Formation

In our approach, we allow agents to form coalitions in order to share reputation information and therefore to improve cooperation. We consider that when an agent joins a coalition, it agrees to share its image score with the rest of the coalition members, but also obtains the image score of the other members of that coalition. We impose that any agent can only belong to a unique coalition at a time, since we consider that coalitions somehow compete in the game. Note that agents belonging to a coalition are not necessary neighbors.

Each coalition has an image score that depends on the average image score of its members. Let Coa_j represent the coalition j with coalition member agents indexed by i. The size of the coalition $|Coa_j|$ gives the number of agents in the coalition. The *coalition score*, CS_j, is specified as follows:

$$CS_j = ln|Coa_j| \cdot \frac{\sum_{i \in Coa_j} s_i}{|Coa_j|} \tag{12.1}$$

where s_i is the image score of member agent i. We include a scaling factor $ln|Coa_j|$ to model that larger-sized coalitions have more influence to attract agents to join those coalitions, as the amount of information they may share is bigger. At the same time, if a cooperative agent joins a bigger coalition it has more chances to be identified as a cooperator (if behaving as a donor) in successive encounters, and obtaining more donations in average than if it was isolated or in a smaller coalition. Thus, the benefits for the agent are double: a higher probability to obtain donations, and better options to rewire.

In Procedure 28 we present the rules for coalition dynamics, which we adapt from other approach that considers direct interactions [2]. The decision to join a coalition is based on simple rules, as in [1], which precludes modeling sophisticated agents that can learn about the rules to form coalitions.

Firstly, if an agent that belongs to a coalition is isolated from its coalition mates, i.e., none of its neighbors belongs to its coalition, then it becomes independent (line 4). We do this since we consider that each agent in a coalition must have at least one connection to another coalition member to transmit/receive information. Otherwise, it checks the payoff of its neighbors to see if its payoff P_i has been the worst in its neighborhood (line 6). If this is the case, it searches among its neighbors the agent or the coalition with the best reputation (s_j and CS_j, lines 7 and 8) to join them depending on the value of CS_j with respect to s_j (lines 10 and 12, respectively).

12.3.4 Changing the Strategy

At the beginning of the game, each agent is randomly assigned a strategy. However, depending on the payoffs it is obtaining, it may change it after every round in order to

Procedure 28: Rules for coalition formation and independence

```
Function ChangeCoalition ()
Coa_i = GetCoalition (a_i);
if ((Coa_i ≠ ∅) & a_i.Isolated (Neighbors, Coa_i)) then
|   a_i.GetIndependence();
else
|   if WorstPayoff (Neighbors, P_i) then
|   |   (a_j, s_j) = BestIndScore (IndepNeighbors);
|   |   (Coa_j, CS_j) = BestCoaScore (CoaNeighbors);
|   |   if (CS_j ≥ s_j) then
|   |   |   JoinCoalition (Coa_j);
|   |   else
|   |   |   CreateCoalition (a_i, a_j);
|   |   end
|   end
end
EndFunction
```

increase its benefits. When agents are neighbors (directly connected in the network), we consider that they know each other's image score, as well as their payoff and their strategy in the previous game. Thus, we assume that agents have access to local information about reputation, payoff and strategies from their neighbors, since they can directly observe them. With that information, an agent changes its strategy to copy the one with the highest payoff in its neighborhood, if higher than its own. We refer to this method as 0 and denote it as *Imitate the best neighbor*.

The previous paragraph describes the main method that agents will use along this chapter for deciding a new strategy after every game round. Nevertheless, there are many other possibilities for deciding the new strategy to play. Next we introduce some alternative imitation[1] strategies that we will study in Sect. 12.4.7. The numbers in the list will be used to refer to them:

1. *Imitate the best strategy*: this method considers the strategy that, in average, has obtained the higher payoff among the strategies used by the neighbors.
2. *Imitate the most popular strategy*: in this case, the most popular strategy, i.e., the one used by more neighbors, is selected.
3. *Infection*: the most successful agent in a neighborhood substitutes the strategies of its neighbors by its own successful strategy. We denote this procedure as infection, considering that this strategy infects its neighbors' strategies.
4. *Random selection*: in this case, every agent randomly selects a new strategy.
5. *Imitate probabilistically the best neighbor*: this method is similar to the main method 0 (i.e., *Imitate the best neighbor*), but imitation is probabilistically selected depending on the payoff achieved by every neighbor.

[1]This chapter uses an imitation approach (memetics) using evolutionary game theory to discover the most successful strategy in a given scenario.

6. *Imitate probabilistically the best strategy*: this case is also similar to method number 1 (i.e., *Imitate the best strategy*), but using a probabilistic distribution.
7. *Imitate probabilistically the most popular strategy*: this method uses a probabilistic distribution based on the popularity of the different strategies in the neighborhood. Again, it is a probabilistic version of method 2 (i.e., *Imitate the most popular strategy*).

12.3.5 Network Topologies

We place agents in a complex network since they provide a realistic model of the topological features found in many nature, social and technological networks [10] (e.g., computer and social networks). For our simulations, we mainly focus on small-world and scale-free network topologies, since they model the most common networks appearing in our human societies and in nature. Nevertheless, we shall also consider spatial and random networks at the end of the chapter. We briefly introduce next these topologies, and redirect to the interested reader to Sect. 3.5 from Chap. 3.

12.3.6 Rewiring

In most real-world network interactions, relationships are not static, i.e., agents can change the individuals that they are linked to. We denote this change in the network topology as rewiring. By using rewiring agents can modify their neighborhood if they are not satisfied with their neighbors.

As a difference with [5], where one agent is randomly chosen to change its neighbors, in our model, we specify a neighborhood measure of satisfaction to decide if an agent wishes to change it or not. In Eq. 12.2 we define the probability of rewiring for an agent i, which depends on the aggregate image score of all the neighbors, i.e., it depends on the average neighborhood reputation.

$$p_i^{rew} = \frac{(10 - \frac{\sum_{j=1}^{F}(s_j+5)}{F})}{10} \tag{12.2}$$

where s_j is the image score of each of the neighbors of a_i and F is the number of neighbors (friends) that the agent a_i has. Observe that $s_j \in [-5, 5]$, thus the maximum difference between scores is 10. Once agent a_i computes this probability, then it samples a Bernoulli distribution to decide if rewiring or not (Procedure 26, line 10). If agent a_i decides to rewire, then it leaves its neighbor with the lowest image score, and joins the neighbor with the highest one in its coalition. The reason for this is that, as we stated above, we consider that coalitions are communities that share reputation information, so agents can benefit from it to change their neighbors. We point out that this rewiring procedure only happens if the agent with the lowest image score does not become isolated, i.e., we do not allow disconnected nodes in our network.

12.4 Results

In this section, we present the performance of our mechanism, using the final strategy selected by the agents, after the simulation has converged, as a measure of the cooperation level achieved by the population. Firstly, we present the empirical setting for our simulations. Afterwards, we analyze how coalitions and rewiring allow for the emergence of cooperation, the differences on results depending on the initial topology, the differences between random or selected rewiring, the effect of alternative strategy dynamics, the dependance on the initial conditions, and finally, the effect of random changes in the agents' strategies (i.e., a mutation effect).

12.4.1 Experimental Settings

We have performed a in-depth experimental study, setting the number of agents N to 1600. Each run is composed of a set of iterations in which several agents play the donation game. Each agent plays in average 1.25 times in each round, i.e., 2000 times two agents are selected randomly, one as a donor and another one as the recipient. The number of rounds varies in each particular run depending on the simulation convergence and stability. We consider convergence when there are no changes in the strategy of the agents during 25 consecutive iterations. Finally, unless otherwise stated, the parameters used for building the networks are a von Neumann neighborhood for the spatial network (i.e., four neighbors), $W_{1600}^{4;0.1}$ for the small-world network, $S_{1600}^{4;-2}$ for the scale-free network and $R_{1600}^{0.0025}$ for the random network. These values have been chosen to compare the different topologies in a fair condition trying to have an average of four neighbors per agent.

Next, we will provide some results from a micro-simulation point of view, i.e., showing some individual runs that can be considered as stereotypes for a particular scenario. Then, we will consider a macro-simulation point of view, averaging the results over multiple runs.

12.4.2 Emergence of Cooperation (Micro-analysis)

In this part, we first review in more detail the results from [9], concerning the effects over cooperation when we use coalitions and the rewiring mechanism in the donation game. Here, we provide a micro-analysis, considering representative executions.

In Fig. 12.1 we see the results of a typical simulation when we do not use coalitions nor rewiring in a scale-free network (see its node degree in Fig. 12.2). In the histogram we represent the distribution of agents with a certain strategy when the simulation has converged. We see that all the agents end up playing $k \geq 0$. This means that

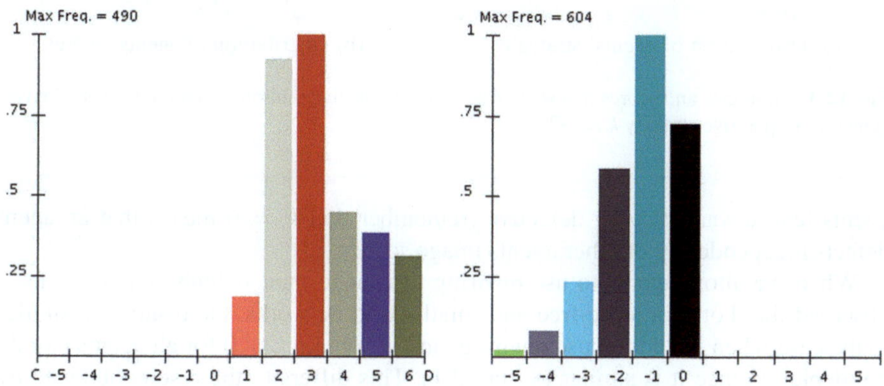

(a) Distribution of agents' strategies. (b) Distribution of agents' scores.

Fig. 12.1 Distribution of agents' strategies and scores with no coalitions, no rewiring, in a scale-free network after convergence

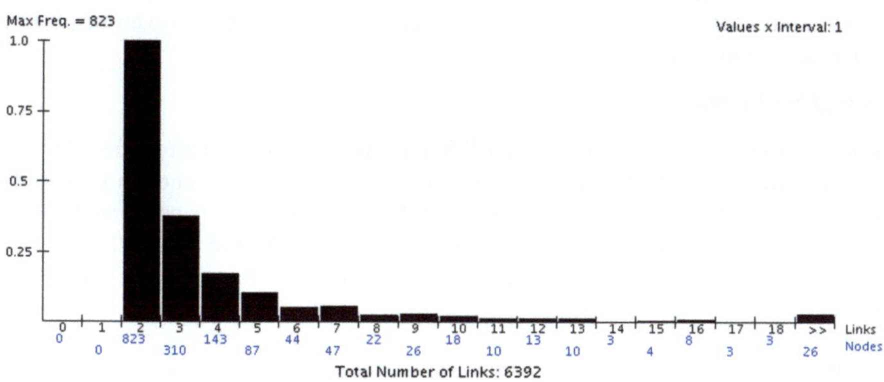

Fig. 12.2 Node degree for a scale-free network. The horizontal axis shows in black the number of links per node, and in blue the number of nodes that have such value. Observe the long tail, where there are 26 nodes with more than 18 links each. The most popular value is 2 happening in 823 nodes, and used to normalize the figure

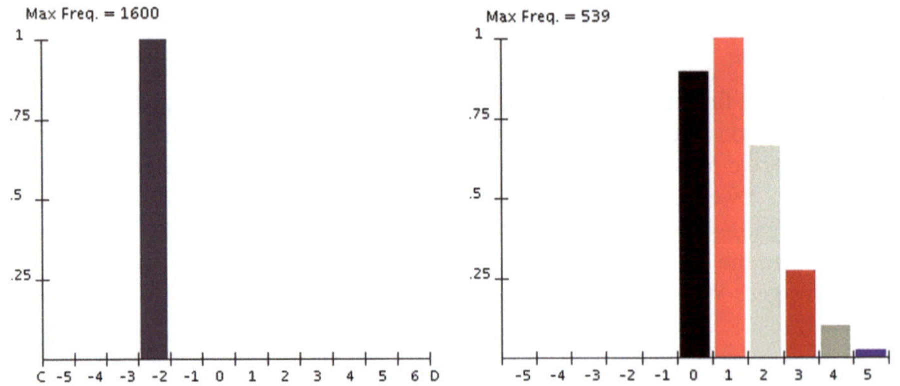

(a) Distribution of agents' strategies. (b) Distribution of agents' scores.

Fig. 12.3 Strategies and scores in a scale-free network, using coalitions but not rewiring. Convergence to cooperative strategy $k = -2$

agents lean toward playing defective (remember that $k = 6$ means that an agent defects independently of other agents image score).

When we allow agents to use rewiring to change their neighborhood, we have observed that both for scale-free and small-world networks, the results are similar to the case when we do not use rewiring, and we obtain $k > 0$ for all agents (we do not depict it since it is similar to Fig. 12.1). This differs to the results obtained by Fu et al. [5], where they successfully use rewiring to improve cooperation among agents. However, here we propose a different environment, where even if agents are connected to others, they can play with any agent in the population. In fact, as they have no information about other agents' reputation, since there are no coalitions for information sharing, the rewiring is done randomly, and it might even result in a worse neighborhood.

Using Coalitions

Now, we endow agents with our coalition formation mechanism alone. We find that in scale-free networks, allowing them to join coalitions is enough to achieve cooperation. In Fig. 12.3 we present the final distribution of agents per strategy when convergence is reached, as well as the resulting scores. We see that in this case all agents converge to strategy $k = -2$. Moreover, we have observed that in different simulations the results vary from one strategy to another, but being $k \leq 0$ in almost all the cases.

The evolution of strategies along the generations can be seen in Fig. 12.4. We see how $k = -2$ becomes more popular since the beginning of the simulation, followed by strategy $k = 0$ that finally disappears.

However, when using small-world networks, we have observed that agents converge to a single strategy, which is not cooperative ($k > 0$). In Fig. 12.5a we show an example where all agents do not converge and several strategies remain available.

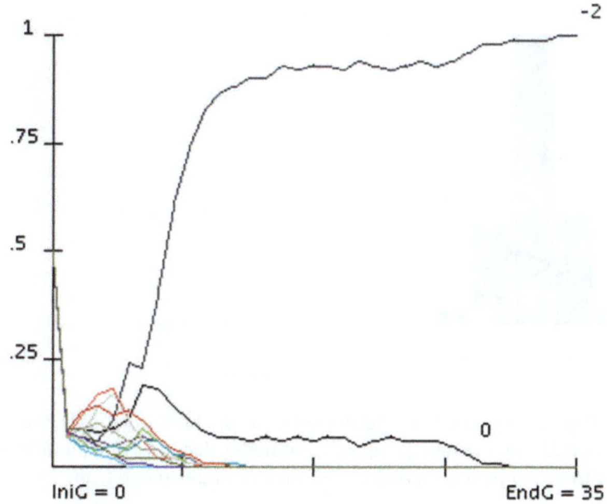

Fig. 12.4 Evolution of the different strategies after 35 generations

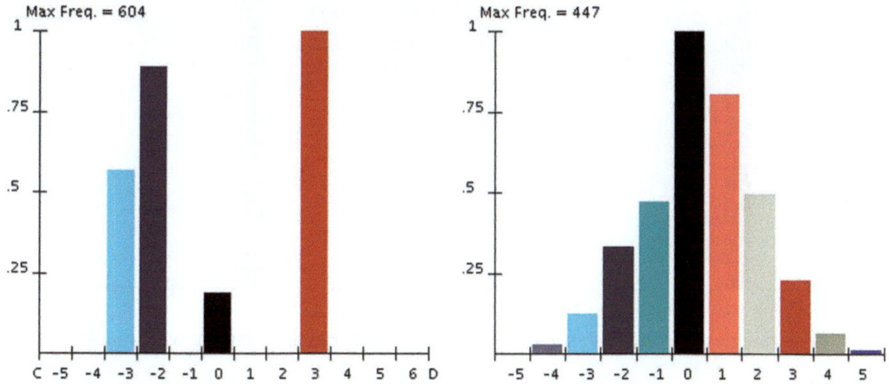

(a) Distribution of agents' strategies. (b) Distribution of agents' scores.

Fig. 12.5 Strategies in small-world, with coalitions but without rewiring

In Fig. 12.5b we see the scores they produce, given some of the strategies are cooperative and one defective. In Fig. 12.6 we see the node degree representation of the network. We can observe that in this type of network there is not a long tail, as it happened in the scale-free case, and this has a relevant influence as stated next.

Using Coalitions and Rewiring

Now, we study if cooperation is improved when we add rewiring to the coalition formation mechanism. In Fig. 12.7 we present the distribution of strategies after one simulation when using our coalitions and rewiring, starting from initial scale-free and small-world topologies (Fig. 12.7a, b, respectively). We see that in both cases, all agents end up a using a cooperative strategy ($k \leq 0$). Moreover, not only agents

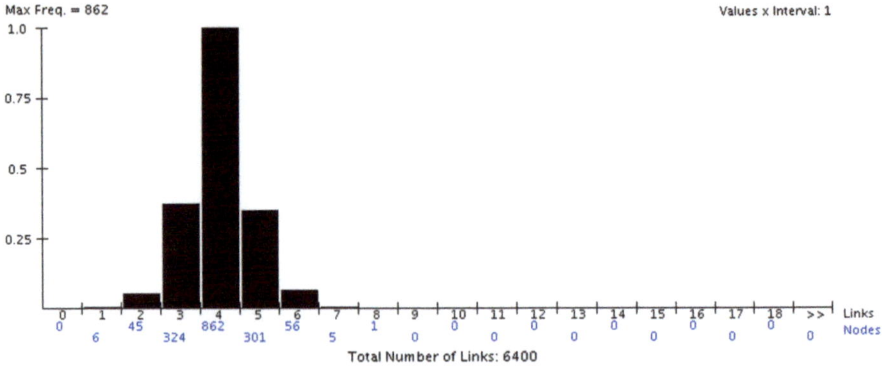

Fig. 12.6 Node degree for a small-world network. In the horizontal axis, we show in black the number of links per node, and in blue the number of nodes that have that number of links. The more popular value is 4 links happening in 862 nodes, and all nodes range from 2 to 6 links

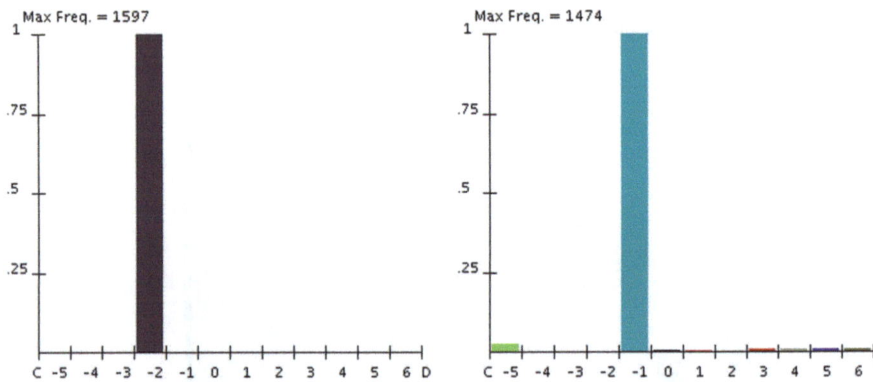

(a) Strategies starting with scale-free. (b) Strategies starting with small-world.

Fig. 12.7 Strategies obtained after simulating with scale-free and small-world initial topologies, using coalitions and rewiring

converge to a cooperative strategy, but we have observed that in every simulation all agents converge to the same cooperative strategy in a particular run.[2]

Moreover, if we now take a look at Fig. 12.8 we see the evolution among the different strategies in both cases, we can observe how only one cooperative strategy remains as the most popular one. However, we notice that the simulation in the scale-free case converges in less than 20 generations; while in small-world case it needs nearly a hundred generations to become stable. The key to understand the effect of rewiring, in the small-word case, comes from the analysis of the node degree distributions for both cases.

[2]Note that the particular strategy may not be the same in different runs.

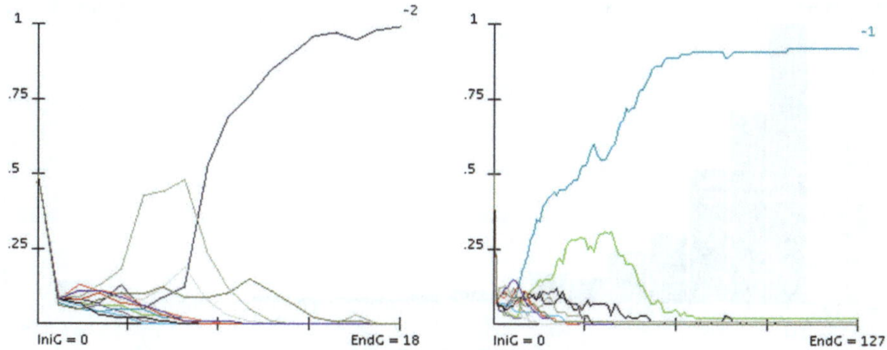

(a) Strategies starting with scale-free. (b) Strategies starting with small-world.

Fig. 12.8 Strategies obtained after simulating with a scale-free and a small-world initial topologies, using coalitions and rewiring

The use of rewiring means that starting from distributions similar to the ones appearing in Figs. 12.2 and 12.6, respectively, they can change their links and we end up in a different topology. Figure 12.9 shows the resulting node degree distributions for the scale-free and the small-world cases, using both coalition and rewiring. The reason why rewiring, together with coalitions, helps to support cooperation in the small-world case is that the topology changes to become something closer to a scale-free one. Although, both distributions shown in Fig. 12.9 do not exactly fit with the scale-free distribution shown in Fig. 12.2 we can see how they have evolved into a network with several nodes having a huge number of connections (41 nodes with more than 18 links in Fig. 12.9a and 58 nodes in Fig. 12.9b). This means that both distributions, even changing they initial structure, have evolved to another ones sharing a heavy tail distribution.

12.4.3 Emergence of Cooperation (Macro-analysis)

In order to see that our combined mechanism allows the emergence of cooperative strategies ($k \leq 0$), now we are going to analyze the results obtained when averaging over a set of runs in every scenario.

In Fig. 12.10 we present the average results after 100 different runs, without using coalitions nor rewiring. We represent the strategy distribution considering all simulations, keeping in mind that in each simulation we can obtain a set of coexisting strategies. We can see how the results starting with a scale-free topology,[3] or with a small-world one are pretty similar. Note how all the strategies chosen by the agents in all simulations are defective, i.e., $k > 0$. Besides, the simulations converge in a few dozens of generations, always less than 50. The standard deviation for these results

[3]Note: a new network is generated for each run.

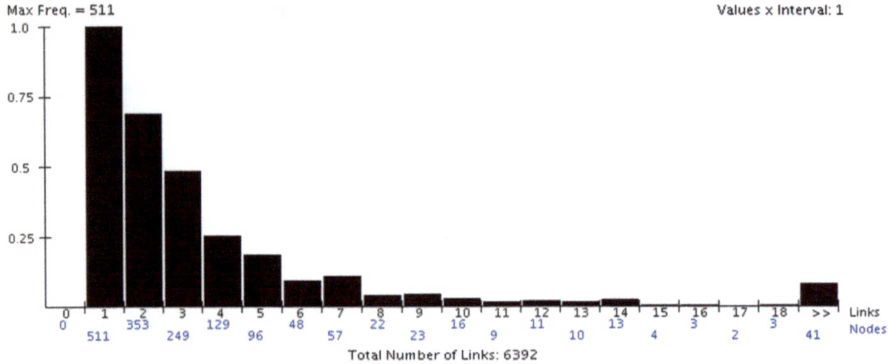

(a) Node degree distribution obtained starting with scale-free.

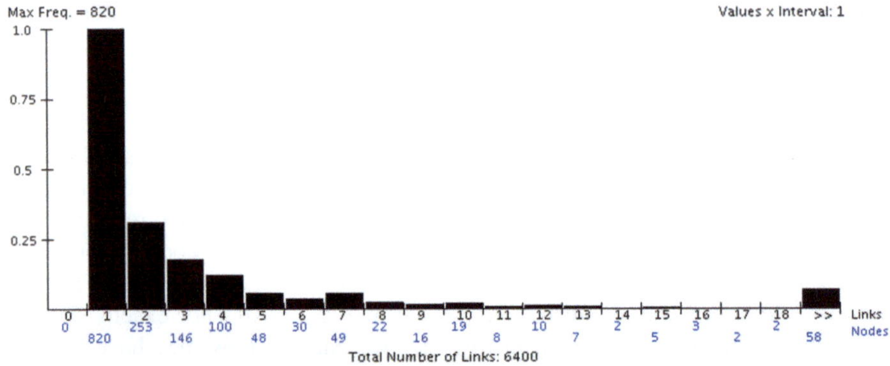

(b) Node degree distribution obtained starting with small-world.

Fig. 12.9 Node degree distributions obtained after two simulations starting with a scale-free and a small-world topologies, and using coalitions and rewiring. Note how the small-world distribution has evolved into one with a long tail

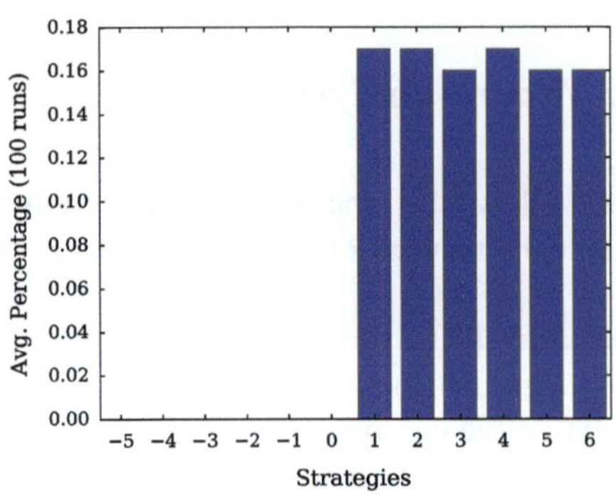

(a) Scale-free topology.

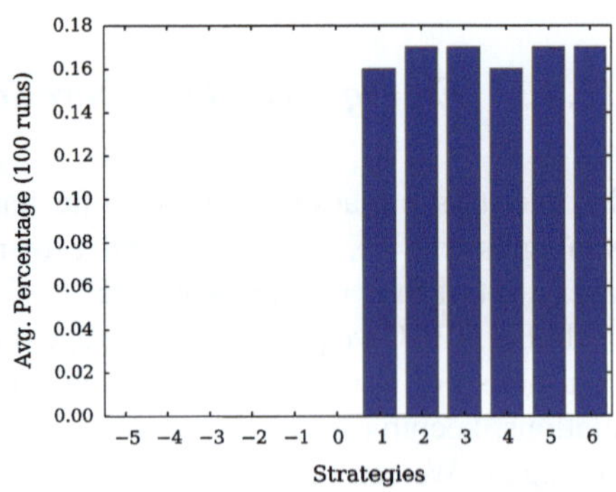

(b) Small-world topology.

Fig. 12.10 Averaged results obtained after 100 runs using scale-free and small-world topologies. All simulations converge in less than 50 generations

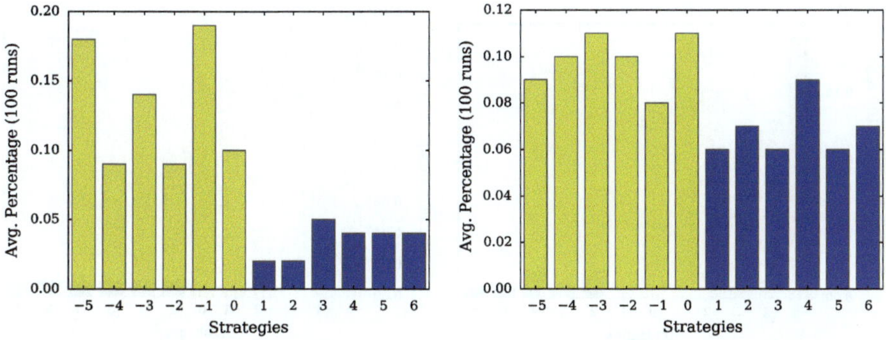

(a) Scale-free topology. (b) Small-world topology.

Fig. 12.11 Averaged results obtained after 100 runs starting with scale-free or small-world topologies using only coalitions

are in the interval $[0.06, 0.08]$ for $k > 0$ in the scale-free case, and in the interval $[0.03, 0.04]$ for $k > 0$ in the small-world one.

In Sect. 12.3.6 we introduced the possibility to do partner-switching, i.e., rewiring; meaning that agents can decide to change their links if they are not very happy with a particular neighbor. So, what happens if an agent decides to disconnect its worst neighbor and reconnect to the best one in its neighbors' neighborhood? The result is that if we add rewiring alone there is no practical effect, i.e., the results are practically the same than in Fig. 12.10. But, we now wonder what happens when we add coalitions. Figure 12.11 shows the results when we add the possibility to participate in coalitions. We see how in the scale-free case, the cooperative strategies $k \leq 0$ become much more popular (around 80%) than defective ones; while in the small-world case cooperation becomes more popular than defection, but not up to a similar level (59%) than in the scale-free case. All simulations end in less than 100 generations in the scale-free case, but do not end (cyclic behaviors) in the small-world one. Concerning the standard deviation, it is approximately equal to the average values $\forall k$; and for both topologies.

Finally, Fig. 12.12 shows the results obtained when we add coalitions and rewiring at the same time. Now we observe very similar results, either for the scale-free topology with a 93% level of cooperation and the small-world topology, where cooperative strategies achieve a 96% of cooperation. All simulations end around 100 generations in the scale-free case, but do not end (cyclic behaviors) in the small-world one. The standard deviation is approximately equal to the average values $\forall k$ and for both topologies. However, given the results shown by the figure, the popularity of cooperative strategies is guaranteed in all cases.

We will see the coherence of these results in the next section, where we explore spatial and random topologies, that are somehow similar to the small-world one, i.e., they share a similar number of links per node, having no initial hubs.

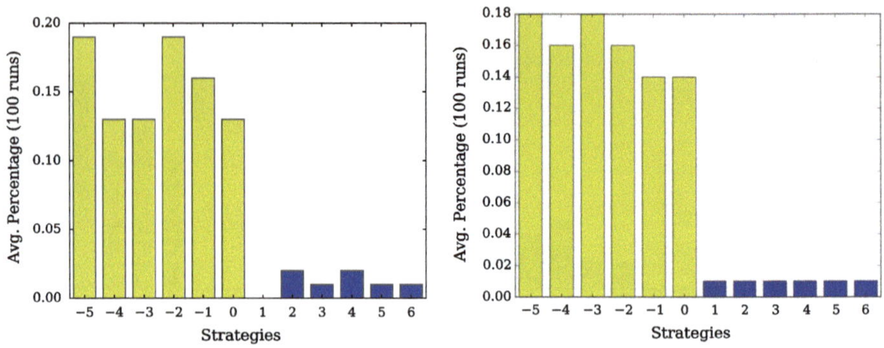

(a) Scale-free topology. (b) Small-world topology.

Fig. 12.12 Averaged results obtained after 100 runs starting with scale-free or small-world topologies including coalitions and rewiring. Standard deviations approximately equal to average values $\forall k$

12.4.4 Regular (SP) Versus Random Networks (RN)

We have seen how coalitions in scale-free networks, together with rewiring in small-world ones help cooperation strategies to emerge in the donation game. However, what would happen if we select a different initial topology? For instance, we may consider pure regular spatial topologies or even random networks. Here we analyze what happens with our model for these two cases.

In the case of regular or spatial topologies, we have selected a two dimensional toroidal grid of 40×40 cells, where the nodes are connected with the four closer neighbors (up, down, left and right, i.e., the von Neumann neighborhood) and using a torus world. Initially, defective strategies completely dominate the results, and including rewiring alone does not change the scenario at all. When adding coalitions, cooperation shifts from 0 to 17%, while when combining coalitions and rewiring cooperation reaches 95%. The pattern observed for the node degree distribution is also the same than in the previous topologies, i.e., there is a need of high connected nodes (hubs) for cooperation to emerge.

In the case of random networks, we have decided to create our random topology connecting every node randomly with other four nodes in the network, to keep the average number of connections around four (as in all the previous topologies used). In this case, simulations show that coalitions have an effect similar to small-world networks (see Fig. 12.13), cooperation moves from 0 to 60%, and when combining coalitions and rewiring cooperation reaches a 96%. This suggest that random networks,[4] as scale-free ones, are appropriate for cooperation emergence when we just use coalitions.

[4]Again, the standard deviation is approximately equal to the average values $\forall k$. This means that in the right figure (rewiring plus coalitions) cooperation is guaranteed.

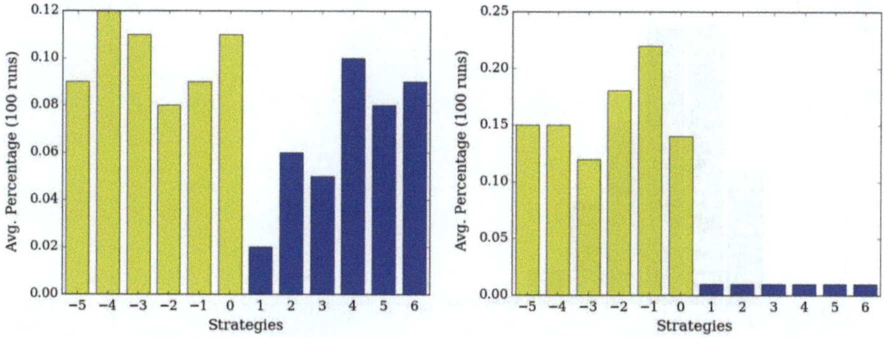

(a) Using only coalitions. (b) Using coalitions and rewiring.

Fig. 12.13 Averaged results obtained after 100 runs starting with a random topology including coalitions (left) together with rewiring (right). In the left case, we get cyclic behaviors, while in the right one simulation usually ends before 500 generations

Figure 12.14 shows the initial distribution of a random network, and the final one obtained after doing a simulation, using coalitions and rewiring. Note how the distribution has evolved into one with a long tail, very similar to what happened in the small world case.

Finally, Fig. 12.15 shows the evolution of the clustering coefficient (cc) along 100 generations, when using rewiring, starting from a spatial network (left) and a random network (right). If we compare these two figures with Fig. 12.17 we realize how the pattern is quite similar to what happens with the small-world topology. Again, the stable value of *cc* fits around 0.05, and this is necessary but not enough to achieve full cooperation, which is obtained when coalitions are present.

12.4.5 Topology Influence

We have further investigated the effects of adding rewiring to the coalition formation mechanism. We have observed that in scale-free networks, hubs (agents with a higher number of connections) have a strong influence over the rest of agents, and also more information than them. This eases the process of convergence to one single coalition, where all the agents use the same cooperative strategy. This happens even when we only use our coalition formation mechanism. But, when we introduce rewiring the process of convergence is even faster. This happens because, as we allow agents to choose their neighbors, hubs are the most successful ones, making their own influence even higher, and also the influence of the coalition where they belong.

In Fig. 12.7 we saw an example with the evolution of strategies along iterations. Therefore, starting with a scale-free topology the convergence to cooperation is much faster, so the time required for the convergence varies depending on the topology.

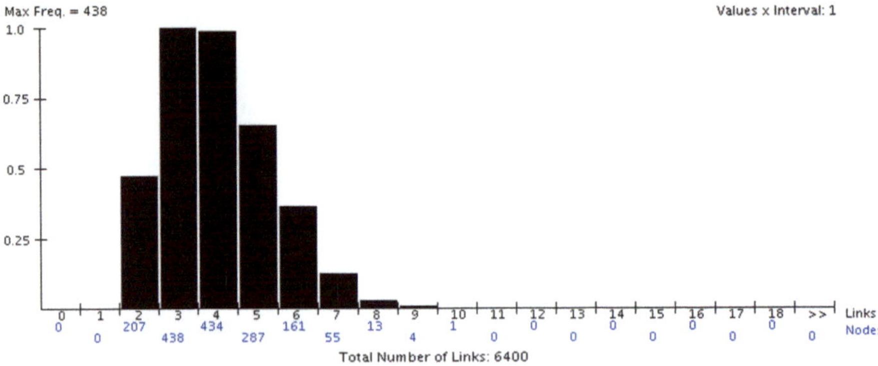

(a) Initial node degree distribution of a random network.

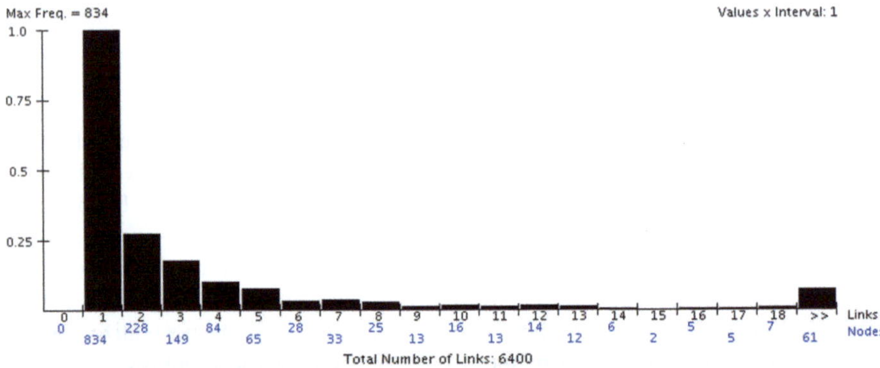

(b) Final node degree distribution obtained starting with a random network.

Fig. 12.14 Node degree distributions of a random network before and after simulating with coalitions and rewiring

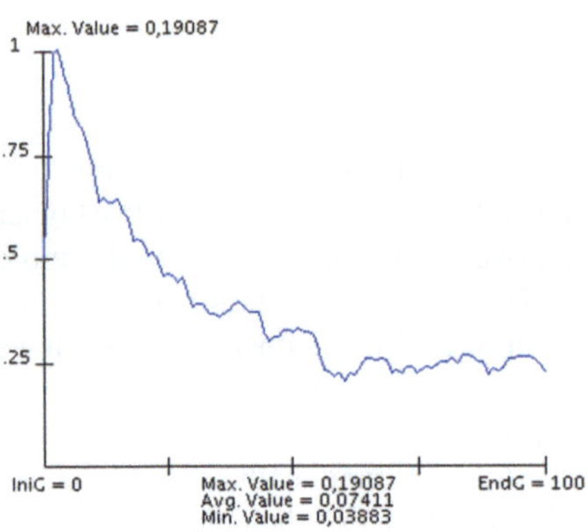

(a) Evolution of *cc* starting with a SP.

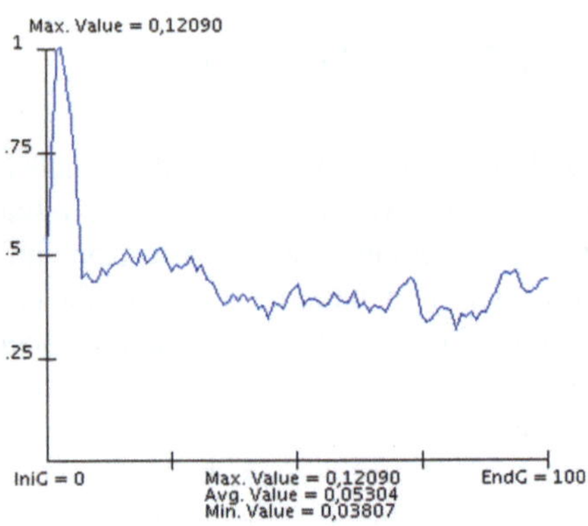

(b) Evolution of *cc* starting with a RN.

Fig. 12.15 Evolution of the clustering coefficient (cc) along 100 generations, when using rewiring, starting from a spatial (SP) and a random network (RN), respectively

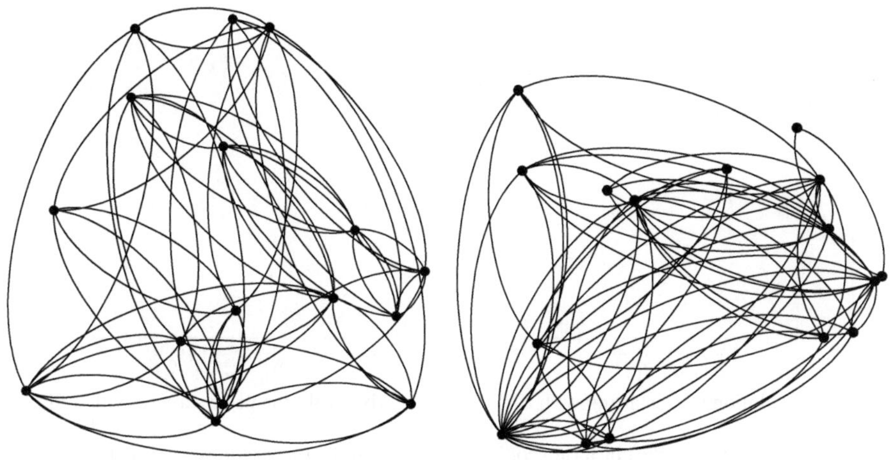

(a) Initial small-world network. (b) Final network.

Fig. 12.16 Initial and final networks obtained, starting with a small-world, after using coalitions and rewiring (16 nodes). Initially all nodes have around 4 links, while in the final network there are nodes with many more links (hubs)

On the one hand, the faster convergence with scale-free is due to the strong influence that hub agents have over the rest of the population. By definition, a hub has a higher number of neighbors than the average agent; so it has more information to play (remember that agents know the image score of their direct neighbors), which increases its chances to donate appropriately, and to receive donations in return to its image score. This puts hub agents in an excellent influence position, since obtaining the highest benefits make other agents to copy their strategy. Moreover, agents with a small neighborhood, but connected to a hub, promptly join the hub to form a coalition, thus less and bigger coalitions are formed faster. Besides, as hubs usually belong to bigger coalitions, it becomes very popular to rewire to them. This causes that they increase even more their individual and coalition influence.

On the other hand, in small-world networks each agent has a similar number of neighbors (as in regular or random networks), so agents have more or less the same level of influence. However, with the use of rewiring, agents with highest score start having more neighbors than the others, which increases their influence. Afterwards, more agents imitate them, and the coalitions they belong start to grow faster, allowing to finally reach a single cooperative strategy. Figure 12.16 shows an example of a final configuration when we start with a small-world network topology (here we used only 16 nodes to ease its display). We see that agents self-reorganize in a structure, where some of them have much more links than the others. Thus, as in the scale-free case, bigger and more influential coalitions (regarding their image score and size) are formed.

Another interesting element to measure what happens in the networks when allowing them to rewire is to use the *clustering coefficient (cc)*, that measures the fraction

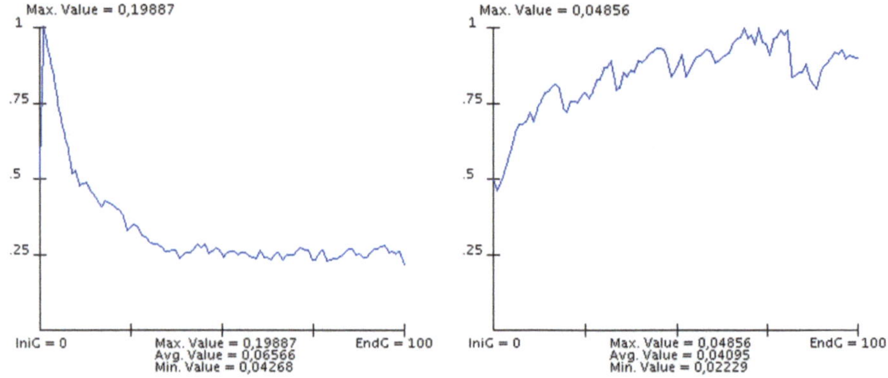

(a) Evolution of *cc* starting with a SW. (b) Evolution of *cc* starting with a SF.

Fig. 12.17 Evolution of the clustering coefficient (cc) along 100 generations, when using rewiring, and starting from a small-world (SW) and a scale-free (SF) network, respectively

of closed paths of length two divided by the total number of paths with length two, with $cc \in [0, 1]$. If $cc = 1$ then we have perfect transitivity, but if $cc = 0$ there is no closed triads (see Sect. 3.4 in Chap. 3). Figure 12.17 presents the evolution of the *cc* in the two topologies under analysis here. In the small-world case (left), we see how it initially starts in a value around 0.2 and evolves to a much lower value around 0.05. In the scale-free case (right), the *cc* starts in a much lower value (around 0.02), and slightly grows to reach again a value around 0.05. The key here is to observe that both nets evolve to have long tails (i.e., hubs) and $cc \simeq 0.05$. However, these two conditions are not enough to achieve full cooperation, as the use of coalitions is necessary too. Nevertheless, now we can understand why scale-free networks are able to reach cooperation without rewiring: the reason is that the scale-free topology has some initial characteristics that previously we have identified as relevant: a long tail distribution and a value of *cc* close to 0.05.

We must also consider the influence of the neighborhood size in the cooperation. Figure 12.18 shows the results obtained changing the average neighborhood size of the network, considering the four topologies; and using coalitions together with rewiring. We can see that increasing the neighborhood size reduces cooperation, and we also point out that curiously the cooperation reduction is lower in the spatial and small-world topologies, which are more regular/homogeneous than the other two. Our explanation is that when the neighborhood size increases, the influence of hubs becomes smaller, and it takes more time for a single coalition to appear. Besides, the rewiring effect is less effective, as an agent that rewires to another still has high probabilities to get some influence from the agent refused. Altogether, it makes the emergence of the desired cooperation more difficult.

Summarizing, we have found that by using coalitions and rewiring cooperation emerges with lower values of the neighborhood size, and this happens mainly by two reasons:

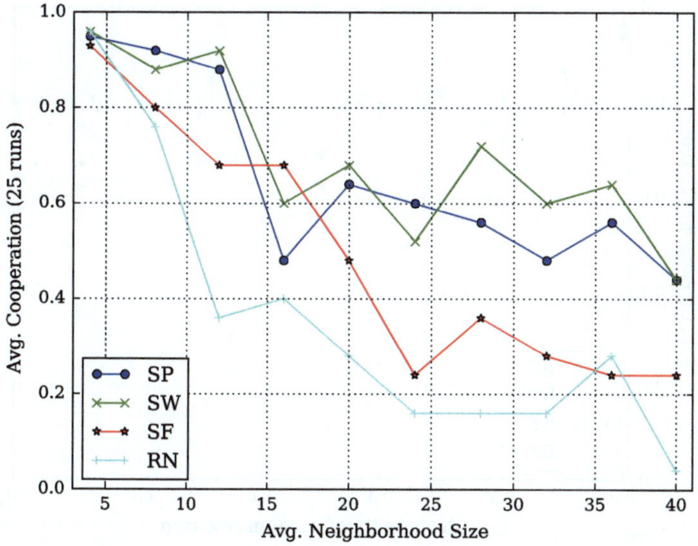

Fig. 12.18 Average cooperation depending on the neighborhood size, using coalitions and rewiring

- Firstly, because usually one single super coalition is formed. As an agent has information not only about its neighbors, but also about its coalition mates, this results in agents having more information about the image score of the whole population as the simulation evolves.
- Secondly, as an agent can change its neighborhood, it can discover and join other agents with higher image score. This allows an agent to donate with higher probability, also increasing its image score, and therefore its chances for obtaining a donation next time it becomes a recipient.

Finally, we compare our results with the ones obtained in [8], which is the basis for our work. The comparison cannot be done under an equivalent framework, as that paper presents a panmictic scenario, and genetics are used to evolve the most popular strategies in the population. Nevertheless, in such work, the scenario with a public image score obtained the strategy $k = 0$ as a final result; but in a second scenario, where agents have a limited view of others' image score, agents tend to be defective ($k > 0$). In our case, our coalitions and rewiring mechanisms allow to achieve cooperation even in scenarios with limited information.

12.4.6 Random Versus Selected Rewiring

The rewiring process, described in Sect. 12.3.6, only performs the partner-switching selecting the best nodes in the neighbors' neighborhood; but perhaps using a random

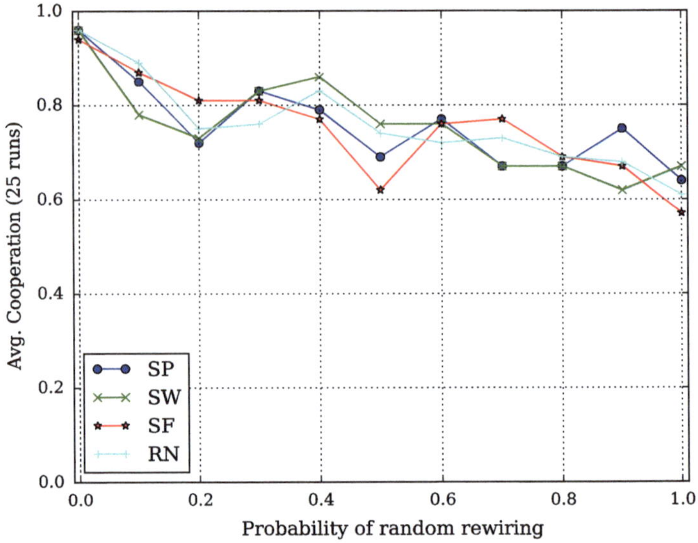

Fig. 12.19 Values of cooperation when rewiring to the best neighbor of a neighbor, or when doing a random rewiring

component in the rewiring could be positive. Here we explore this possibility seeing how it affects the emergence of cooperation.

Figure 12.19 shows the results when we introduce a certain probability to rewire to random nodes, instead of rewiring to the best neighbor, which is the method agents have used in our previous simulations. In the left side of the figure, we rewire always to the best node of the neighbors' neighborhood; while in the right side of the picture we rewire always randomly. We can observe how a random rewire drops the percentage of cooperation around 60% in all topologies. The reason is that a random rewire makes more difficult the creation of hubs, as any node can be selected. This, in average, increases the number of links for nodes with a low number of connections, i.e., it makes more difficult to achieve an optimum topology for the donation game.

12.4.7 Alternative Strategy Dynamics

The strategy dynamics selected previously are based in evolutionary imitation, where an agent basically imitates the best strategy found in its neighborhood. But we can imagine many other dynamics for deciding the strategy in the next iteration. Here, we consider some popular alternatives.

Now we are going to enumerate all the methods we are going to compare, according to the description introduced in Sect. 12.3.4:

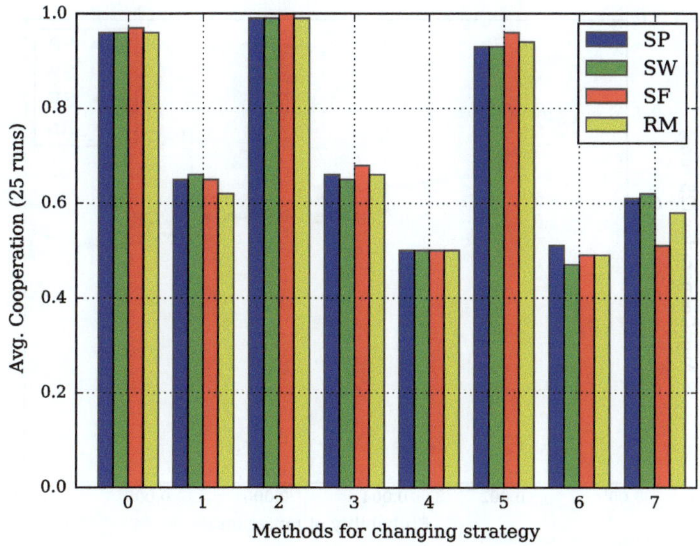

Fig. 12.20 Values of cooperation when using several methods for changing the agent's strategy, using coalitions and rewiring, and starting with different topologies

0. *Imitate the best neighbor.*
1. *Imitate the best strategy.*
2. *Imitate the most popular strategy.*
3. *Infection.*
4. *Random selection.*
5. *Imitate probabilistically the best neighbor.*
6. *Imitate probabilistically the best strategy.*
7. *Imitate probabilistically the most popular strategy.*

Figure 12.20 presents the results averaging over 25 runs. The vertical axis shows the cooperation level (i.e., the aggregate distribution of strategies for $k <= 0$) obtained averaging over 25 runs for each particular topology. In the horizontal axis we show the different methods (according to previous enumeration). We can see that the best results, are obtained with methods 2, 0 and 5 (in this order). Method 2 imitates the most popular strategy, i.e., it imitates strategies based on how popular they are, without taking into account their payoff. Method 5 is very similar to method 0, but imitates probabilistically the best node. Finally, note that method 4 has the expected behavior in all networks, as it produces a 50% of cooperation; as it uses random selection of strategies. Note also how methods 6 and 7 are the worst, and perform quite close to the random results obtained by method 4.

Therefore, it seems that for these external dynamics, a direct or probabilistic imitation of the best neighbor, together with imitating the most popular strategy are the most successful approaches. Note that these methods are very common strategies in human societies.

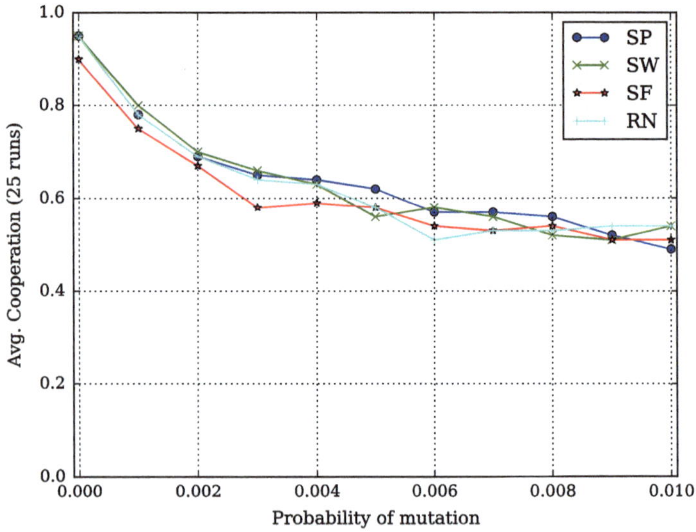

Fig. 12.21 Values of cooperation when using coalitions and rewiring, and changing the mutation probability

12.4.8 Mutation

Mutation is a powerful mechanism to introduce diversity among the agents in a simulation, and to avoid a quick convergence to a local optima. In previous simulations, agents have chosen among the available strategies in the neighborhood, but without introducing new strategies or changing randomly their previous decisions. Now, we analyze what happens with cooperation when the agents can change randomly their own selected strategies. To do that, we introduce a probability of mutation, that allow an agent to randomly select a new strategy among the whole set.

Figure 12.21 shows the results over the different topologies, when introducing a certain mutation probability. All the topologies decrease the value of cooperation, which stabilizes around 50% for upper values of the mutation probability. We can see that even in this challenging conditions, by means of coalitions and rewiring, we are able to stabilize cooperation in half of the agents in the network.

12.4.9 Dependance on the Initial Conditions

In our previous simulations we assumed that the agents have initially a uniform probability for selecting among the twelve possible strategies $k \in [-5, 6]$. Here we analyze what happens when the number of defective strategies $k \in [1, 6]$ is initially increased. Figure 12.22 shows the results, using the four types of networks, when

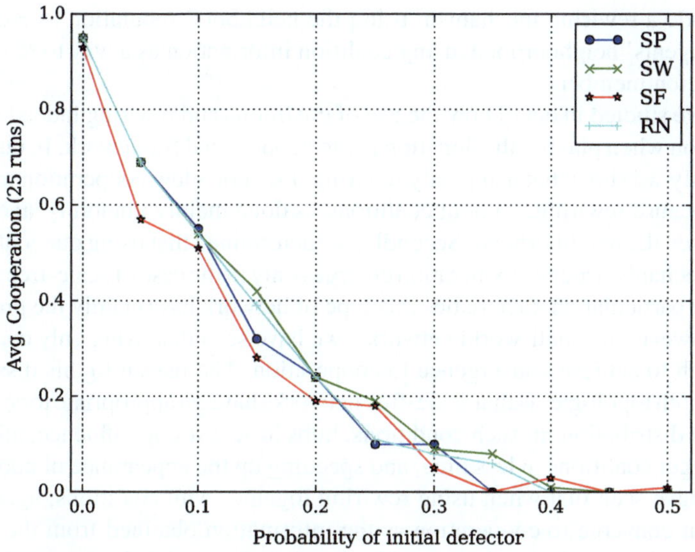

Fig. 12.22 Values of cooperation when using coalitions and rewiring, and increasing the number of defectors, i.e., $k \in [1, 6]$, at the beginning of the simulation

increasing the probability of being a defector at the beginning of the simulation. If a defector is initially selected (according to the probability shown in the horizontal axis in Fig. 12.22), then we randomly choose its strategy in the defection interval $[1, 6]$. Otherwise, we select randomly the strategy in the whole interval $[-5, 6]$. As we see, all the networks evolve in a similar pattern and cooperation disappears in all cases when the initial number of defectors gets over 0.4. This remarkable result means that the indirect reputation model is strongly influenced by the initial ratio of defective strategies in the population.

12.5 Conclusions

In this chapter, we have analyzed in depth an indirect reputation scenario previously introduced in [9], where agents are connected in a social network; but where any agent may interact with any other in the population. Agents' neighbors are their close related contacts from which agents obtain information, as strategy, payoff and reputation. Thus, the neighborhood of an agent models a social group with a friendly exchange of information. Besides the agent direct neighborhood, we also model a second set of contacts using coalitions (to describe mates in a club or in a team), as a way to share information about the environment where agents play. In real life, coalitions are useful to share information, and we also consider a measure of the coalition image score, in order to decide what coalition an agent should join. We also include a rewiring mechanism using the neighbors' reputation as a criteria to improve agents' neighborhood, using coalition information as a way to rewire to the best coalition members.

We have studied in detail how the use of coalitions and rewiring indeed improves cooperation when playing the donation game in our social framework. In our simulations, firstly we show that using only rewiring does not allow cooperation to emerge. This is because rewiring, without coalitions, is done mainly randomly, and that can even worsen the neighborhood. Secondly, we determined that using our coalition formation mechanism alone, cooperation emerges only in the case of scale-free networks due to the particular characteristics this type of network has (mainly the presence of hubs). However, in small-world networks, we have seen that using only coalitions is not enough to achieve convergence to cooperation. The reason to this dissimilarity between both topologies is that scale-free networks have an appropriate topology with a long tail distribution. In such conditions, hubs have a strong influence; allowing to create bigger coalitions in less time, and speeding up the appearance of cooperation. We also have seen that when using rewiring together with coalitions, agents in the population converge to cooperation as the information obtained from the coalition mates allow them to change their neighborhood, ending in topologies quite similar to scale-free networks. This effect has also been considered using regular and random networks, obtaining the very similar results to the ones shown with small-world ones.

Along the chapter, we have evaluated, in more detail than our previous work [9], the positive effects that grouping and social networking have over cooperation in complex networks with indirect reciprocity. However, we also have evaluated the strength of this result under different scenarios. We have seen how selected rewiring, directed to the best potential neighbors is more adequate under the scenarios seen here. We have also considered alternative strategy dynamics, using several strategies for imitating or spreading information about the best decisions to take. We have found that, together with imitating the best neighbor, the most successful methods have been to imitate the most popular strategy, and to imitate probabilistically the best neighbor (which is a probabilistic variation of our basic method). We also allow agents to mutate randomly their strategies along the simulation, but in this case, the model is robust and able to stabilize cooperation around a 60%. Finally, we have also considered how increasing the initial number of defectors affects cooperation, and we have seen how the scenario is particularly sensible about having a uniform distribution at the beginning. In fact, when we slightly increase the initial rate of defecting strategies in the population, cooperation is significantly reduced.

To sum up, this chapter has provided a comprehensive understanding on how cooperation can emerge in networked indirect reciprocity scenarios driven by coalitions. We can conclude that in all the scenarios analyzed, the use of coalitions plus rewiring have shown to be particularly effective as a way to enhance cooperation in the donation game.

Future work can consider the use of learning, (i.e., introspection) instead of memetics (imitation), but in such case the model will move away from the evolutionary game theory approach we have followed in this chapter. However, a mix of both approaches will be also interesting to evaluate.

Notes

This chapter provides an extension of the work presented in [9]. In the present chapter, we have performed the simulations considering a more complex scenario; increasing the number of agents and the strategy dynamics, including mutation and the dependency on initial conditions.

Acknowledgements This work was partially supported by the European Regional Development Fund (ERDF) together with the Galician Regional Government under agreement for funding the Atlantic Research Center for Information and Communication Technologies (AtlantTIC).

References

1. Axelrod, R.: The Complexity of Cooperation: Agent-Based Models of Competition and Collaboration, 1st edn. Princeton University Press, Princeton (1997)
2. Burguillo, J.C.: A memetic framework for describing and simulating spatial prisoner's dilemma with coalition formation. In: Proceedings of the 8th International Conference on Autonomous Agents and Multiagent Systems, pp. 441–448 (2009)
3. Chong, S.Y., Yao, X.: More choices and reputation in multi-agent interactions. IEEE Trans. Evol. Comput. **11**(6), 689–711 (2007)
4. Fu, F., Wu, T., Wang, L.: Partner switching stabilizes cooperation in coevolutionary prisoner's dilemma. Phys. Rev. E **79**(3), 036101 (2009)
5. Fu, F., Hauert, C., Nowak, M.A., Wang, L.: Reputation-based partner choice promotes cooperation in social networks. Phys. Rev. E **78**(2), 026117+ (2008)
6. Griffiths, N., Luck, M.: Changing neighbours: improving tag-based cooperation. In: Proceedings of the 9th International Conference on Autonomous Agents and Multiagent Systems: Volume 1, AAMAS '10, pp. 249–256 (2010)
7. Nowak, M.A.: Five rules for the evolution of cooperation. Science **314**, 1560–1563 (2006)
8. Nowak, M.A., Sigmund, K.: Evolution of indirect reciprocity by image scoring. Nature **393**, 573–577 (1998)
9. Peleteiro, A., Burguillo, J.C., Chong, S.Y.: Exploring indirect reciprocity in complex networks using coalitions and rewiring. In: Proceedings of the 2014 International Conference on Autonomous Agents and Multi-Agent Systems, pp. 669–676. International Foundation for Autonomous Agents and Multiagent Systems (2014)
10. Reka, A., Barabási, A.: Statistical mechanics of complex networks. Rev. Mod. Phys. **74**, 47–97 (2002)
11. Salazar, N., Rodriguez-Aguilar, J.A., Arcos, J.L., Peleteiro, A., Burguillo-Rial, J.C.: Emerging cooperation on complex networks. In: The 10th International Conference on Autonomous Agents and Multiagent Systems - Volume 2, AAMAS '11, pp. 669–676 (2011)
12. Sandholm, T., Larson, K., Andersson, M., Shehory, O., Tohmé, F.: Coalition structure generation with worst case guarantees. Artif. Intell. **111**(1–2), 209–238 (1999)
13. Shehory, O., Kraus, S.: Task allocation via coalition formation among autonomous agents. In: Proceedings of the 14th international joint conference on Artificial intelligence - Volume 1, IJCAI'95, pp. 655–661 (1995)

Chapter 13
A Coalitional Game of Life

Juan C. Burguillo

Abstract This chapter introduces the interesting Conway's Game of Life, describing first its simple rules; and then some static, oscillating and moving patters that emerge from the game iteration depending deterministically on its initial configuration. We also describe some Life properties concerning the complexity that emerges from simple local interactions, its capabilities for self-replicating patterns and several variants of this game described in the Life literature. Afterwards, a coalitional version of Life (CoaLife) is introduced as an upper layer, keeping the basic rules used by Life; and using new ones to create, modify or release coalitions along the game execution. Finally, we consider an Iterated Prisoner's Dilemma (IPD) extension for CoaLife (IPD-CoaLife), as another upper layer over CoaLife, where alive cells play the IPD with their alive neighbors. As an illustrative example, we provide a comparison between an IPD-Life (without coalitions) and the IPD-CoaLife, where the latter shows a better cooperation level. The aim of this last chapter is to show how the combination of coalitions and classical game theory models allows to explore competition and cooperation among cells in complex and rich environments such as the Game of Life.

13.1 Introduction

The Game of Life (GoL), also known as Life Game or simple Life, as will be denoted in this chapter, is the most popular cellular automata. Life was created by the British mathematician John Horton Conway, and published by Martin Gardner at Scientific American [9] in 1970 in his mathematical game column. John von Neumann created the cellular automata model in order to construct a hypothetical machine that could build copies of itself. Neumann found a complex mathematical model for such a machine with 29 states and several rules working on a rectangular grid. Conway was interested in such a problem, and Life appeared as a successful model that simplifies

J. C. Burguillo (✉)
Department of Telematics Engineering, School of Telecommunications Engineering,
University of Vigo, 36310 Vigo, Spain
e-mail: J.C.Burguillo@uvigo.es

J. C. Burguillo, *Self-organizing Coalitions for Managing Complexity*, Emergence, Complexity and Computation 29, https://doi.org/10.1007/978-3-319-69898-4_13

von Neumann's model. Since then, Life became a very popular and interesting game for the computer science community, and also for a broader audience. Many papers, books and simulators have been published in the last decades, and a detailed review of many of them, together with an introduction and a complete description of Life features can be found at [1].

This chapter introduces the interesting Conway's Game of Life, its properties and its capabilities for self-replicating patterns. The main contributions include a coalitional version of Life (CoaLife), introduced as an upper layer that keeps the basic rules used by Life; but using new ones to create, modify or release coalitions along the game execution. We also consider an Iterated Prisoner's Dilemma (IPD) extension for CoaLife (IPD-CoaLife), as another upper layer over CoaLife, where alive cells play the IPD with their alive neighbors. The main aim of this last chapter is to show how the combination of coalitions and classical game theory models allows to explore competition and cooperation among cells in complex and rich environments such as the Game of Life.

The rest of the chapter is organized as follows. Section 13.2 introduces the basic rules of the Game of Life, and Sect. 13.3 describes some simple and remarkable patters observed. Afterwards, Sect. 13.4 discusses some of the relevant properties shown by this game. Section 13.5 introduces CoaLife, a coalitions version of Life, described as an upper layer, and Sect. 13.6 creates another upper layer over Life, where cells play the Iterated Prisoner's Dilemma among them. Finally, Sect. 13.7 concludes the chapter.

13.2 Life Rules

Life is a 2D cellular automata game, where the neighborhood of a cell is composed by the cell itself and the 8 closer ones surrounding it (known as Moore neighborhood), as shown in Fig. 13.1. In Conway's Life cells can be only in two states: dead or alive, and they evolve in a two dimensional grid using three very simple rules:

Fig. 13.1 A cell (green) with the 8 surrounding neighbor cells (blue)

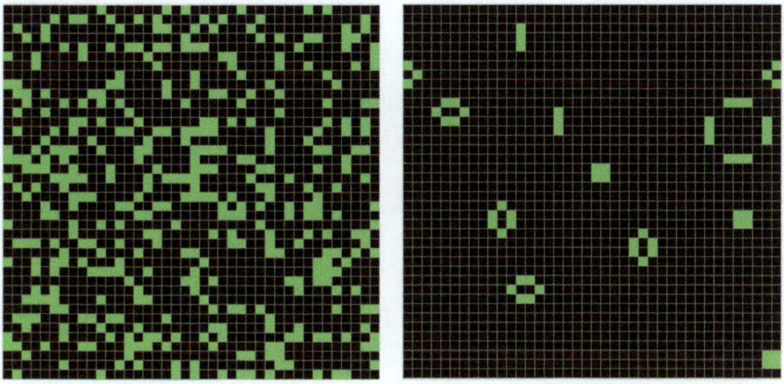

(a) Initial state (b) Ending state

Fig. 13.2 An execution of Life starting from a toroidal grid with 1600 cells (40×40), and randomly generating initial alive cells with a probability of 0.25. Alive cells appear in green, while dead ones in black

1. *Birth*: a dead cell gets back to live if it has three neighbors.
2. *Death*: an alive cell dies if it has less than two cells around (isolation) or more than three (overpopulation).
3. *Survive*: alive cells stay alive with two or three cells around.

The initial configuration of the grid is usually generated randomly, and then generations evolve by applying the above rules simultaneously to every cell in the lattice. Births and deaths of cells happen synchronously, and time passes discretely as a tick. Given there is no random influence in the rules, each generation is a pure result of the preceding one, and the whole computation depends deterministically on such initial configuration.

Figure 13.2 shows a typical execution of Life. In the ending state we can appreciate some static patterns and oscillators, which are commented in the next section.

13.3 Patterns

Along last decades, many interesting patters have been discovered in Life. The first ones were found without the use of computers, like static patterns or basic oscillators. Afterwards, with the use of computing power, many other patterns were found including those ones that are able to move across the grid, denoted as spaceships, being the glider the most popular one. In this section we show some frequent and interesting patterns representing alive cells in green over a white grid.

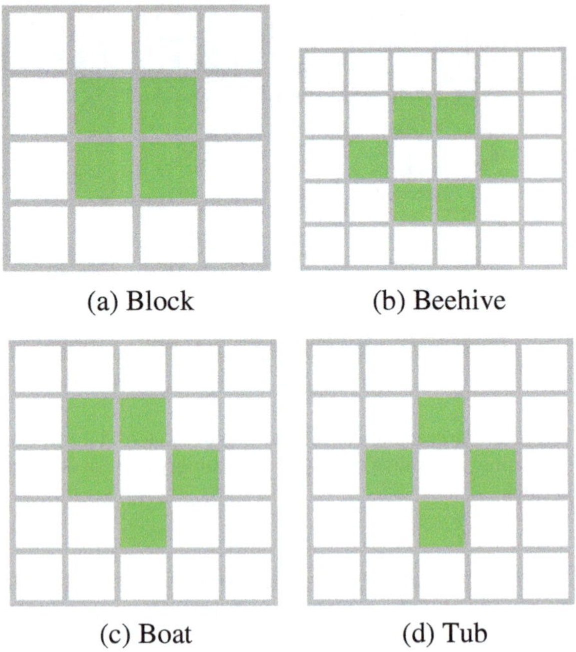

Fig. 13.3 Some usual static patterns

13.3.1 *Static Patterns*

Stable patterns, also called *still life*, do not change without external interference, and can be used as building parts to construct more complex patterns. The stable state of these patterns enables them to store information, to stop the movement of other patterns or to keep other unstable patterns stable. Figure 13.3 shows four of the most common static patterns: block, beehive, boat and tub.

13.3.2 *Oscillators*

These patterns cycle over a specific period, repeating their pattern infinitely. The basic oscillators have periods of two or three ticks, but complex oscillators have been discovered with higher order periods. Oscillators can be used for bumping stable patterns to start chain reactions. Figures 13.4, 13.5 and 13.6 show the blinker, the toad and the beacon, which are three oscillators with a period of two.

Fig. 13.4 Blinker oscillator with period 2

(a) Blinker phase 1 (b) Blinker phase 2

Fig. 13.5 Toad oscillator with period 2

(a) Toad phase 1 (b) Toad phase 2

Fig. 13.6 Beacon oscillator with period 2

(a) Beacon phase 1 (b) Beacon phase 2

13.3.3 Spaceships

Spaceships are patters that shift along the grid, going back to the initial configuration after a certain period, i.e., they are moving oscillators. Spaceships are very relevant patterns in Life, and mainly responsible for the great complexity and the interest created by this game. Conway realized that gliders were self-replicant configurations that can transport information from one place of the grid to another. Because of the locality of Life rules, no information can travel through the grid at a greater rate than one cell per unit time, so this velocity is said to be the cellular automaton *speed of light* and denoted as c.

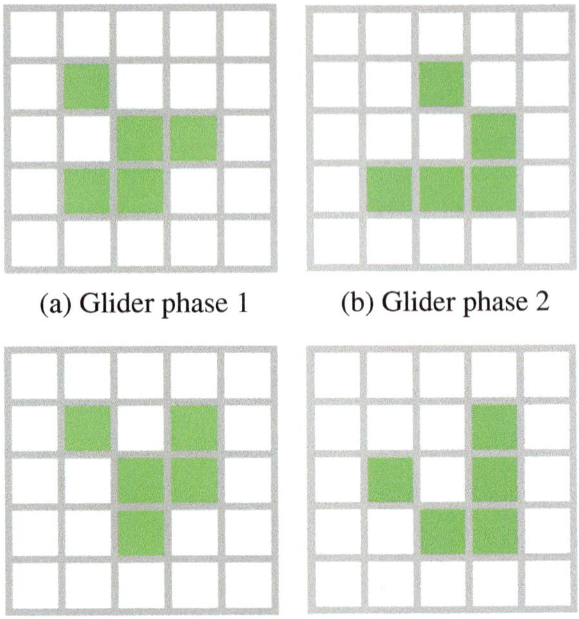

(a) Glider phase 1 (b) Glider phase 2

(c) Glider phase 3 (d) Glider phase 4

Fig. 13.7 A glider moves diagonally from the up-left corner to the bottom-right one

Figure 13.7 shows the most popular example, which is the glider, with period of four, that shifts one diagonal cell every four ticks, i.e., it moves at one-quarter the speed of light.

13.3.4 Guns

Guns are patterns that produce other patterns, like the *Gosper glider gun* discovered in 1970 by Bill Gosper, and that produces a new glider every 30 generations. Since then, researchers and Life players have discovered multiple new patterns, some similar to the basic patters mentioned here and others much more complex (see [12] for many more examples).

13.4 Properties

Life is not a classical player game, it is referred as a zero-player game, given that starting from an initial configuration the next generations are completely determined by the rules of the game. Therefore, a player (or Life *designer*) interacts with the game by creating an initial configuration, and observing how it evolves.

Conway's Life has attracted much interest since it was published, because it provides an example of emergence and self-replicating patterns. Gliders are relevant as they interact with other patterns in interesting ways, and can be used to simulate counters or to construct logic gates such as AND, OR and NOT. This means that any problem that can be computed algorithmically can be computed by Conway's Game of Life [1, 7, 13]. In fact, Life is one of the simplest CA to exhibit universal computation and universal construction, meaning that with an appropriate initial distribution of the cells, it is equivalent to a Turing Machine [10].

13.4.1 Algorithmic Complexity

When running the game, usually we see a combination of static patterns, oscillators and spaceships; and we may wonder if Life is decidable, i.e., if there is an algorithm that can tell whether, if starting from an initial pattern then a later one is ever going to appear. This is another formulation of the halting problem, discussed when talking about Turing Machines in Sect. 2.3 of Chap. 2; so, from that discussion, we can realize that such algorithm cannot exist.

13.4.2 Emergence and Self-replication

Starting from random initial patterns on the grid, we may find new populations of patterns changing, moving, emerging and disappearing as generations passed by. In a very few cases the whole population eventually dies out, while this is not so frequent, as initial configurations usually produce stable patterns, oscillators and even spaceships, being the glider the most common one due to its simplicity. Conway's Life is a nice example of emergence properties, for instance, when starting from an random configuration a glider emerges traveling along the 2D world.

13.4.3 Bounding an Unbounded Life

Like the tape in a Turing machine, Life grid is infinite in its mathematical model; but computers do have a finite memory, so a problem appears at the boundaries of finite grid arrays. There has been several solutions to cope with such limitations:

- The simplest solution is to consider that outside the array all cells are permanently dead. This solution is easy to code, but destroys moving patters, when they reach the edges of the grid world.
- A still simple, but more powerful solution is to consider that the left and the right edges of the world are connected, and also the top and bottom ones, yielding a

connected toroidal array. The result is that moving patters shift from one side of the square to the other. However, the result still may become inadequate when a set of patterns grow too much, but at least there are no edge defects.

- A more sophisticated alternative is to use dynamic storage allocation to create dynamic arrays for growing patterns. An alternative is to use vectors for representing alive cells. However, the limitation of this solution comes when the population exceeds the size of the computer memory. Besides, counting live neighbors needs a hash-table lookup and search operations that slow down the simulation speed. These problems can be partially managed by complicating the code, and using even more sophisticated data structures and algorithms; being Hashlife a relevant example created by Bill Gosper in the 1980s.

13.4.4 Variants of Life

Conway explored several rules and neighborhoods, but his criterion was that the rule should neither lead to populations which quickly died out, nor which expanded without end from limited beginnings. We can use a formal notation to describe Life rules, like $B3/S23$, where $B3$ means that a cell is born when it has exactly 3 neighbors, and $S23$ means that it survives having 2 or 3 neighbors; otherwise the cell dies. Therefore, in general, the list of numbers after the B, is what is required for a dead cell to be born; while the second list after the S is what an alive cell needs to survive to the next generation.

Cellular automata on a 2D grid, that can be described in this way are known as Life-like cellular automata. For instance, the rule $B36/S23$ describes another popular life-like automata known as Highlife, where replicators are very common. The only difference with the original Life is that in Highlife cells born when they have exactly 3 or 6 alive cells around.

Many other rules have been considered by researchers, although the majority of them produce universes that are either too chaotic or too empty to be relevant. Other models change the dimensions, the geometry and/or the rules of the CA universe. One dimensional square (1D) variations (known as elementary cellular automata) and 3D square variations have been developed, and there are also 2D hexagonal and 2D triangular versions of life [6]. The game can also be modified in order to introduce more than the two basic states, with more complicated transition rules (see [1]).

13.5 A Coalitional Game of Life

In this section we present a coalitional version of the game of life (CoaLife). The idea is to keep the basic rules and properties of Life, while adding an extra upper layer to allow cells to interact, join or leave coalitions.

13.5.1 Rules of CoaLife

The design of CoaLife took in mind the simpleness considered by Conway when designing Life; so CoaLife rules are also very simple. For the the birth and dead of new cells, it keeps the previous B3/S23 rules of Life. Of course, they can be changed in any way, and also the game geometry or neighborhood, without affecting the coalition rules that are presented next.

The new coalition rules only affect how coalitions born, extend and/or disappear; and they can be described as:

1. *New Birth*: If a new cell will born in the next generation, it joins the biggest coalition in its neighborhood. *Optional: If there is no such coalition (or there are more than one with equal popularity) then the new born cell creates its own coalition with its independent neighbors as new coalition members.*
2. *Keeping Alive*: If a cell will keep alive there are two cases: (i) if the cell is independent it captured by the biggest coalition in its neighborhood; and (ii) if the cell already belongs to a coalition, but it is surrounded by two or more cells from another coalition, then such coalition captures the cell.
3. *Dying*: if an alive cell will die in the next generation, and it is a coalition member; then it leaves its coalition.

In the first rule there is an optional part, that deals with the apparition of new coalitions. This part is needed when cells are not assigned to coalitions in the initial configuration of the grid, so new coalitions must emerge at sometime. The second rule allows coalitions to capture independent cells (only one coalition cell is required) or coalition[1] cells (more than one cell from another coalition is required). Finally, the third rule is oriented to have a memoryless reborn, avoiding zombie cells getting back to life, and remembering coalitions they belong to in past lives.

Algorithm 29 presents a more formal but intuitive description of the previous rules in pseudo-code. Lines 4–9 correspond to the first rule, being line 8 the optional part, depending on the playing mode. Lines 10–15 correspond to the second rule. Finally, lines 16–18 correspond to the third rule.

13.5.2 CoaLife Scenarios

The reason for having an optional part of the first rule is that, depending on the scenario, we may need new coalitions to appear from scratch. Concerning such optional rule, we may have three possible scenarios to consider:

[1]This rule can be simplified allowing only to capture independent cells. However, the results have been more interesting when allowing coalitions to capture other coalition cells.

Procedure 29: void *ChangeCoalitions* (boolean *CreateCoalitions*)

```
foreach c_i ∈ Cells do
    N_i = GetNeighbors(c_i);
    C_b = BiggestCoalition(N_i);

    if (WillBorn(c_i)) then
        if (C_b ≠ ∅) then
            C_b = JoinCoalition(c_i, C_b);

        else if CreateCoalitions then
            C_i = NewCoalition(c_i, GetIndependentCells(N_i));
        end

    else if (KeepsAlive(c_i)) then
        if (isIndependent(c_i) & (C_b ≠ ∅)) then
            C_b = JoinCoalition(c_i, C_b);

        else if (isCoaMember(c_i) & (size(C_b) > 1)) then
            C_b = JoinCoalition(c_i, C_b);
        end

    else if (WillDie(c_i) & isCoaMember(c_i)) then
        LeaveCoalition(c_i, C_i);
    end
end
```

1. *Only initial coalitions*: some cells start linked to some coalitions, but new coalitions cannot appear along the simulation run. The optional part of the rule is not applied in this case.
2. *Initial and new coalitions*: some cells start linked to some coalitions, and new coalitions may appear along the run. Here, the optional part of the rule is applied.
3. *Only new coalitions*: cells do not start linked to any coalition, but new coalitions may appear along the run. The optional rule part is also applied in this case.

The first scenario is intended to pass from a zero-player game, as in the case of Life, to an n-player game; where players setup initial configurations, for a certain part of the automata world they own. They can try to avoid invasions from other players (coalitions) by using static patters or oscillators; or to invade other regions by means of spaceships. In this case, after players configure their initial coalitions state, then the game can be run without intervention to see what is the winning coalition. A possible modification of this first scenario would be to allow each player to modify some of its cells state after every generation as it is allowed in certain proposals [2–4], that usually allow players to add/delete own cells and even cells from other players. Nevertheless, adding cells individually make the game uncomfortable to play given its parallel nature, difficult to predict by human players, and that is a reason why the n-player version of Life has not been successful in the past. Another attempt to

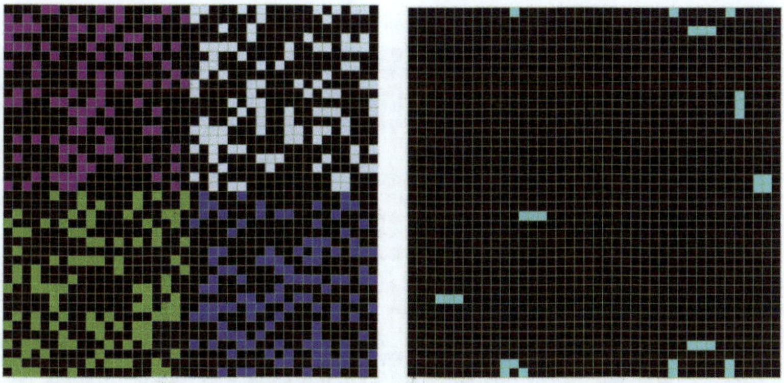

(a) Initial state with four coalitions (b) Final state with only one winner

Fig. 13.8 An execution of CoaLife with four coalitions, starting from a toroidal grid with 1600 cells (40 × 40), and randomly generating initial cells with $P_{alive} = 0.25$

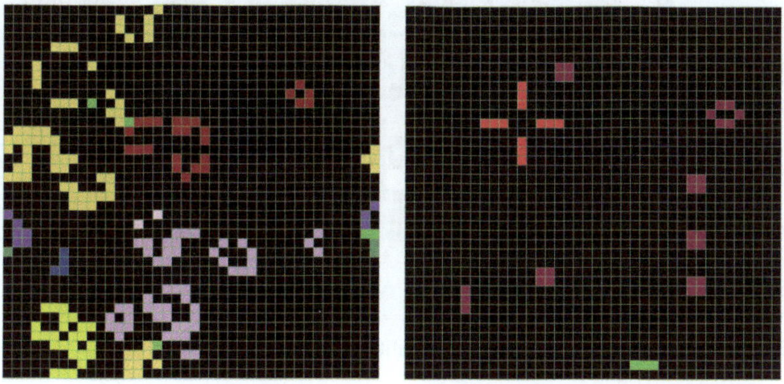

(a) Intermediate state (b) Final state with two coalitions

Fig. 13.9 Another run with intermediate and final states. Independent cells appear in green

obtain a 2-player Life version was done in [11], where two players were considered and represented by white or black cells, and Life rules for birth and survival were adapted depending on the differences between the number of white or black cells in a cell neighborhood. However, the most interesting possibility to play a n-player version of Life is to do it by means of computers, which can simulate a few new rounds and decide where to place a new cell or delete an existing one.[2]

The second and third scenarios allow the emergence of new coalitions to see how they interact with the others depending on the initial configuration. These two scenarios are more intended to observe how new coalitions emerge, and interact dynamically along the CoaLife game, and they are again zero-player scenarios.

---

[2]Remember that there is no algorithm to predict future Life configurations, so the only chance for a computer based algorithm is to run the game to evaluate what is the best action to play and modify a present configuration.

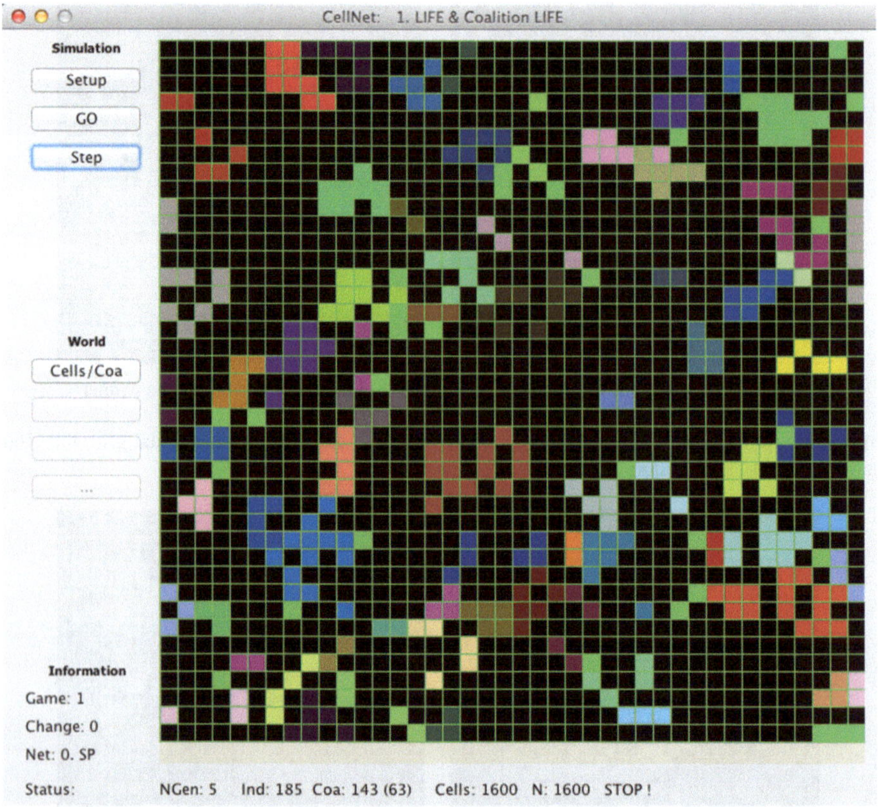

Fig. 13.10 CellNet simulator with multiple coalitions and independent cells

13.5.3 Running CoaLife

Although Life was intended for an infinite square lattice, the limited space resources available in any physical computer force us to take decisions about what to do when alive cells reach the borders of the grid. The implementation of CoaLife presented in this chapter uses a torus world, which is a simple but powerful solution to avoid edge effects.

Figure 13.8 shows two states of a CoaLife execution. On the left we have the initial configuration, where the grid was divided in four parts corresponding to each of the four initial coalitions. The right part of the figure shows a final state when only one coalition remains. The grid contains $40 \times 40 = 1600$ automata, and the initial cells were generated randomly with a probability of 0.25.

Figure 13.9 shows on the left hand side an intermediate state with some independent cells in green. The final state contains cells belonging to two coalitions and an oscillator. Independent cells appear in green color. In a case like this, we can assume that the winner is the coalition with more cells.

Finally, Fig. 13.10 shows a snapshot of the CellNet simulator running CoaLife in a multiple coalition scenario, i.e., it allows the emergence of new coalitions. In fact, the status bar shows that, after 5 generations, there are 185 independent cells and also 143 cells in 63 coalitions within a grid of 1600 potential cells.

13.6 An Iterated Prisoner's Dilemma for CoaLife

In this section we build a new upper layer over the two previous ones, i.e., Life and CoaLife, to provide a competitive/cooperative environment with coalitions and independent cells in order to play the iterated version of the famous Prisoner's Dilemma (IPD) (see Sect. 7.7 in Chap. 7).

The roots of the model introduced here come from [8], which is based in Axelrod's seminal work [5]. Nevertheless, we consider a simplified version of the model, given the aim of this chapter is to show how we can add multiple layers over the complex and interesting scenarios provided by the Life game using the coalition approaches introduced along the previous chapters of this book.

13.6.1 IPD-Life and IPD-CoaLife Rules

Here we describe the rules for the IPD-CoaLife in two parts. First we describe the basic rules to play an IPD version of the Life game (IPD-Life):

- We consider only the two pure basic strategies of such game, i.e., to be a defector (D) or to be a cooperator (C).
- When starting the simulation, each cell initializes its strategy and tax randomly. Afterwards, when a cell is born, it imitates the strategy of the alive neighbor with the highest payoff.
- Along each iteration, every alive cell play with its alive neighbors the Prisoner's Dilemma game.

Second, we describe the competitive/cooperative version to play the IPD with coalitions over the Life game (IPD-CoaLife):

- Coalitions are considered as groups or tribes managed by a leader. Belonging to a coalition means that all the cells in the coalition play cooperative among themselves, and play the leaders strategy against other coalitions or independent cells. But nothing is free, and given such benefits provided to the coalition members, the coalition leader imposes a tax over the income of coalition members.
- When, due to Life rules, a coalition leader dies, then the heiress cell[3] belonging to the coalition inherits it, and becomes the new leader with its own new external strategy and internal tax.

[3] The successor will be the cell that joined the coalition before the rest of the members.

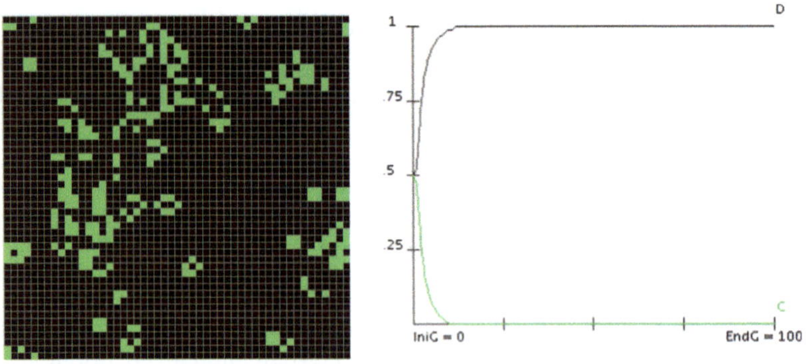

(a) Grid state after 100 iterations (b) Strategy evolution

Fig. 13.11 IPD-Life game: grid and strategy evolution along 100 iterations

- If a new born cell is captured by a coalition, then it imitates the leader tax; otherwise it generates randomly a new tax.
- A coalition member that has the worst payoff in its neighborhood may rebel, depending on a certain rebellion probability (P_{reb}) configured as a game parameter.

The reason to divide the description of the rules in two parts is to compare the behavior of the IPD over Life using coalitions or not.

13.6.2 Running the IPD-Based CoaLife

In this last section we present some snapshots of the evolution of a game simulation considering the IPD-Life, which does not use coalitions, and the IPD-CoaLife, which uses coalitions. The payoff matrix considered is a classical one with the values Temptation ($T = 5$), Reward ($R = 3$), Punishment ($P = 1$) and Suckers payoff ($S = 0$). These payoffs fulfill the two basic properties for the IPD: $T > R > P > S$ and $2R > S + T$.

Figure 13.11 shows the IPD-Life version of the game. On the left appears a snapshot of the grid after 100 iterations, while on the right appears the strategy evolution. Given the iterations among nearby cells are not continuous, given the nature of the Life game, where cells born and die frequently; it is natural to have an environment where defection is the ruling strategy, as can be seen in the right figure. This outcome is the rule for this game given these conditions.

Figure 13.12 presents the IPD-CoaLife, again showing on the left an intermediate snapshot of the grid after 100 iterations, and on the right figure appears the strategy evolution. In this case, the simulation started with four initial coalitions and no more coalitions were allowed to emerge. Looking at the figures we observe that cooperation emerges among cells in this particular simulation. This is a usual outcome of the game, but for a certain percentage of runs the cooperation level can be lower.

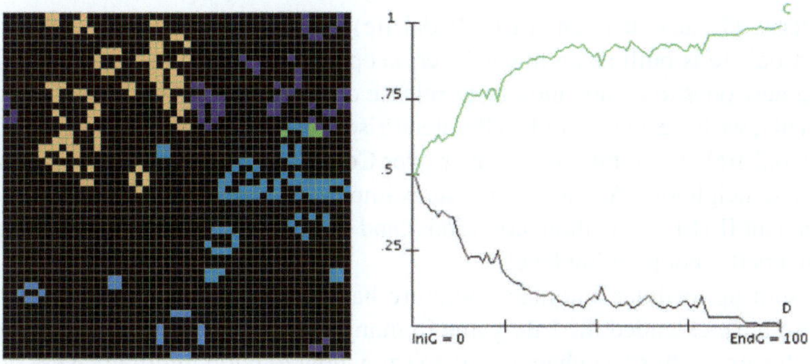

(a) Grid state after 100 iterations (b) Strategy evolution

Fig. 13.12 IPD-CoaLife game: grid and strategy evolution along 100 iterations

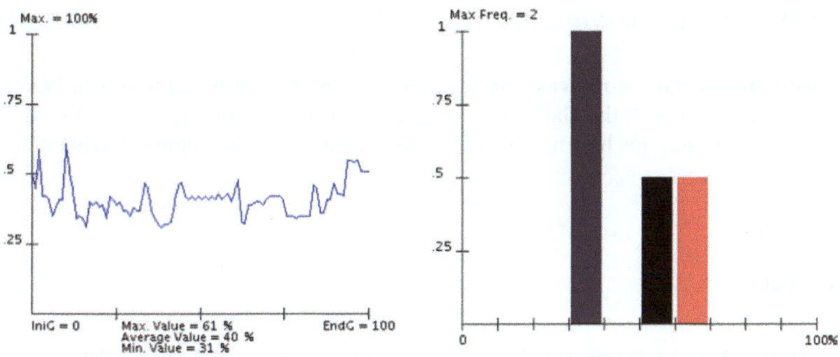

(a) Tax per cell evolution (b) Tax histogram

Fig. 13.13 Taxes per cell along 100 iterations, and taxes in the 100th iteration

Finally, Fig. 13.13 shows the average tax per cell evolution along the first 100 iterations, and also the tax histogram in the 100th iteration containing the taxes of the four coalitions. In the IPD-CoaLife, taxes usually move around 50% if we setup an initial number of coalitions (as in this case) or if we allow new coalitions to emerge.

13.7 Conclusions

This chapter has introduced the interesting Conway's Game of Life, describing first its simple rules and some static, oscillating and moving patters that emerge along the game iteration depending deterministically on its initial configuration. We have also described some Life properties concerning the complexity that emerges from simple local interactions, its capabilities for self-replication of patterns, and several game variants described in the Life literature.

Afterwards, an extension of Life (CoaLife) has been introduced to support coalitions. CoaLife is built as an upper layer, keeping the basic rules used by Life, and adding new ones to create, modify or release coalitions along the game execution.

Finally, we have considered an Iterated Prisoner's Dilemma extension for CoaLife (IPD-CoaLife), as another upper layer over CoaLife, where alive cells play the IPD with alive neighbors. As an illustrative example, we provided a simple comparison between an IPD-Life (without coalitions) and the IPD-CoaLife, where the latter has shown a better cooperation level.

To sum up, cellular automata literature has considered many rules, topologies, and also have extended the Life game in many ways. Within the framework of this book, the aim of this last chapter was to show how to build multiple layers over the Life game, providing support first for coalitions, and then for the classical IPD game. The combination of these two elements gives the possibility to explore competition and cooperation among cells in complex, emergent and rich frameworks such as the one provided by the Game of Life.

Acknowledgements This work was partially supported by the European Regional Development Fund (ERDF) together with the Galician Regional Government under agreement for funding the Atlantic Research Center for Information and Communication Technologies (AtlantTIC).

References

1. Adamatzky, A. (ed.): Game of Life Cellular Automata. Springer, Berlin (2010)
2. A Multiplayer Version of Conway's Game of Life. http://lifecompetes.com. Accessed Sept 2017
3. A Two Player Version of Conway's Game of Life. http://www.math.cornell.edu/~lipa/mec/lesson6.html. Accessed Sept 2017
4. A War Version of Conway's Game of Life. http://gameoflifetotalwar.com. Accessed Sept 2017
5. Axelrod, R.M.: The Complexity of Cooperation: Agent-Based Models of Competition and Collaboration. Princeton University Press, New Jersey (1997)
6. Bays, C.: The game of life in non-square environments. Game of Life Cellular Automata, pp. 319–330. Springer, Berlin (2010)
7. Berlekamp, C., Conway, J.H., Guy, R.K.: Winning Ways for Your Mathematical Plays. Academic Press, Cambridge (1982)
8. Burguillo-Rial, J.C.: A memetic framework for describing and simulating spatial prisoner's dilemma with coalition formation. In: Proceedings of the 8th International Conference on Autonomous Agents and Multiagent Systems. International Foundation for Autonomous Agents and Multiagent Systems, Vol. 1, pp. 441–448 (2009)
9. Gardner, M.: Mathematical games: the fantastic combinations of John Conway's new solitaire game Life. Sci. Am. **223**(4), 120–123 (1970)
10. Goucher, A.P.: Universal computation and construction in GoL cellular automata. Game of Life Cellular Automata, pp. 505–518. Springer, Berlin (2010)
11. Levene, M., Roussos, G.: A two-player game of life. Int. J. Mod. Phys. C **14**(02), 195–201 (2003)
12. McIntosh, H.V.: Conways Life. Game of Life Cellular Automata, vol. 1, pp. 17–34. Springer, Berlin (2010)
13. Rendell, P.: Turing universality of the game of life, Collision-Based Computing, pp. 513–539. Springer, Berlin (2002)

Appendix
CellNet: A Hands-On Approach for Agent-Based Modeling and Simulation

CellNet is a free to use open-source Java-based software, developed by Juan C. Burguillo,[1] and licensed under the GNU Lesser General Public License (LGPL). CellNet has its origins in 2003, as a research resource created by the author to study evolutionary game theory and cellular automata simulations. Since then, it has been used in a number of research works by the author, Ph.D. students and some research colleagues from other Universities.

Installing instructions, documentation, and related material can be accessed via the website: https://sites.google.com/view/cellnet-sim

A.1 CellNet Basic Features

CellNet was created at a time were most of the simulators were too complex or do not provide enough tools to comfortably manage cellular automata networks and evolutionary game theory simulations in Java with user-friendly interfaces.

CellNet works in two modes: (i) using a graphical user interface (GUI) for doing micro-simulations or (ii) using a batch mode for doing macro-simulations. CellNet also provides support for:

- Visualizing the whole set of cells and their state along each simulation iteration.
- Visualizing the simulation results in real time at each iteration. A set of graphical windows are provided for every relevant simulation result.
- Importing network data to reuse particular network structures to run experiments.
- Exporting network data, to save a particular network structure. The format used for the exported files is compatible with popular network analyzers such as Pajek[2] or Gephi.[3]

[1]The present importing/exporting package has been developed by Ana Peleteiro.

[2]http://mrvar.fdv.uni-lj.si/pajek/.

[3]https://gephi.org.

J. C. Burguillo, *Self-organizing Coalitions for Managing Complexity*, Emergence, Complexity and Computation 29, https://doi.org/10.1007/978-3-319-69898-4

A.2 CellNet Role in This Book

CellNet allows a hands-on approach to simulating and modifying most of the coalition-based experiments presented in this book:

- *Running Micro-simulations*: most of the experiments described in the research chapters (except the ones from Chap. 10, which were programmed in Netlogo) can be directly run in a one-shot mode, selecting them directly from the main menu of the simulator. The different game simulation parameters can be reviewed and modified from the experiment option window, or from a general configuration window. No programming skills are needed to run the simulator in this mode, so the reader can explore the contents of the book, repeat some experiments or perform new ones by just selecting different parameter values.
- *Running Macro-simulations*: having very basic general programming skills allows the reader to configure some batch files to execute a set of experiments, involving multiple runs, to analyze the average results provided by such set of game simulations.
- *Modifying the algorithms*: readers having standard Java programming skills can redesign their own algorithms, and then test the behaviors obtained by performing new experiments. For this, new algorithms can be created from scratch, or more commonly, the algorithm files already included can be inherited and used as templates. CellNet code includes support for generating different types of complex topologies, using several machine learning techniques, performing evolutionary meta-decisions, generating real-time visualizations and interacting with external network analyzers.

The reader is encouraged to access CellNet at
https://sites.google.com/view/cellnet-sim
and to follow the instructions there to use the full CellNet resources, and get hands-on to explore the use of self-organizing coalitions for agent-based simulation of complex systems!

Index

J. C. Burguillo, *Self-organizing Coalitions for Managing Complexity*, Emergence, Complexity and Computation 29, https://doi.org/10.1007/978-3-319-69898-4

MIX
Papier aus verantwortungsvollen Quellen
Paper from responsible sources
FSC® C105338

If you have any concerns about our products,
you can contact us on
ProductSafety@springernature.com

In case Publisher is established outside the EU,
the EU authorized representative is:
Springer Nature Customer Service Center GmbH
Europaplatz 3, 69115 Heidelberg, Germany

Printed by Libri Plureos GmbH
in Hamburg, Germany